Developing Skills and Competencies for Digital and Green Transitions

Patricia Ordóñez de Pablos
The University of Oviedo, Spain

Mohammad Nabil Almunawar
School of Business and Economics, University of Brunei Darussalam, Brunei

Muhammad Anshari
School of Business and Economics, University of Brunei Darussalam, Brunei

A volume in the Advances in Educational Technologies and Instructional Design (AETID) Book Series

IGI Global
PUBLISHER of TIMELY KNOWLEDGE

Published in the United States of America by
IGI Global
Business Science Reference (an imprint of IGI Global)
701 E. Chocolate Avenue
Hershey PA, USA 17033
Tel: 717-533-8845
Fax: 717-533-8661
E-mail: cust@igi-global.com
Web site: http://www.igi-global.com

Library of Congress Cataloging-in-Publication Data

Names: Ordóñez de Pablos, Patricia, 1975- editor. | Almunawar, Mohammad Nabil, 1960- editor. | Anshari, Muhammad, 1977- editor.
Title: Developing skills and competencies for digital and green transitions / edited by Patricia Ordóñez de Pablos, Mohammad Nabil Almunawar, Muhammad Anshari.
Description: Hershey, PA : Information Science Reference, [2024] | Includes bibliographical references and index. | Summary: "The book will explore how green education and advanced information technologies can be applied in the education sector to create more resilient, climate neutral and green economies and societies. Chapters will discuss how green education and digital technologies can boost the creation of new jobs, build circular economies and societies and increase the inclusiveness of citizens. Impact: Thanks to its innovative topics (education, circular economy, digital tools, green economy, climate neutral economy) and approaches, geographical focus (international focus, with special attention to Asian region) and methodologies (empirical studies, case studies and comparative studies), the book will be a reference for students (undergraduate and postgraduate), academics, experts on IT, CEOs, policy makers and other relevant stakeholders around the world"-- Provided by publisher.
Identifiers: LCCN 2023016889 (print) | LCCN 2023016890 (ebook) | ISBN 9781668490891 (h/c) | ISBN 9781668490938 (s/c) | ISBN 9781668490907 (ebook)
Subjects: LCSH: Environmental protection--Vocational guidance. | Core competencies. | Computer literacy.
Classification: LCC TD170.2 .D48 2024 (print) | LCC TD170.2 (ebook) | DDC 363.70023--dc23/eng/20230824
LC record available at https://lccn.loc.gov/2023016889
LC ebook record available at https://lccn.loc.gov/2023016890

This book is published in the IGI Global book series Advances in Educational Technologies and Instructional Design (AETID) (ISSN: 2326-8905; eISSN: 2326-8913)

British Cataloguing in Publication Data
A Cataloguing in Publication record for this book is available from the British Library.
All work contributed to this book is new, previously-unpublished material.

For electronic access to this publication, please contact: eresources@igi-global.com.

Advances in Educational Technologies and Instructional Design (AETID) Book Series

Lawrence A. Tomei
Robert Morris University, USA

ISSN:2326-8905
EISSN:2326-8913

MISSION

Education has undergone, and continues to undergo, immense changes in the way it is enacted and distributed to both child and adult learners. In modern education, the traditional classroom learning experience has evolved to include technological resources and to provide online classroom opportunities to students of all ages regardless of their geographical locations. From distance education, Massive-Open-Online-Courses (MOOCs), and electronic tablets in the classroom, technology is now an integral part of learning and is also affecting the way educators communicate information to students.

The **Advances in Educational Technologies & Instructional Design (AETID) Book Series** explores new research and theories for facilitating learning and improving educational performance utilizing technological processes and resources. The series examines technologies that can be integrated into K-12 classrooms to improve skills and learning abilities in all subjects including STEM education and language learning. Additionally, it studies the emergence of fully online classrooms for young and adult learners alike, and the communication and accountability challenges that can arise. Trending topics that are covered include adaptive learning, game-based learning, virtual school environments, and social media effects. School administrators, educators, academicians, researchers, and students will find this series to be an excellent resource for the effective design and implementation of learning technologies in their classes.

COVERAGE

- Adaptive Learning
- Game-Based Learning
- Virtual School Environments
- Curriculum development
- E-Learning
- K-12 Educational Technologies
- Hybrid Learning
- Instructional Design
- Digital Divide in Education
- Collaboration Tools

IGI Global is currently accepting manuscripts for publication within this series. To submit a proposal for a volume in this series, please contact our Acquisition Editors at Acquisitions@igi-global.com or visit: http://www.igi-global.com/publish/.

The Advances in Educational Technologies and Instructional Design (AETID) Book Series (ISSN 2326-8905) is published by IGI Global, 701 E. Chocolate Avenue, Hershey, PA 17033-1240, USA, www.igi-global.com. This series is composed of titles available for purchase individually; each title is edited to be contextually exclusive from any other title within the series. For pricing and ordering information please visit http://www.igi-global.com/book-series/advances-educational-technologies-instructional-design/73678. Postmaster: Send all address changes to above address.

Titles in this Series

For a list of additional titles in this series, please visit: http://www.igi-global.com/book-series/

Empowering Students Through Multilingual and Content Discourse
Stacie Lynn Finley (Missouri State University, USA) Pamela Correll (Missouri State University, USA) Cathy Pearman (Missouri State University, USA) and Stephanie Huffman (Missouri State University, USA)
Information Science Reference • © 2023 • 340pp • H/C (ISBN: 9798369305430) • US $215.00

Perspectives on Indigenous Pedagogy in Education Learning From One Another
Sheila Cote-Meek (York University, Canada) and Taima Moeke-Pickering (Laurentian University, Canada)
Information Science Reference • © 2023 • 279pp • H/C (ISBN: 9781668434253) • US $215.00

Implementing Rapid E-Learning Through Interactive Materials Development
Mohammad Issack Santally (University of Mauritius, Mauritius) Yousra Rajabalee (Mauritius Institute of Education, Mauritius) and Ravi Rajputh (University of Mauritius, Mauritius)
Information Science Reference • © 2023 • 232pp • H/C (ISBN: 9781668449400) • US $215.00

Perspectives on Enhancing Learning Experience Through Digital Strategy in Higher Education
Afzal Sayed Munna (University of Sunderland in London, UK) Vipin Nadda (University of Sunderland in London, UK) Theo Ammari Allahyari (University of Sunderland in London, UK) Giuseppe Cantafio (University of Sunderland in London, UK) and Sahidi Bilan (University of Sunderland in London, UK)
Information Science Reference • © 2023 • 325pp • H/C (ISBN: 9781668482827) • US $215.00

Inclusive Phygital Learning Approaches and Strategies for Students With Special Needs
Efthymia Efthymiou (Zayed University, UAE)
Information Science Reference • © 2023 • 357pp • H/C (ISBN: 9781668485040) • US $215.00

IGI Global
PUBLISHER of TIMELY KNOWLEDGE

701 East Chocolate Avenue, Hershey, PA 17033, USA
Tel: 717-533-8845 x100 • Fax: 717-533-8661
E-Mail: cust@igi-global.com • www.igi-global.com

Editorial Advisory Board

List of Reviewers

Iris-Panagiota Efthymiou, *LABHEM University of Pireaus, Greece*

Table of Contents

Detailed Table of Contents

Digital competence is one of the most fundamental competences pre-service teachers should have. Training programs have been carried out to support digital competence. It is also important to examine how pre-service teachers plan to use the knowledge and skills that they have gained from the digital competence training in different contexts. Thus, this case study examined the pre-service teachers' perspectives on personal, pedagogical, and professional digital competence in digital competence training. The study was conducted with 30 pre-service teachers. As a result, the digital competence training supported pre-service teachers to develop a perspective for reaching their personal, pedagogical, and professional goals through activating their information and data literacy, communication, and collaboration, digital content creation, safety, and problem-solving competence areas.

Online learning has become a widely adopted mode of education, particularly during the COVID-19 pandemic. In general, individuals with disabilities face challenges when using non-technology components for studying. This chapter

has a direct relationship with training. The professional quality and the perceived satisfaction vary from levels to levels in a healthcare organization. Hence, conducting job profiling is important to identify skills across the hierarchy. This research article aims to explore the Employability Skills required by the professionals working in the healthcare sector at Medical Service Vertical, further, the utility of each skill is identified and then their relative importance is studied at each hierarchical level. Digital technology and sustainable manpower can boost the creation of new jobs.

This research aims to evaluate current and desired conditions, develop innovative learning methods, and measure their effectiveness in enhancing Thai communication competencies among teachers in the 21st century. The core competencies include: Principles of the Thai language, listening, speaking, reading, and writing; the use of the language; communication etiquette; Thai language aesthetics; moral and communication ethics. The results showed that teachers' knowledge and skill competencies were at a moderate level, but their desire for improvement was high. Data analysis evidenced the effectiveness of the Thai language learning innovation; indeed, the posttest score of the teachers' Thai communication competency assessment was statistically significantly higher (70%) than the benchmark (p < .05).

Digital skills and competencies are becoming increasingly important in today's rapidly changing labor market in Indonesia. The adoption of new technologies and the increasing digitalization of work processes have created a significant demand for individuals who possess the necessary digital skills and competencies to succeed in the modern workplace. Many industries in Indonesia are rapidly adopting digital technologies to improve productivity and efficiency. This means that workers lacking digital skills and competencies are disadvantaged in the job market and may struggle to find employment. As more tasks become automated, workers who possess digital skills and competencies will be better positioned to adapt to new job requirements and remain employable. The growth of the digital economy in Indonesia is creating new job opportunities in fields such as e-commerce, digital marketing, and data analytics. Workers with digital skills and competencies will be better positioned to take advantage of these opportunities and succeed in these industries.

The purpose of the study reported on in this chapter was to investigate the extent to which educators were developing skills and competencies for their transitions to using digital or Information and Communication Technologies (ICTs) at a specific school in the greater Johannesburg area, Gauteng province, South Africa. Against the background of Developing Skills and Competencies for Digital and Green Transitions, the objectives of this study also included establishing to what extent the school was ready to implement such technologies in teaching and learning and hence be characterized as an e-school.

The issues of mismatching between the feedback and the strategies have deteriorated most businesses in their transitions. The purpose of this chapter is thus to provide an instant digital marketing transition from consumers' feedback to providers' strategies. The three theories of action regulation, actor-network, and motivation are used to direct the solutions and recommendations whereby the feedback has been clustered by the digital Linguistic Inquiry and Word Count into 12 words and transited to 12 marketing strategies. Findings indicated that the feedback words related to "body", "health", "risk", "work", "friend", "social", "adjectives", "family", "biological" "drives", "affect", and "positive emotion" have been transited to physical handling, fact finding, packaging, product planning, distribution, display, advertising, personal selling, branding, servicing, pricing, and promotion, respectively. Solutions and recommendations have been discussed.

This study examines emerging trends in organizational learning and their profound impact on modern businesses. In an era marked by rapid technological developments, globalization, and changing customer expectations, businesses must constantly adapt to remain competitive. Organizational learning as a strategic tool plays a crucial role

This study aims to identify the areas in which Enterprises need to focus for improving their level of readiness and develop strategies and plans to adopt Industry 4.0 technologies successfully. 219 samples were collected using snowball sampling from Marathwada District in Maharashtra, India. factor analysis was conducted using SPSS and different factors acting as barriers to implementation of Industry 4.0 technologies were identified.

Green supply chain concept is used to reduce environmental degradation and emissions of air, water, and waste by incorporating green practices into business operations. Growing impacts of global warming, climate change, waste, and air pollution problems have prompted experts all over the world to think more environment friendly and find the best possible approach for "Green" solutions. green supply chain management is one of the factors that motivates organizations to be more sustainable. This study focuses on the green supply chain management in the auto parts industry in South Indian. Data from three green initiatives: recyclable packaging, green warehouse management and milk run approach for logistics is taken and compared with non-green approaches. It is found that there is significant reduction in costs by adopting the green approaches. With environmental issues growing all the time, green supply chain deserves to be a long-term community concern in developing countries.

In this paper, an attempt has been made to look into the effects of external shock generated by Pandemic on the industrial output growth in India by analysing the macroeconomic time series data generating process. The authors have carried out a number of tests concerning the stochastic behaviour and nature of trend of the annual industrial output growth at national level as well as for some major states for the period 1989-90 to 2020-21 locating possible structural break in the series. The authors observe that average annual growth rate of industrial production has decreased significantly in every sector in the lockdown period as compared to pre lockdown period. Testing of unit root hypothesis confirms the presence of stochastic

Preface

INTRODUCTION

Digital literacy is critical to develop inclusive and resilient societies and economies. Educational technologies and digital skills and competences are key for the green economic transition. Education and training systems are crucial to support the transformation of societies and economies in the digital age and boost these skills and competences. Students, workers and citizens need to develop digital skills and competences to live and work in the digital economy. All relevant stakeholders in the field of education must foster active dialogue and exchange on digital literacy and green education as well as focus on priorities to support an innovative digital education system (Almunawar *et al.*, 2022, 2023; European Commission, 2023a,b,c).

In the case of the European Union, the Digital Education Action Plan (2021-2027) supports the transformation of the traditional education and training systems to adapt to the new labour market and digital era. Additionally it aims to provide digital and inclusive education for EU students and citizens, with special focus on digital, green skills and competences (European Commission, 2023 d).

The book aims to foster international dialogue of approaches, methodologies and tools for the transformation of education in the digital era, with special focus on green education and sustainability. It will broaden the visibility of new digital technologies for education, create an international scientific platform for collaboration on digital and green education and accelerate innovation in educational technologies.

OBJECTIVES

The book explores how green education and advanced information technologies can be applied in the education sector to create more resilient, climate neutral and green economies and societies. Chapters discuss how green education and digital technologies can boost the creation of new jobs, build circular economies and societies and increase the inclusiveness of citizens.

Thanks to its innovative topics (education, circular economy, digital tools, green economy, climate neutral economy) and approaches, geographical focus (international focus, with special attention to Asian region) and methodologies (empirical studies, case studies and comparative studies), the book will be a reference for students (undergraduate and postgraduate), academics, experts on IT, CEOs, policy makers and other relevant stakeholders around the world.

STRUCTURE AND CONTENTS OF THE BOOK

The books present a collection of 13 chapters. The first section of the book *Developing Skills and Competencies for Digital and Green Transitions*,

Section I: Digitals skills and competences for the new labour market", has a collection of 7 chapters.

Chapter 1 titled **"**Does Context Matter? Examining Personal, Pedagogical, and Professional Digital Competence**"** (by Ayça Çebi, Canan Çolak, İlknur Reisoğlu, & Tuğba Bahçekapılı Özdemir) highlights "Digital competence is one of the most fundamental competences pre-service teachers should have. Training programs have been carried out to support digital competence. It is also important to examine how pre-service teachers plan to use the knowledge and skills that they have gained from the digital competence training in different contexts. Thus, this case study examined the pre-service teachers' perspectives on personal, pedagogical, and professional digital competence in digital competence training. The study was conducted with 30 pre-service teachers. As a result, the digital competence training supported pre-service teachers to develop a perspective for reaching their personal, pedagogical, and professional goals through activating their information and data literacy, communication, and collaboration, digital content creation, safety, and problem-solving competence areas".

Chapter 2 titled **"**Equitable and Inclusive Online Learning- A Framework for Supporting Students with Disabilities: Online Learning for Disabled Learners" (by Jayapriya J & Vinay M) discusses Online learning has become a widely adopted mode of education, particularly during the COVID-19 pandemic. In general, individuals with disabilities face challenges when using non-technology components for studying. This chapter proposes a framework for equitable and inclusive online learning practices that support students with disabilities. The framework is based on a review of current research and best practices for online learning and disability accommodations. The framework emphasizes a collaborative, student-centered approach to online learning that acknowledges the unique needs and experiences of students with disabilities. Depending on the disabilities, the framework is divided into two phases namely: Prevalent Learning, and Discrete Learning. The former comprised components: Accessibility, Accommodation, and Engagement, and later has components like Methodology, Evaluation. The framework proposed provides a roadmap for addressing the challenges faced by students with disabilities in online learning environments".

Chapter 3 titled "Youth Aspirations Towards Industry 4.0 Job Requirements - The Example of Serbian Labor Market" (by Mihajlo Dusan Djukic & Dejana Pavlovic) analyzes "A push towards digitalization and a solid inflow of foreign investments has provided a basis to create new jobs and facilitate a smoother transition of youth into the Serbian labor market. Although still inordinately high when compared to its EU counterparts, the unemployment rate youth has decreased over the last decade following improved macroeconomic trends. Further positive outcomes necessitate targeted, labor market policies focused on strengthening employability - of interest to research are changes in the external environment (particularly Industry 4.0 requirements) and deficiencies caused by rigidity in formal education. Taking into consideration youth career aspirations and capacities, a Principal Component Factor Analysis was applied to data collected from the survey of youth (18-29 years old, n=125). The results help provide conclusions on youth attitudes related to learning preferences, additional skills required and the self-estimation of their digital-skill capacities. The analysis extracts five factors relevant for developing effective active labor market measures".

Chapter 4 titled "Profiling Industry-Relevant Employability Skills A Study on Healthcare Industry in India" (by Swati sisodia & Sumaira Jan) discusses "The demand for skilled and qualified manpower in India is likely to remain elevated and robust in the coming years, both nationally and internationally. To encourage, continuous development of employees' regular competency mapping of skills with changing job roles is essential to create a productive and expert employee. Competitiveness is closely related to the concept of employee performance, which has a direct relationship with training. The professional quality and the perceived satisfaction vary from levels to levels in a healthcare organization. Hence, conducting job profiling is important to identify skills across the hierarchy. This research article aims to explore the Employability Skills required by the professionals working in the healthcare sector at Medical Service Vertical, further, the utility of each skill is identified and then their relative importance is studied at each hierarchical level. Digital technology and sustainable manpower can boost the creation of new Jobs".

Chapter 5 titled "Learning Innovations for Developing Teachers' Competencies in Thai Communication in the 21st Century" (by Rungarun Rojrattanadamrong Chaisri, Suppawan Satjapiboon, & Adisorn Rueangkitchanan) analyzes "This research aims to evaluate current and desired conditions, develop innovative learning methods, and measure their effectiveness in enhancing Thai communication competencies among teachers in the 21st century. The core competencies include: Principles of the Thai language, listening, speaking, reading, and writing; the use of the language; communication etiquette; Thai language aesthetics; moral and communication ethics. The results showed that teachers' knowledge and skill competencies were at a moderate level, but their desire for improvement was high. Data analysis evidenced the effectiveness of the Thai language learning innovation; indeed, the posttest score of the teachers' Thai communication competency assessment was statistically significantly higher (70%) than the benchmark".

Chapter 6 titled "Digital Skills and Competencies for a Changing Labor Market in Indonesia" (by Hani Gita Ayuningtias) examines "Digital skills and competencies are becoming increasingly important in today's rapidly changing labor market in Indonesia. The adoption of new technologies and the increasing digitalization of work processes have created a significant demand for individuals who possess the necessary digital skills and competencies to succeed in the modern workplace. Many industries in Indonesia are rapidly adopting digital technologies to improve productivity and efficiency. This means that workers lacking digital skills and competencies are disadvantaged in the job market and may struggle to find employment. As more tasks become automated, workers who possess digital skills and competencies will be better positioned to adapt to new job requirements and remain employable. The growth of the digital economy in Indonesia is creating new job opportunities in fields such as e-commerce, digital marketing, and data analytics. Workers with digital skills and competencies will be better positioned to take advantage of these opportunities and succeed in these industries".

Chapter 7 titled "Developing Skills and Competencies for Educators' Transitions to Using Digital Technologies: An e-School in Johannesburg, Gauteng, South Africa" (by Peter Kisembo & Leila Goosen) discusses "The purpose of the study reported on in this chapter was to investigate the extent to which educators were developing skills and competencies for their transitions to using digital or Information and Communication Technologies (ICTs) at a specific school in the greater Johannesburg area, Gauteng province, South Africa. Against the background of *Developing Skills and Competencies for Digital and Green Transitions*, the objectives of this study also included establishing to what extent the school was ready to implement such technologies in teaching and learning and hence be characterized as an e-school".

Section II "Digital technologies and competitiveness in the green economy" has 6 chapters.

Chapter 8 titled "Developing Consumers' Competencies for Digital Marketing Transitions" (by Xuan Tran) examines "The issues of mismatching between the feedback and the strategies have deteriorated most businesses in their transitions. The purpose of this chapter is thus to provide an instant digital marketing transition from consumers' feedback to providers' strategies. The three theories of action regulation, actor-network, and motivation are used to direct the solutions and recommendations whereby the feedback has been clustered by the digital Linguistic Inquiry and Word Count into 12 words and transited to 12 marketing strategies. Findings indicated that the feedback words related to "body", "health", "risk", "work", "friend", "social", "adjectives", "family", "biological" "drives", "affect", and "positive emotion" have been transited to physical handling, fact finding, packaging, product planning, distribution, display, advertising, personal selling, branding, servicing, pricing, and promotion, respectively. Solutions and recommendations have been discussed".

Chapter 9 titled "New Trends in Organizational Learning and Its Impact on Business" (by Ayşe Meriç Yazıcı) discusses "Firms need to be able to understand and capture the significant changes in teaching and learning environment in organizational education and envision the tech-embedded potential to build and advance competencies in the age of digitalization and automation when adaptative learning organizational structures promoting a diverse workforce becomes a norm. Some of the new trends in organizational learning include artificial intelligence, online learning, blended learning, microlearning, continuous learning, cloud learning, gamification, adaptive learning, technological learning, m-learning, modular learning, and smart working. This study aims to advocate the adoption of various organizational learning approaches hoping to shed light on the new learning trend in businesses".

Chapter 10 titled "Volume Growth Advocacy and Sustainability Options in Sikkim Tourism in Post-pandemic Scenario" (by Supriya Dam) highlights "This study explores a dichotomy between the volume growth advocacy and survival options available to stakeholders in the post-pandemic scenario, applying sustainability principles in tourism in Sikkim. Pandemic inflicted respite from the volume growth advocacy model provided much-needed time to restore sanity, given the ecological fragility of the state. Scenario analysis of actual space on offer for tourism and host-guest ratio suggests the destination has embraced tourists beyond the acceptable limits. Per capita, municipal solid waste generation per day has left a lasting impact on land and masses. Mushrooming of concrete structures across urban centers exposes poorly planned arrangements for tourism facilities over sustainability considerations. Again, the projected demand for accommodation infrastructure and adjoining services is inadequate to meet spiraling demand. The study advocates for micro tourism besides practicing high-value and low-impact activities and spot-specific plans to strike a chord between volume growth advocacy and sustainability options".

Chapter 11 titled "Study of Micro and Small Enterprise's Readiness in Implementing Industry 4.0: A study in Marathwada District of Maharashtra, India" (by Saibal Kumar Saha, Janhavi Ravindra Deshmukh, & Ramakrishnan N) examines "Industry 4.0 aims tp transform the development of global value chains and the development of a digital revolution, with intelligent machines capable of communicating via wireless connections and a connection thought system, resulting in autonomous decision-making. Although large sized firms are adopting Industry 4.0, the small and micro enterprises are facing great difficulties in adopting them. This study aims to identify the areas in which Enterprises need to focus for improving their level of readiness and develop strategies and plans to adopt Industry 4.0 technologies successfully. 219 samples were collected using snowball sampling from Marathwada District in Maharashtra, India. factor analysis was conducted using SPSS and different factors acting as barriers to implementation of Industry 4.0 technologies were identified".

Chapter 12 titled "Application of Green Logistics in Supply Chain of Auto Parts: A South Indian Scenario" (by Saibal Kumar Saha, Rashmi S, & Siva Kumar Pujari) mentions "Green supply chain concept is used to reduce environmental degradation and emissions of air, water, and waste by incorporating green practices into business operations. Growing impacts of global warming, climate change, waste, and air pollution problems have prompted experts all over the world to think more environment friendly and find the best possible approach for "Green" solutions. green supply chain management is one of the factors that motivates organizations to be more sustainable. This study focuses on the green supply chain management in the auto parts industry in South Indian. Data from three green initiatives: recyclable packaging, green warehouse management and milk run approach for logistics is taken and compared with non-green approaches. It is found that there is significant reduction in costs by adopting the green approaches. With environmental issues growing all the time, green supply chain deserves to be a long-term community concern in developing countries".

Finally, Chapter 13 titled "Impact of Covid-19 Pandemic on Industrial Recession in India and the Emerging Structural Breaks - Testing Unit Root Hypothesis" (by Asim Kumar Karmakar & Tushar Das) highlights in this chapter, an attempt has been made to look into the effects of external shock generated by Pandemic on the industrial output growth in India by analysing the macroeconomic time series data generating process. We have carried out a number of tests concerning the stochastic behaviour and nature of trend of the annual industrial output growth at national level as well as for some major states for the period 1989-90 to 2020-21 locating possible structural break in the series. We observe that average annual growth rate of industrial production has decreased significantly in every sector in the lockdown period as compared to pre lockdown period. Testing of unit root hypothesis confirms the presence of stochastic trend in the logarithmic values of industrial output series both at the national and at the state level. The total production declined sharply during the lockdown and the structural break towards the fall in industrial production appeared before the lockdown. Hence, it seems that the lockdown by the pandemic may not be the primary source of industrial recession in India".

REFERENCES

Almunawar, M. N., Islam, Z., & Ordóñez de Pablos, P. (2022). *Digital Transformation Management: Challenges and Futures in the Asian Digital Economy*. Routledge. doi:10.4324/9781003224532

Almunawar, N., Ordóñez de Pablos, P., & Anshari, M. (2023a). *Sustainable Development and the Digital Economy: Human-centricity, Sustainability and Resilience in Asia*. Routledge. doi:10.4324/9781003388753

Almunawar, N., Ordóñez de Pablos, P., & Anshari, M. (2023b). *Digital Transformation for Business and Society: Contemporary Issues and Applications in Asia*. Routledge.

European Commission. (2023a). Green Transition. Accessed on July 27, 2023 https://reform-support.ec.europa.eu/what-we-do/green-transition_en

European Commission. (2023b). A European Green Deal Striving to be the first climate-neutral continent. Accessed on July 27, 2023 https://commission.europa.eu/strategy-and-policy/priorities-2019-2024/european-green-deal_en

European Commission. (2023c). Digital Education Action Plan (2021-2027). Accessed on July 27, 2023 https://education.ec.europa.eu/focus-topics/digital-education/action-plan

European Commission. (2023d). Communication from the Commission to the European Parliament, The Council, The European Economic And Social Committee And The Committee Of The Regions Digital Education Action Plan 2021-2027 Resetting Education And Training For The Digital Age. Accessed on July 27, 2023 https://eur-lex.europa.eu/legal-content/EN/TXT/?uri=CELEX%3A52020DC0624

Section 1

Digital Skills and Competences for the New Labour Market

Chapter 1
Does Context Matter?
Examining Personal, Pedagogical, and Professional Digital Competence

Ayça Çebi
Trabzon University, Turkey

Canan Çolak
Giresun University, Turkey

İlknur Reisoğlu
https://orcid.org/0000-0002-6485-254X
Recep Tayyip Erdogan University, Turkey

Tuğba Bahçekapılı Özdemir
https://orcid.org/0000-0003-3339-7756
Trabzon University, Turkey

ABSTRACT

Digital competence is one of the most fundamental competences pre-service teachers should have. Training programs have been carried out to support digital competence. It is also important to examine how pre-service teachers plan to use the knowledge and skills that they have gained from the digital competence training in different contexts. Thus, this case study examined the pre-service teachers' perspectives on personal, pedagogical, and professional digital competence in digital competence training. The study was conducted with 30 pre-service teachers. As a result, the digital competence training supported pre-service teachers to develop a perspective for reaching their personal, pedagogical, and professional goals through activating their information and data literacy, communication, and collaboration, digital content creation, safety, and problem-solving competence areas.

DOI: 10.4018/978-1-6684-9089-1.ch001

INTRODUCTION

Nowadays, education systems are in a transformation due to social changes created by digital technologies. In this process, the roles and competences of teachers have been redefined. The rapid spread of digital technologies calls for teachers to use these technologies in their professional lives (Guillén-Gámez et al., 2021). Thus, teachers need to be digitally competent and use digital technologies effectively in learning environments (Tondeur et al., 2018). On the other hand, some environmental challenges arise as a result of technological transformations such as climate change and loss of biodiversity (Bianchi et al., 2022). Thus, teachers need to be aware of the environmental impact of digital technologies and their use as a digitally competent citizen. Also teachers should be role models for their students in using technology safely to protect the environment, such as knowing the environmental impact of energy use and carbon emissions caused by online video viewing or creating digital material to be shared online platforms regarding the sustainable use of digital devices at school and home (Vuorikari, et al., 2022).

Digital competence (DC) is essential for teachers to deal with challenging situations such as ensuring safety in the digital world, assessing the legal consequences of actions carried out through digital tools, or dealing with cyberbullying behaviors of students (Tomczyk, 2019). In addition, teachers need DC to create solutions to solve complex problems that they encounter in the learning environment. The transformation of education during the coronavirus outbreak can be a notable example of this. In fact, the Digital Education Action Plan 2021-2027 announced by the European Commission highlighted that the COVID-19 crisis revealed the importance of having digital competences for digital transformation (European Commission, 2020). Researchers similarly noted that the digital competences of pre-service teachers have become even more important in the pandemic because of the required qualifications during transition to online learning. Beyond the pandemic, pre-service teachers need to develop new skills to sustain their profession in the digital learning environments and to improve the quality of the teaching and learning process by using digital technologies (Damşa et al., 2021). On the other hand, teachers with insufficient digital competencies may have difficulties in using technology to support students with diverse needs (Montenegro-Rueda & Fernández-Batanero, 2022). Therefore, it is important to (re)think about the preparation of future teachers (Flores & Swennen, 2020). Pre-service teachers need to have personal DC in order to participate in a democratic society. Accordingly, pedagogical and professional DC is important for pre-service teachers to design their learning-teaching, assessment, and evaluation activities, to select and develop digital content, and communicate and interact with other stakeholders in the educational process (Gudmundsdottir et al., 2020). Because, using digital technologies in an educational context requires specific competencies that differ from other contexts (Pettersson, 2018). Previous research has revealed that pre-service teachers are not sufficiently trained in terms of pedagogical and professional DC in teacher training programs (Gudmundsdottir & Hatlevik, 2018; Lund et al., 2014). It is stated that pre-service teachers do not have sufficient practical experience in the pedagogical use of digital technologies in the classroom environment, in the evaluation of the use of digital technologies in accordance with learning objectives, and in the use of digital technologies in the realization of administrative procedures at school. It is also noted that pre-service teachers do not acquire sufficient knowledge and experience in the development of content in different formats, behavioral norms in the use of digital media, and collaborative work through technology (Gudmundsdottir & Hatlevik, 2018). Accordingly, there is a need to examine different types of DC of pre-service teachers (Tondeur et al., 2018).

On the other hand, most of the studies are conducted with teachers or pre-service teachers by considering demographic variables such as attitudes towards DC, gender, and discipline (Ottenbreit-Leftwich et al., 2018; Tondeur et al., 2018). In addition, these studies generally focus on how DC training should be delivered (Reisoğlu & Çebi, 2020; Røkenes & Krumsvik, 2016), DC definitions (Lund et al., 2014), perceptions of DC (Alarcón et al., 2020; Guillén-Gámez et al., 2021; Ramírez-Montoya et al., 2017), DC conceptions (List et al., 2020), and DC training outcomes (Brevik et al., 2019; Gudmundsdottir et al., 2020; Ramírez-Montoya et al., 2017; Reisoğlu & Çebi, 2020). However, few studies adequately examine how pre-service teachers plan to use the competencies they acquire from the DC training in relation to personal, pedagogical, and professional DC contexts. In this regard, this study aims to explore the perspectives of pre-service teachers in a DC training program regarding personal, pedagogical, and professional DC.

BACKGROUND

Competence is much more than knowledge and skills; it is the ability to meet demands by mobilizing psychosocial resources in a given context (OECD, 2005). It is also defined as personal development, social inclusion, and the knowledge, skills and attitudes needed for employment (Guillén-Gámez et al., 2020). Digital competence is more than using a digital software or device; it involves the use of a wide range of complex skills such as cognitive, psychomotor, sociological and emotional in the use of digital environments (Ilomäki et al, 2011). Along with this, digital competence not only enables the use of the opportunities offered by digital technologies, but also helps to cope with their disadvantages (Napal-Fraile et al., 2018).

To date, frameworks and models have emerged to determine and develop the digital competence of educators, such as DigiLit Leicester (Fraser et al., 2013), Professional Digital Competence Framework for Teachers (INTEF, 2017), Digital competence building blocks (Janssen et al., 2013), Three pillar models of professional digital competence (Ottestad et al., 2014), Digital competence (Krumsvik, 2014). One of the most recent and comprehensive frameworks among them is DigComp (Siddiq et al., 2016), which was developed by the European Commission Joint Research Center's Prospective Technological Studies Institute to better understand and develop digital competence in Europe (Vuorikari et al., 2016). Unlike other frameworks that consider 21st century skills, DigComp addresses metacognitive skills (Janssen et al., 2013). It also provides information on how digital competence is reflected in curricula, teaching and academic studies (Ferrari, 2012). This framework consists of five different digital competence areas which are "information and data literacy", "communication and collaboration", "digital content creation", "safety" and "problem solving" (Ferrari, 2012; Ilomäki et al., 2016). Each digital competence domain includes competencies related to thinking, acting, learning and teaching with technology, as well as technical skills (Tømte, 2013). Teacher professional development programs in various European countries have adopted DigComp for the purpose of structuring teachers' digital competencies (Vuorikari et al., 2016).

Pre-service teachers are expected to transform and improve their classroom practice and support their professional development by activating their knowledge, skills, and attitudes regarding using digital technologies (Hall et al., 2014). In addition to making use of digital technologies for their personal needs, they are asked to be role models for the use of these technologies and to consider how they can benefit from the potential of digital technologies (Krumsvik, 2008). Eventually, pre-service teachers should have personal, pedagogical, and professional DC at a level that can produce solutions to the situations they face (Lund et al., 2014). Starkey (2020) states that although personal, pedagogical, and professional DC areas overlap at some point, they aim to serve the different aspects of teacher education. Thus, this study considers the DC in three dimensions- personal, pedagogical, and professional competences.

Personal Digital Competence

Personal DC includes using digital technologies to solve problems encountered in daily life contexts. It involves DC which citizens must develop for a successful life in a digital society (Caena & Redecker, 2019). According to Ilomäki et al. (2016), "*personal digital competence consists of the skills and practices required to use new technologies in a meaningful way and as a tool for learning, working and leisure time, understanding the essential phenomena of digital technologies in society as well as in one's own life, and the motivation to participate in the digital world as an active and responsible actor.*" (p. 671).

However, personal DC does not include the knowledge and skills required for integrating digital technologies into learning-teaching processes and ensuring professional development for teachers (Fraillon et al., 2014). Personal DC can form the basis for integrating technology with pedagogical methods and strategies, providing institutional communication and collaboration, reflecting on teaching and learning, and professional development. For example, pre-service teachers who can benefit from digital technologies for communication purposes in daily life may not have any problems communicating with colleagues, students, and parents in their professional life. Therefore, it is stated that professional and pedagogical competences related to the teaching profession cover more complex knowledge and skills than personal DC (Instefjord & Munthe, 2016).

Pedagogical Digital Competence

Pedagogical DC includes complex and holistic competences for using digital technologies with pedagogical judgments in the educational context (Krumsvik, 2008). According to Olofsson et al. (2017), pedagogical DC is related to the ability of teachers to plan, conduct, continually evaluate, and revise learning-teaching activities supported by digital technologies. It includes creating and managing learning, teaching, assessment, evaluation, and classroom management activities in line with learning objectives on different topics (Gudmundsdottir & Hatlevik, 2018; Ottestad et al., 2014) and providing students access to learning resources (Lund et al., 2014). Pre-service teachers are expected to evaluate the potential of digital resources and integrate them with their learning objectives to achieve optimum learning outcomes (Brevik et al., 2019).

Professional Digital Competence

Professional DC involves the knowledge and skills of pre-service teachers that are outside the teaching activities but related to the teaching profession (Gudmundsdottir et al., 2020). It includes the use of digital technologies to communicate effectively with colleagues, students, families, and others (Gudmundsdottir & Hatlevik, 2018). Professional DC also indicates individual or group reflections about teaching practices, evaluating, and developing the effectiveness and suitability of teaching strategies (Caena & Redecker, 2019). Similarly, teachers' continuous use of opportunities to improve themselves in the use of digital technology for educational purposes was identified as professional DC (Gudmundsdottir et al., 2020; Gudmundsdottir & Hatlevik, 2018).

Purpose of The Study

This study was conducted to understand the perspectives that pre-service teachers gained in terms of personal, pedagogical, and professional DC as they were participating in a DC training. It is because if pre-service teachers gain a multi-faceted perspective rather than focusing on the technological dimension of DC, they can integrate more comprehensive digital activities into learning and teaching practices, and enable students to make critical reasoning about digital technologies (List et al., 2020). Studies have been carried out to determine the personal, pedagogical, and professional competences of pre-service teachers. However, it has not been investigated what kind of perspectives DC training brings to pre-service teachers for these three components. For this purpose, this research seeks to address the following research question:

What are pre-service teachers' perspectives on (a) personal DC? (b) pedagogical DC? (c) professional DC regarding the knowledge and skills they gained from the DC training?

METHOD

This research was carried out as a case study that examined the DC perspectives of pre-service teachers from various disciplines with an in-depth and holistic perspective. "*The case study offers a means of investigating complex social units consisting of multiple variables of potential importance in understanding the phenomenon.*" (Merriam, 2009, p.50). The case in this study is limited to the perspectives of pre-service teachers from the training program implemented to improve their DC.

Context

The DC training program carried out in this study was part of a funded project. Participants gave written informed consent to participate in the research as a part of the project activity. The grant covered accommodation costs, daily expenses as well as material purchases required for DC activities. In order to improve pre-service teachers' current DC, face-to-face training was structured based on the DigComp framework (Vuorikari et al., 2022). The framework was developed primarily to define and exemplify the knowledge and skills that citizens should have in the digital age. This framework consists of five DC areas which are "information and data literacy", "communication and collaboration", "digital content creation", "safety" and "problem-solving" (Vuorikari et al., 2022).

The training continued for six days during the semester break. Within the scope of the training program, a total of 51-hour training was provided for pre-service teachers through five modules: information and data literacy, communication and collaboration, digital content creation, safety, and problem-solving. Trainers prepared the training modules in cooperation with the researchers of the study in accordance with the personal, pedagogical, and professional DC. Each module started with theoretical information about five DC areas mentioned in the DigComp framework. Pre-service teachers were given information on how to integrate this theoretical information into their teaching with the help of sample classroom activities. Afterward, pre-service teachers were divided into groups of three, with at least one person in each group having experience with digital technologies. In each module, it was tried to ensure that pre-service teachers put their theoretical knowledge into practice by working in cooperation with their group members and exchanging ideas about how they can use this information in their own teaching areas. In addition, each group developed different types of digital learning materials (infographic, animation, augmented reality, and interactive video) using web 2.0 tools for a learning objective they determined.

Participants

The participants of this study consisted of pre-service teachers attending the DC training. Participants were selected among the pre-service teachers who voluntarily applied to the DC training by considering the maximum diversity in terms of the university and the department they are studying. Within the scope of the project budget, a total of 30 pre-service teachers in their 3rd and 4th years in Turkey took part in the study. The ages of the participants ranged between 20 and 35 (M=22.6; SD=3.23). Table 1 shows the demographic information of the participants.

Pre-service teachers studying at the Computer Education and Instructional Technology (CEIT) department are more experienced than the others due to the education they receive in digital technologies and their use in education. Students of CEIT department acquire knowledge and experience in using different digital technologies in education and daily life during their four-year education. In this study, pre-service teachers studying at the CEIT department are included in order to work in collaboration with other pre-service teachers of other departments.

Data Sources and Collection

Data collection tools of the study consist of participant diaries and focus group interview forms. Thus, collecting data from the participants in different ways has supported the credibility of the findings by providing method triangulation (Patton, 1999) and richness of data (McMillan & Schumacher, 2014).

Table 1. Demographic information of participants

Participant	Department	Gender	Participant	Department	Gender
P1	CEIT	Female	P16	Social Sciences Teaching	Female
P2	CEIT	Female	P17	Early Childhood Education	Female
P3	CEIT	Female	P18	Early Childhood Education	Female
P4	CEIT	Male	P19	Science Teaching	Female
P5	CEIT	Male	P20	Science Teaching	Female
P6	CEIT	Female	P21	Science Teaching	Female
P7	CEIT	Female	P22	Science Teaching	Female
P8	CEIT	Male	P23	Special Education	Female
P9	CEIT	Female	P24	Special Education	Female
P10	CEIT	Male	P25	Primary School Teaching	Female
P11	Mathematics Teaching	Male	P26	Primary School Teaching	Female
P12	Mathematics Teaching	Male	P27	Primary School Teaching	Female
P13	Mathematics Teaching	Female	P28	Primary School Teaching	Male
P14	Mathematics Teaching	Female	P29	Turkish Language Teaching	Female
P15	Social Sciences Teaching	Female	P30	Turkish Language Teaching	Male

Firstly, the purpose and the process of the research were introduced to the participants. Then, the participants were informed about what their participation would involve in the research, and they were asked to give their consent. Open-ended questions were prepared for the participant diaries according to the daily topics in the training program. The questions such as "*What are your thoughts on the usability of infographics in education?*" and "*What difficulties did you face and how did you overcome them when developing augmented reality applications?*" were included in the diaries for participants to express their perspectives in detail. Participant diaries were collected through Google forms during the training.

The focus group interview form was developed by the researchers. Two experts working on DCs reviewed the questions in the protocol for clarity and relevance. The questions such as "*How do you evaluate the activities in terms of developing your DC skills?*", "*How do you plan to use the information you have obtained in the activities in your professional life and development?*", and "*How do you intend to use the information you have obtained in your teaching?*" were included in the interview form. Six participants took part in each focus group interview, and these interviews, which lasted an average of 75 minutes, were held after the whole training was completed. A total of five focus group interviews were recorded and later transcribed for data analysis. Interviews followed a standard protocol and question schedule.

Data Analysis

The data obtained through the participant diaries and focus group interviews were first read by the researchers. In the first step, the data were categorised into personal, pedagogical, and professional contexts. Content analysis was employed in this step to analyze the data according to the specified dimensions (Elo & Kyngäs, 2008). Thus, personal, pedagogical, and professional DC emerging from the data were determined. In the second step, for detailed investigation, each of the DC areas was analyzed according to DigComp framework competency areas as information and data literacy, communication and collaboration, digital content creation, safety, and problem-solving. The themes and sub-themes obtained from all these analysis results were summarised, and the intended use of pre-service teachers' DC was exemplified. This process is shown in Figure 1.

Figure 1. Data analysis process

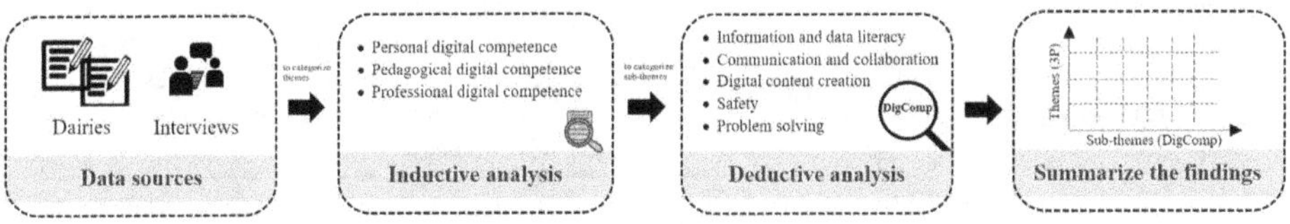

In the analysis of data, the MAXQDA qualitative data analysis program was used (VERBI Software, 2021). Participant diaries and focus group interview data were analyzed separately by two researchers. In addition, some data collected in both types were analyzed by all researchers, and this process continued until a consensus was reached on the codes. To complete the analysis process, the themes, sub-themes, and codes obtained were checked.

FINDINGS

As a result of the content analysis, the perspectives of the participants regarding DC were gathered under personal, pedagogical, and professional dimensions. Each dimension was also categorized according to the information and data literacy, communication and collaboration, digital content creation, safety, and problem-solving competences areas of DigComp.

Personal Digital Competence

When DC is considered in a personal context, pre-service teachers' views are categorized under the themes of information and data literacy, communication and collaboration, digital content creation, safety, and problem-solving. In line with these categories, awareness of digital technologies, knowledge, and self-confidence in using various web 2.0 tools, overcoming digital problems, coping with safety threats, and developing various strategies to reach the information sought was evaluated within the context of personal DC. Personal DC and examples are presented in Table 2 within the scope of DigComp DC areas.

Table 2. Personal DC and intended uses

	Competence areas	Competences	Intended use
Personal DC	**Information and data literacy**	Realizing the ethical use of information	Realizing citing and paying attention to copyrights when information is used.
		Questioning the accuracy of information	Examining the identity of the information (place of publication, date, author, etc.), accessing the primary source, accessing scientific resources, reading comments, using reliable websites
		Using information seeking strategies	Using keywords while searching information, using search operators-filters, using search engines, etc.
	Communication and collaboration	Realizing virtual classroom applications	Data storage, facilitating collaboration, accessing data through various technologies, avoiding viruses
		Using cloud technologies	
	Digital content creation	Realizing digital content creation with various web 2.0 tools	Gaining awareness of web 2.0 tools
		Building digital content creation confidence with various web 2.0 tools	Creating web 2.0 tool usage confidence
	Safety	Realizing digital threats	Identifying safety threats such as fraud, cyberbullying, digital piracy
		Ensuring digital safety	Securing personal data, knowing their rights, knowing strategies to deal with digital threats, taking measures to protect personal data ın digital environments (using virtual cards, not clicking on links, using strong passwords, etc.)
	Problem-solving	Realizing digital problems	Solving the problems encountered in digital environments by themselves (research, trial and error), seeking help from others and consulting with experts
		Developing strategies to solve digital problems	

In terms of information and data literacy competence area, the participants were aware of the need to question the accuracy of information obtained from digital environments and the ethical considerations while using this information. In addition, they gained knowledge of using various strategies such as using keywords, filtering, and using databases while accessing digital information. The comments such as "*We learned about copyright, using free content, and citing properly. I think, in line with this information, my contents will be more qualified and owned by me*" (Diary-P2), and "*...by searching from many sources, confirming the accuracy of the information*" (Diary-P29) support the related findings.

In the communication and collaboration competence area, the participants gained awareness and usage knowledge of various cloud technologies to facilitate cooperation, archive data, access this data through different digital tools, and avoid viruses. This finding is supported by the following comment: "*It is a storage tool that I use frequently in my daily life. I think storing my important information online is more reliable than storing it via a USB or CD*" (Diary-P10). In addition, it was determined that the participants gained awareness about various virtual classroom applications. It was stated that: "*I had heard of Edmodo, but I had never used it. I had a chance to use it here, and I liked it very much...*" (Interview-P2).

Considering the digital content creation competence area, the participants gained awareness in terms of using web 2.0 tools in the development of various digital content such as interactive video, animation, and infographics. In addition, participants gained confidence in developing digital content using web 2.0 tools. These findings are supported by participant comments such as "*I did not know how to use the interactive application and I was not aware of the existence of such applications before. I learned it here for the first time and learned incredibly beautiful apps. I learned the use of applications like Powtoon, Visme, Edpuzzle, Easly, Kahoot, and Nearpad.*" (Diary-P16). "*I think I can improve because before participating in the training, I only knew MorpaCampus that we used in mathematics, for example, they used Powtoon there ... I was wondering how they are developing it. I found it easier after the training*" (Interview-P12).

In the safety competence area, the participants gained digital safety competences in ensuring the safety of personal data, knowing digital rights and responsibilities, developing strategies to deal with digital threats, and taking measures to protect data in digital environments. The participants gained awareness about digital threats such as fraud, cyberbullying, and viruses. At the same time, they learned about the measures to ensure their safety in digital media and online environments. In other words, the participants realised that the disclosure of personal data is a digital safety vulnerability, and they learned about the legal measures or solutions that can be taken in accordance with the Personal Data Protection Law. This finding is in line with the comments such as "*I realized that I had not been taking many precautions.*

I had shared a picture of my identity with a relative. I thought it was okay since no one could see it but learned that someone else could access the messages or pictures we sent" (Diary-P21), and *"...When I reached my phone after class, I first turned off the permissions of the applications. I turned off camera permission, microphone permission, and permission to access my album. Now, I pay attention to these as much as I can"* (Interview-P30).

In the problem-solving competence area, the participants developed strategies to solve digital problems such as watching videos, seeking technical support, getting help from nearby others, and using the help menu to solve possible hardware or software problems. This finding is supported by the participants' comments such as *"I watch videos to solve problems. I use YouTube, Expert TV or Udemy."* (Diary-P7), and *"... if not, I call my friends who are software developers. They help me."* (Interview-P1).

Pedagogical Digital Competence

For which purpose and how pre-service teachers intend to use their DC in the context of education was examined and described as pedagogical DC. Pedagogical DC of pre-service teachers and their purposes of use regarding five competence areas are presented in Table 3.

Considering the area of information and data literacy competence, the participants stated that they would pay particular attention to questioning the accuracy of the information in the evaluation of student assignments, the use of information search strategies, and the ethical presentation of the information while questioning the accuracy and reliability of the information in the content they would prepare for their students. The following comments are in accordance with the findings: *"... At least, we should teach children how to access the right information. For example, if I explain what a safe source is while giving homework, it will be information for them, and they will have access to reliable and correct information"* (Interview-P15), and *"I am thinking of reaching educational resources by using the methods that I have learned to access correct and reliable information"* (Diary-P30).

Regarding the communication and collaboration competence area, it is seen that the competences of the participants to use virtual classrooms and cloud applications come to the fore. In the pedagogical context, the participants intend to use their competences to share classroom resources and activities, increase teacher-student communication, make lessons attractive, support extracurricular activities, facilitate evaluation, make announcements, provide feedback, encourage group work, collect assignments, make online exams, and create a discussion environment. These findings are supported by *"If virtual classroom applications are used in social media style as in Edmodo, students can share information with each other. They can help each other with their homework. Students can share notes with each other. Teachers can also share exam results on this platform with their students"* (Diary-P17), and *"... it would be nice to implement a virtual classroom....we can communicate using one platform such as the virtual classroom. Currently, WhatsApp groups are used to communicate with parents. It is not clear to parents where to check students' homework. I can give them feedback. Before the exam, for example, I can prepare a quiz. I put on my presentations, etc. There is no confusion about where the children will get information. Collectively, everyone can look there. I think I can use it in that way"* (Interview-P15).

Table 3. Pedagogical DC and intended uses

	Competence areas	Competences	Intended use
Pedagogical DC	**Information and data literacy**	Using information searching strategies	Assessing student works and preparing educational content
		Questioning the accuracy of information	
		Ethical use of information	
	Communication and collaboration	Using virtual classroom applications	Sharing classroom resources and activities, increasing teacher-student communication, making subjects attractive, complementing extracurricular activities, facilitating evaluation, making announcements, providing feedback, encouraging group work, collecting assignments, making virtual exams, creating a discussion environment.
		Using cloud technologies	
	Digital content creation	Selecting web 2.0 tools for their intended purpose	Developing content suitable for teaching objectives
		Developing augmented reality applications	Increasing interaction in classroom, attracting attention, ensuring permanence, embodying, assessment and evaluation, supporting the subject, providing feedback, complementing extracurricular activities, addressing individual differences, making the classes enjoyable and creating a safe learning environment.
		Developing interactive videos	
		Developing animations	
		Developing infographic	
	Safety	Knowing the safety measures in digital environments	Informing students about safety measures in digital environments
	Problem-solving	Realizing digital problems	Solving the problems that may arise while performing educational applications in digital environments.
		Developing strategies to solve digital problems	

Considering the digital content creation competence area, the participants stated that they could choose suitable web 2.0 tools while developing various digital content such as interactive video, animation, and infographics and develop digital content using these tools. Participants stated that they would use the digital content developed through web 2.0 tools to increase the interaction in the classroom, attract attention, ensure the retention of the information learned, embody the abstract information, evaluate, to support the subject, provide feedback, to complement extracurricular activities, to address individual differences, to make the classes enjoyable, and to create a safe learning environment. The following participant comments support these findings: "*My discipline is mathematics teaching, and my students cannot visualize solid objects. With augmented reality, I could show 3D versions of those shapes, especially for the areas and volumes of prisms*" (Diary-P11), and "*Especially we want to get feedback from the students at the end of the day. Edpuzzle or Kahoot will be very nice for this purpose ...*" (Interview-P29), and "*Animation should be used in education because students now want to see different materials in the lessons, and I think that the animation will attract more attention when we look at the age group of the audience we are going to teach. I can develop animations that are in parallel with my goal*" (Diary-P6).

Regarding the safety competence area, the participants stated that they would inform their students about the safety measures in digital environments. This finding is supported by the comments such as "*Because we are IT pre-service teachers, our students always come and ask us questions... There may be clickbait, there may be Internet fraud, and they always ask us about similar issues. I think I am now more conscious of these issues. In other words, I think I can raise my students' awareness more in that regard. I can design banners for them. They can hang it in their class. If you encounter something like this, do this. If you receive a message from unknown people in your email, do not open it. I can design such a poster or infographic*" (Interview-P9).

When pedagogical DC in the problem-solving competence area is examined, it is determined that the participants can realize the digital problems they may encounter while performing educational activities in digital environments and develop a strategy to solve these problems. Some participant comments on these findings are as follows: "*I think that I will do research on problem-solving first, then I will evaluate what I find on the Internet and perform the most appropriate solution. I will evaluate it on different web pages I can trust*" (Interview-P9), and "*If we use the programs that record simultaneously while working online, our information will not be lost in case of Internet outage*" (Diary-P13).

Professional Digital Competence

In the professional DC dimension, pre-service teachers' knowledge, and skills, which are out of their teaching activities but related to the teaching profession, were discussed. The professional DC of the pre-service teachers and their intended purposes of using these competences are presented in Table 4.

Table 4. Professional DC and intended uses

	Competence areas	Competences	Intended use
Professional DC	Information and data literacy	Using information search strategies	Accessing up-to-date content
		Questioning the accuracy of information	
	Communication and collaboration	Using virtual classroom applications	Sharing professional knowledge among teachers, ensuring teacher-parent communication
		Using cloud technologies	Digital archiving (storing exam documents and professional materials), collaborating with stakeholders (colleagues, parents)
	Safety	Knowing the safety measures in digital environments	Informing colleagues

In the information and data literacy competence area, it was determined that the participants would use information search strategies and question the accuracy of the information they obtained to reach the current content related to their profession. This finding is supported by the comments such as "*At least I will set an example for my students by questioning the accuracy of the information I will use in my teaching with the methods I have learned here, and transferring it to my students in that way*" (Interview-P30) and "*I will use information seeking strategies to produce content that I intend to use in my own class and to follow up-to-date content*" (Diary-P26).

Regarding the communication and collaboration competence area, it was determined that the participants thought to benefit from virtual classroom applications to provide professional information sharing among teachers and teacher-parent communication. These findings are supported by the comments such as "*...virtual classrooms can actively be involved in teacher-parent interaction. It can be used in terms of sharing child observation forms, obtaining information from parents, and presenting the values that students need to gain in the home environment to parents in this way*" (Diary-P18), and "*In the groups, I created for teachers, I can* *share information about the training to be held, etc. and I can share it on social networks developed for education"* (Diary-P4). In addition, it was determined that the participants believed to benefit from cloud technology to archive digital materials or documents and cooperate with stakeholders. Related to this finding, a participant stated that: *"For example, I can use Google documents for simultaneous work with another teacher friend working in another city or for preparing my plan"* (Diary-P16).

In the safety competence area, the participants reported that they would inform their colleagues about safety measures in digital environments. The following comment supports our finding: *"We can share what we have learned about safety with our colleagues to raise awareness among them"* (Interview-P11).

DISCUSSION

Pre-service teachers' perspectives regarding personal, pedagogical, and professional DC are examined in this study in terms of information and data literacy, communication and collaboration, digital content creation, safety, and problem-solving competences. Starkey and Yates (2021) criticise the DC frameworks for being developed for in-service teachers and for focusing on digital skills regardless of the educational context. In this direction, this study exemplifies the use of a DC framework to prepare pre-service teachers for the digital age.

Personal Digital Competence

In terms of personal DC, pre-service teachers developed perspectives on all DC areas as indicated in the DigComp framework such as questioning the accuracy, and reliability of information obtained from the digital environment, using various cloud technologies to facilitate cooperation, using web 2.0 tools for developing various digital contents, taking measures to protect personal data in digital environments, and developing strategies to solve technical problems. Similarly, Reisoğlu and Çebi (2020) found that thanks to DC training, pre-service teachers gained knowledge and experience in developing content in simple form and improved in using information searching strategies. Also, they found that pre-service teachers improved their knowledge and skills in safety policies for protecting personal information and for recognizing risks and dangers in digital environments. However, when sufficient DC training was not given to pre-service teachers, the pre-service teachers considered themselves to have a low level of DC in the areas of digital content creation (Galindo-Domínguez & Bezanilla, 2021; Gutiérrez-Porlán & Serrano-Sánchez, 2016; Hinojo-Lucena et al., 2019), safety (Gutiérrez-Porlán & Serrano-Sánchez, 2016), and problem-solving (Esteve-Mon et al., 2020). Similarly, the low level of awareness of pre-service teachers about digital safety and need for support in this direction was emphasized (Alnasib, 2023; McGarr & McDonagh, 2021). In this respect, this study can be considered effective in supporting the personal DC of pre-service teachers in the areas of digital content development, safety, and problem-solving.

In terms of personal DC, Çebi et al. (2022) found DC training effective to gain awareness in managing digital identity, re-elaborating digital content, and gaining experience in using digital technologies creatively and innovatively. However, in this study, pre-service teachers indicated insufficient or no comments regarding interaction and collaboration through digital technologies, creating and managing digital identities, digital citizenship, integrating and re-elaborating digital content, and using digital technologies creatively. This result can be explained by inadequate practice due to the time limitation in the DC training (Reisoğlu & Çebi, 2020). On the other hand, pre-service teachers' experiences as a result of the widespread use of digital citizenship applications and frequent use of digital technologies in their daily lives for communication, cooperation, and creating a digital identity may have affected this result. Thus, it will be beneficial to plan digital competency training in such a way that pre-service teachers gain new experience in these knowledge and skills.

Pedagogical Digital Competence

In terms of pedagogical DC, regarding the information and data literacy area, pre-service teachers indicated their views to pay particular attention to questioning the accuracy and reliability of the information in the content they will prepare for their students. In the communication and collaboration competence area, pre-service teachers intend to use their gained knowledge to benefit from virtual classroom and cloud technologies to create effective learning environments. In the digital content creation competence area, pre-service teachers developed a point of view on using digital content to enrich the learning process. Regarding the safety competence area, pre-service teachers gain a better understanding in informing their students about the safety measures in digital environments. In the problem-solving competence area, pre-service teachers realise the digital problems they may encounter in digital learning environments and develop a strategy to solve these problems. Yoon (2022) emphasized that pre-service teachers' information evaluation strategies and digital content development skills are low. Similarly, Tárraga-Mínguez et al. (2021) draw attention to the limited DCs of pre-service teachers to enrich the learning-teaching processes and the need to develop these competences. In this direction, Andreasen et al. (2022) claimed that pre-service teachers need opportunities to actively engage in their learning process to apply pedagogical DC in their classroom teaching. According to Drotner (2020), digital content creation is an essential skill for teachers in the 21st century. López-Belmonte et al. (2019) pointed out that having greater competence in the creation of digital content contributes to making greater use of active and innovative methodologies in the classroom. Zhao et al. (2021) drew attention to the need for future teachers to develop their pedagogical DC since the outbreak of the COVID-19 pandemic and the emergence of new models of teaching-learning processes. In this respect, DC training that includes theoretical and practical activities related to digital content creation, and managing activities related to the teaching-learning process is crucial.

On the other hand, on pedagogical DC, pre-service teachers did not express views regarding selecting digital content, accessing learning resources, organising individual or group learning activities through digital technologies, giving feedback to students, and ensuring active participation of students. Reisoğlu and Çebi (2020) pointed to the fact that the pre-service teachers may not have expressed their views on these issues due to the lack of sufficient pedagogical knowledge about learning-teaching methods and the limited training period. Yoon (2022) indicated that pre-service teachers' have low-level DCs in choosing ready-made digital content, using digital technologies to organize individual and group learning activities, evaluation, organizing learning-teaching activities and ensuring active participation of students. In this perspective, it is important to provide experiences for pre-service teachers to develop these skills related to pedagogical DC in a given digital competency training. At this point, introducing the pedagogical potential of digital technologies with practical activities could provide ideas for pre-service teachers on how they could use digital technologies for organizing individual or group learning activities, giving feedback to students, ensuring active participation of students (Røkenes & Krumsvik, 2016). Pre-service teachers should be given an opportunity to observe, reflect, and experience how digital technologies could be used in learning and teaching activities (Gill et al., 2015; Yoon, 2022). Pedagogical digital competencies of preservice teachers should be developed for the effective use of digital technologies in the education of students with different needs or individual characteristics (Masalimova et al., 2022).

In general, pedagogical digital competence includes teachers' knowledge and skills about when and how to integrate digital technologies to meet students' social and cognitive needs and learning goals, as well as their ability to make pedagogical-didactic judgments about how different technologies can extend and deepen students' learning (Skantz-Åberg et al., 2022).

Professional Digital Competence

Finally, in professional DC, regarding the information and data literacy competence area, pre-service teachers underlined using information search strategies and the accuracy of the information they have obtained to reach the current content related to their profession. In the communication and collaboration competence area, pre-service teachers are thought to benefit from virtual classroom applications and cloud technologies to communicate with other teachers and parents. Pre-service teachers are thought to benefit from cloud technology to archive digital materials or documents and to cooperate with stakeholders. In the safety competence area, the pre-service teachers reported that they would inform their colleagues about safety measures in digital environments. Damşa et al. (2021) emphasized the need to develop and maintain DC beyond pedagogical boundaries. In this direction, it can be stated that pre-service teachers should not only improve their pedagogical but also their professional DC. Professional digital competence implies that pre-service teachers take advantage of the opportunities offered by digital learning platforms in order to improve their professional development. Professional digital competence can also contribute to the development of pre-service teachers as lifelong learners (Redecker, 2017).

In terms of professional DC, pre-service teachers did not express their views on determining their DC needs, attempting to support their own development in this sense, and reflecting on the activities they carried out. However, studies highlighted the need to take a multi-faceted approach to technology integration, where leadership, cooperation within and between institutions, professional staff development, and systematic change efforts, is essential (Nelson et al., 2019; Tondeur et al., 2012). According to Yoon (2022), pre-service teachers need knowledge about developing and improving instruction in digital education by analysing their digital education activities and should be supported to accurately understand the meaning of professional engagement. This can be provided with practical teaching experience as planned coursework or teaching practice in pre-service teacher education (Starkey, 2020; Starkey & Yates, 2021).

CONCLUSION

In conclusion, DC training is important in terms of helping pre-service teachers to reach their personal, pedagogical, and professional goals or solving the problems they encounter by using the knowledge and skills they have acquired. When pre-service teachers' DCs are evaluated in context, there are many differences among them. Mainly, personal DC includes awareness, usage of knowledge, skills, and strategies related to a DC area; pedagogical DC includes knowledge and skills related to this competence in educational practices; and professional DC includes reflections of this DC area in the teaching profession. Moreover, personal, pedagogical, and professional DCs are connected within the scope of information literacy, communication and collaboration, digital content creation, safety, and problem-solving DC areas. Particularly, in the digital content creation competence area, while being aware that digital content can be produced with various web 2.0 tools is considered as a personal DC, being able to develop digital content with appropriate web 2.0 tools for purposes such as making information transfer fun, enriching the visuality and increasing the interaction is considered as a pedagogical DC. Thus, there is a concurrence among these areas. In this respect, pre-service teachers' personal knowledge of any web 2.0 tools may cause them to use the relevant tool in education and teaching applications or in the process of professional development. In this sense, the personal DC can form the basis for the acquisition of pedagogical and professional competences. The model proposed by Starkey (2020) through a literature review on teacher education in the digital age also addresses similar points. Accordingly, the use of software and hardware tools falls under general DC, the use of these tools for teaching purposes falls under pedagogical DC, and the use of these tools for the professional development of teachers falls under professional DC; and these areas are interconnected. However, the personal DCs of pre-service teachers may not show that they are sufficiently equipped for the use of technology for teaching purposes. Teachers and pre-service teachers should have the knowledge and skills to integrate the most appropriate digital technologies into the teaching-learning processes to meet students' educational needs (Alarcón et al., 2020). In this regard, it is important to provide environments where pre-service teachers can freely practice their personal DC for pedagogical purposes. In addition, it may be useful to include practices that raise awareness of why personal DCs are important in terms of pedagogical and professional competences in teacher training.

On the other hand, the differences among personal, pedagogical, and professional DC may originate from their intended use or context. For example, the safety area is considered personal DC with recognizing digital dangers, taking protective measures in digital tools and online environments for the safety of personal data, knowing the strategies to deal with digital threats, and knowing their rights and responsibilities in the face of digital dangers themes; pedagogical DC with informing students about ensuring digital safety theme, and professional DC with the theme of informing the colleagues about ensuring digital safety. Likewise, not using the strategies to question the accuracy of information acquired from digital environments at a personal level may result in the transfer of unverified information as content from wrong sources during the education process.

As a result, pre-service teachers developed perspectives in information and data literacy, communication and collaboration, digital content creation, safety, and problem-solving areas in personal and pedagogical DC. However, there is no finding related to digital content creation and problem-solving competence areas in the professional DC. The reason for this situation may be that the professional knowledge and skills of pre-service teachers are limited to the teaching practice course in their undergraduate education. Accordingly, experiences should be designed for the development of pre-service teachers related to different DC areas during teacher education. Teacher educators also need to consider the relation of pre-service teachers' personal DC with the pedagogical and professional DC while organizing a training program. In addition to teaching digital competence knowledge and skills theoretically in teacher training, providing concrete experience by supporting pre-service teachers to use these knowledge and skills in their teaching practices is crucial.

This study is limited to participant diaries and interviews to understand the perspectives of pre-service teachers. In future studies, benefiting from long-term observations in examining pre-service teachers' personal, pedagogical, and professional DC may be useful for in-depth analysis. Also, researchers should consider providing experiences in teaching practice regarding preparing pre-service teachers for the digital age. In this study, the DC training was structured based on the DigComp framework. Future studies can benefit from different frameworks to examine the impact of the training regarding personal, pedagogical, and professional DC in pre-service teacher education.

ACKNOWLEDGMENT

The research was supported by the Scientific and Technological Research Council of Turkey 2237-A program for scientific educational activities support under grant [1129B371801312].

REFERENCES

Alarcón, R., del Pilar-Jiménez, E., & de Vicente-Yagüe, M. I. (2020). Development and validation of the DIGIGLO, a tool for assessing the digital competence of educators. *British Journal of Educational Technology*, *51*(6), 2407–2421. doi:10.1111/bjet.12919

Alnasib, B. N. M. (2023). Digital competencies: Are pre-service teachers qualified for digital education? [IJEMST]. *International Journal of Education in Mathematics, Science, and Technology*, *11*(1), 96–114. doi:10.46328/ijemst.2842

Andreasen, J. K., Tømte, C. E., Bergan, I., & Kovac, V. B. (2022). Professional digital competence in initial teacher education: An examination of differences in two cohorts of pre-service teachers. *Nordic Journal of Digital Literacy*, *17*(1), 61–74. doi:10.18261/njdl.17.1.5

Bianchi, G., Pisiotis, U., & Cabrera Giraldez, M. (2022). *GreenComp-The European sustainability competence framework*. Publication Office of the European Union.

Brevik, L. M., Gudmundsdottir, G. B., Lund, A., & Strømme, T. A. (2019). Transformative agency in teacher education: Fostering professional digital competence. *Teaching and Teacher Education, 86*, 102875. doi:10.1016/j.tate.2019.07.005

Caena, F., & Redecker, C. (2019). Aligning teacher competence frameworks to 21st century challenges: The case for the European Digital Competence Framework for Educators (Digcompedu). *European Journal of Education, 54*(3), 356–369. doi:10.1111/ejed.12345

Çebi, A., Özdemir, T. B., Reisoğlu, İ., & Çolak, C. (2022). From digital competences to technology integration: Re-formation of pre-service teachers' knowledge and understanding. *International Journal of Educational Research, 113*, 101965. doi:10.1016/j.ijer.2022.101965

Damşa, C., Langford, M., Uehara, D., & Scherer, R. (2021). Teachers' agency and online education in times of crisis. *Computers in Human Behavior, 121*, 106793. doi:10.1016/j.chb.2021.106793 PMID:36568039

Drotner, K. (2020). Children's digital content creation: Towards a processual understanding of media production among Danish children. *Journal of Children and Media, 14*(2), 221–236. doi:10.1080/17482798.2019.1701056

Elo, S., & Kyngäs, H. (2008). The qualitative content analysis process. *Journal of Advanced Nursing, 62*(1), 107–115. doi:10.1111/j.1365-2648.2007.04569.x PMID:18352969

Esteve-Mon, F. M., Llopis, M. Á., & Adell-Segura, J. (2020). Digital competence and computational thinking of student teachers. [iJET]. *International Journal of Emerging Technologies in Learning, 15*(2), 29–41. doi:10.3991/ijet.v15i02.11588

European Commission. (2020). *Digital education action plan 2021–2027. Resetting education and training for the digital age*. https://ec.europa.eu/education/sites/default/files/document-library-docs/deap-communication-sept2020_en.pdf

Ferrari, A. (2012). DigComp: *A framework for developing and understanding digital competence in Europe*. Luxembourg: Publications Office of the European Union. Retrieved from https://ec.europa.eu/jrc/sites/default/files/lb-na-26035-enn.pdf

Flores, M. A., & Swennen, A. (2020). The COVID-19 pandemic and its effects on teacher education. *European Journal of Teacher Education, 43*(4), 453–456. doi:10.1080/02619768.2020.1824253

Fraillon, J., Ainley, J., Schulz, W., Friedman, T., & Gebhardt, E. (2014). *Preparing for life in a digital age. The IEA International Computer and Information Literacy Study International Report*. Springer Open. doi:10.1007/978-3-319-14222-7

Fraser, J., Atkins, L., & Richard, H. (2013). *DigiLit leicester. Supporting teachers, promoting digital literacy, transforming learning*. Leicester City Council.

Galindo-Domínguez, H., & Bezanilla, M. J. (2021). Digital competence in the training of pre-service teachers: Perceptions of students in the degrees of early childhood education and primary education. *Journal of Digital Learning in Teacher Education, 37*(4), 262–278. doi:10.1080/21532974.2021.1934757

Gill, L., Dalgarno, B., & Carlson, L. (2015). How does pre-service teacher preparedness to use ICTs for learning and teaching develop through their degree program? *The Australian Journal of Teacher Education, 40*(1), 36–59. doi:10.14221/ajte.2015v40n1.3

Gudmundsdottir, G. B., Gassó, H. H., Rubio, J. C. C., & Hatlevik, O. E. (2020). Student teachers' responsible use of ICT: Examining two samples in Spain and Norway. *Computers & Education, 152*, 103877. doi:10.1016/j.compedu.2020.103877

Gudmundsdottir, G. B., & Hatlevik, O. E. (2018). Newly qualified teachers' professional digital competence: Implications for teacher education. *European Journal of Teacher Education, 41*(2), 214–231. doi:10.1080/02619768.2017.1416085

Guillén-Gámez, F. D., Mayorga-Fernández, M. J., Bravo-Agapito, J., & Escribano-Ortiz, D. (2021). Analysis of teachers' pedagogical digital competence: Identification of factors predicting their acquisition. *Technology. Knowledge and Learning, 26*(3), 481–498. doi:10.100710758-019-09432-7

Gutiérrez-Porlán, J., & Serrano-Sánchez, J. L. (2016). Evaluation and development of digital competence in future primary school teachers at the University of Murcia. *Journal of New Approaches in Educational Research, 5*(1), 51–56. doi:10.7821/naer.2016.1.152

Hall, R., Atkins, L., & Fraser, J. (2014). Defining a self-evaluation digital literacy framework for secondary educators: The DigiLit Leicester project. *Research in Learning Technology, 22*. Advance online publication. doi:10.3402/rlt.v22.21440

Hinojo-Lucena, F. J., Aznar-Díaz, I., Cáceres-Reche, M. P., Trujillo-Torres, J. M., & Romero-Rodriguez, J.-M. (2019). Factors influencing the development of digital competence in teachers: Analysis of the teaching staff of permanent education centres. *IEEE Access : Practical Innovations, Open Solutions, 7*, 178744–178752. doi:10.1109/ACCESS.2019.2957438

Ilomäki, L., Kantosalo, A., & Lakkala, M. (2011). What is digital competence? In *Linkedportal*. European Schoolnet., http://linked.eun.org/web/guest/in-depth3

Ilomäki, L., Paavola, S., Lakkala, M., & Kantosalo, A. (2016). Digital competence– an emergent boundary concept for policy and educational research. *Education and Information Technologies*, *21*(3), 655–679. doi:10.100710639-014-9346-4

Instefjord, E. J., & Munthe, E. (2016). Preparing pre-service teachers to integrate technology: An analysis of the emphasis on digital competence in teacher education curricula. *European Journal of Teacher Education*, *39*(1), 77–93. doi:10.1080/02 619768.2015.1100602

INTEF. (2017). *Marco común de competencia digital docente*. Retrieved from http://educalab.es/documents/10180/12809/MarcoComunCompeDigiDoceV2.pdf

Janssen, J., Stoyanov, S., Ferrari, A., Punie, Y., Pannekeet, K., & Sloep, P. (2013). Experts' views on digital competence: Commonalities and differences. *Computers & Education*, *68*, 473–481. doi:10.1016/j.compedu.2013.06.008

Krumsvik, R. J. (2008). Situated learning and teachers' digital competence. *Education and Information Technologies*, *13*(4), 279–290. doi:10.100710639-008-9069-5

Krumsvik, R. J. (2014). Teacher educators' digital competence. *Scandinavian Journal of Educational Research*, *58*(3), 269–280. doi:10.1080/00313831.2012.726273

List, A., Brante, E. W., & Klee, H. L. (2020). A framework of pre-service teachers' conceptions about digital literacy: Comparing the United States and Sweden. *Computers & Education*, *148*, 103788. doi:10.1016/j.compedu.2019.103788

López-Belmonte, J., Pozo-Sánchez, S., Fuentes-Cabrera, A., & López-Núñez, J. A. (2019). Content creation and flipped learning: A necessary pairing for education in the new millennium. *Revista Española de Pedagogía*, *77*(274), 535–555. doi:10.22550/ REP77-3-2019-07

Lund, A., Furberg, A., Bakken, J., & Engelien, K. L. (2014). What does professional digital competence mean in teacher education?. *Nordic journal of digital literacy*, *9*(4), 280-298. https://doi.org/ doi:10.18261/ISSN1891-943X-2014-04-04

Masalimova, A. R., Khvatova, M. A., Chikileva, L. S., Zvyagintseva, E. P., Stepanova, V. V., & Melnik, M. V. (2022). Distance learning in higher education during COVID-19. *Frontiers in Education*, *7*, 822958. doi:10.3389/feduc.2022.822958

McGarr, O., & McDonagh, A. (2021). Exploring the digital competence of pre-service teachers on entry onto an initial teacher education programme in Ireland. *Irish Educational Studies*, *40*(1), 115–128. doi:10.1080/03323315.2020.1800501

McMillan, J. H., & Schumacher, S. S. (2014). *Research in education: Evidence–based inquiry* (7th ed.). Pearson.

Merriam, S. (2009). *Qualitative research: A guide to design and implementation.* Jossey-Bass.

Montenegro-Rueda, M., & Fernández-Batanero, J. (2022). Digital competence of special education teachers: Impact, challenges and opportunities. *Australasian Journal of Special and Inclusive Education*, *46*(2), 178–192. doi:10.1017/jsi.2022.8

Napal-Fraile, M., Peñalva-Vélez, A., & Mendióroz-Lacambra, A. M. (2018). Development of digital competence in secondary education teachers' training. *Education Sciences*, *8*(3), 104. doi:10.3390/educsci8030104

Nelson, M. J., Voithofer, R., & Cheng, S. L. (2019). Mediating factors that influence the technology integration practices of teacher educators. *Computers & Education*, *128*, 330–344. doi:10.1016/j.compedu.2018.09.023

Olofsson, A. D., Lindqvist, M. H., Lindberg, O. J., From, J., Stödberg, U., & Holmgren, T. (2017). Pedagogical digital competence for police teachers in relation to distance-based police education. In *INTED2017 Proceedings* (pp. 4219-4227). IATED. 10.21125/inted.2017.1010

Organisation of Economic Cooperation and Development [OECD] (2005). *Definition and selection of key competencies. executive summary (online resource).* OECD Publishing.

Ottenbreit-Leftwich, A., Liao, J. Y. C., Sadik, O., & Ertmer, P. (2018). Evolution of teachers' technology integration knowledge, beliefs, and practices: How can we support beginning teachers use of technology? *Journal of Research on Technology in Education*, *50*(4), 282–304. doi:10.1080/15391523.2018.1487350

Ottestad, G., Kelentrić, M., & Gudmundsdottir, G. B. (2014). Professional digital competence in teacher education. *Nordic Journal of Digital Literacy*, *9*(4), 243–249. doi:10.18261/ISSN1891-943X-2014-04-02

Patton, M. Q. (1999). Enhancing the quality and credibility of qualitative analysis. *Health Sciences Research, 34*(5 Pt 2), 1189–1208. https://www.ncbi.nlm.nih.gov/pmc/articles/PMC1089059/

Pettersson, F. (2018). On the issues of digital competence in educational contexts – a review of literature. *Education and Information Technologies*, *23*(3), 1005–1021. doi:10.100710639-017-9649-3

Ramírez-Montoya, M. S., Mena, J., & Rodríguez-Arroyo, J. A. (2017). In-service teachers' self-perceptions of digital competence and OER use as determined by a xMOOC training course. *Computers in Human Behavior*, *77*, 356–364. doi:10.1016/j.chb.2017.09.010

Redecker, C. (2017). European framework for the digital competence of educators: DigCompEdu (No. JRC107466). Joint Research Centre.

Reisoğlu, İ., & Çebi, A. (2020). How can the digital competences of pre-service teachers be developed? Examining a case study through the lens of DigComp and DigCompEdu. *Computers & Education*, *156*, 103940. doi:10.1016/j.compedu.2020.103940

Røkenes, F. M., & Krumsvik, R. J. (2016). Prepared to teach ESL with ICT? A study of digital competence in Norwegian teacher education. *Computers & Education*, *97*, 1–20. doi:10.1016/j.compedu.2016.02.014

Siddiq, F., Hatlevik, O. E., Olsen, R. V., Throndsen, I., & Scherer, R. (2016). Taking a future perspective by learning from the past–A systematic review of assessment instruments that aim to measure primary and secondary school students' ICT literacy. *Educational Research Review*, *19*, 58–84. doi:10.1016/j.edurev.2016.05.002

Skantz-Åberg, E., Lantz-Andersson, A., Lundin, M., & Williams, P. (2022). Teachers' professional digital competence: An overview of conceptualisations in the literature. *Cogent Education*, *9*(1), 2063224. doi:10.1080/2331186X.2022.2063224

Starkey, L. (2020). A review of research exploring teacher preparation for the digital age. *Cambridge Journal of Education*, *50*(1), 37–56. doi:10.1080/0305764X.2019.1625867

Starkey, L., & Yates, A. (2021). Do digital competence frameworks align with preparing beginning teachers for digitally infused contexts? An evaluation from a New Zealand perspective. *European Journal of Teacher Education*. Advance online publication. doi:10.1080/02619768.2021.1975109

Tárraga-Mínguez, R., Suárez-Guerrero, C., & Sanz-Cervera, P. (2021). Digital teaching competence evaluation of pre-service teachers in Spain: A review study. *IEEE-RITA*, *16*(1), 70–76. doi:10.1109/RITA.2021.3052848

Tomczyk, Ł. (2019). What do teachers know about digital safety? *Computers in the Schools*, *36*(3), 167–187. doi:10.1080/07380569.2019.1642728

Tømte, C. E. (2013). Educating teachers for the new millennium? Teacher training, ICT and digital competence. *Nordic Journal of Digital Literacy, 10*(Jubileumsnummer), 138-154.

Tondeur, J., Aesaert, K., Prestridge, S., & Consuegra, E. (2018). A multilevel analysis of what matters in the training of pre-service teacher's ICT competencies. *Computers & Education, 122*, 32–42. doi:10.1016/j.compedu.2018.03.002

Tondeur, J., van Braak, J., Guoyuan, S., Voogt, J., Fisser, P., & Ottenbreit-Leftwich, A. S. (2012). Preparing student teachers to integrate ICT in classroom practice: A synthesis of qualitative evidence. *Computers & Education, 59*(1), 134–144. doi:10.1016/j.compedu.2011.10.009

VERBI Software. (2021). MAXQDA 2022 [computer software]. Berlin, Germany: VERBI Software. Available from maxqda.com.

Vuorikari, R., Kluzer, S., & Punie, Y. (2022). DigComp 2.2: The digital competence framework for citizens - With new examples of knowledge, skills and attitudes, EUR 31006 EN, Publications Office of the European Union, Luxembourg.

Vuorikari, R., Punie, Y., Carretero, S., & van den Brande, L. (2016). *DigComp 2.0: The Digital Competence Framework for Citizens. Update Phase 1: The conceptual reference model.* Luxembourg Publication Office of the European Union.

Yoon, S. H. (2022). Gender and digital competence: Analysis of pre-service teachers' educational needs and its implications. *International Journal of Educational Research, 114*, 101989. doi:10.1016/j.ijer.2022.101989

Zhao, Y., Llorente, A. M. P., & Gómez, M. C. S. (2021). Digital competence in higher education research: A systematic literature review. *Computers & Education, 168*, 104212. doi:10.1016/j.compedu.2021.104212 PMID:36568577

ADDITIONAL READING

European Commission/EACEA/Eurydice. (2019). *Digital Education at School in Europe. Eurydice Report.* Publications Office of the European Union.

Redecker, C. (2017). European framework for the digital competence of educators: DigCompEdu (No. JRC107466). Joint Research Centre.

Vuorikari, R., Kluzer, S., & Punie, Y. (2022). DigComp 2.2: The Digital Competence Framework for Citizens - With new examples of knowledge, skills and attitudes, EUR 31006 EN, Publications Office of the European Union, Luxembourg.

KEY TERMS AND DEFINITIONS

Digital Competence: Attitude, knowledge, and skills in using digital technologies for various purposes to successfully participate in a digital society.

Personal Digital Competence: Attitude, knowledge, and skills in using digital technologies for daily activities.

Pedagogical Digital Competence: Attitude, knowledge, and skills in using digital technologies for teaching and learning related activities.

Professional Teacher Competence: Attitude, knowledge, and skills in using digital technologies in a related profession.

Training Programs: Short educational programs aimed developing specific skills or abilities.

Teachers' Perspectives: Teachers' point of view regarding to a specific issue based on their experiences.

Information and Data Literacy: Accessing, using, and managing the digital content in a proper way.

Communication and Collaboration: Communicate and collaborate using digital technologies and participate in digital networks properly.

Digital Content Creation: Creating materials using digital technologies and editing.

Safety: Protecting devices, personal data and privacy, health and well-being, and environment while using digital technologies.

Problem-Solving: Solving problems encountered in digital environments and providing solutions by using digital technologies.

Chapter 2
Equitable and Inclusive Online Learning:
A Framework for Supporting Students With Disabilities

J. Jayapriya
https://orcid.org/0000-0001-6672-6865
Christ University, Bangalore, India

M. Vinay
https://orcid.org/0000-0003-0297-3597
Christ University, Bangalore, India

ABSTRACT

Online learning has become a widely adopted mode of education, particularly during the COVID-19 pandemic. In general, individuals with disabilities face challenges when using non-technology components for studying. This chapter proposes a framework for equitable and inclusive online learning practices that support students with disabilities. The framework is based on a review of current research and best practices for online learning and disability accommodations. The framework emphasizes a collaborative, student-centered approach to online learning that acknowledges the unique needs and experiences of students with disabilities. Depending on the disabilities, the framework is divided into two phases namely: Prevalent Learning, and Discrete Learning. The former comprised components: Accessibility, Accommodation, and Engagement, and later has components like Methodology, Evaluation. The framework proposed provides a roadmap for addressing the challenges faced by students with disabilities in online learning environments.

DOI: 10.4018/978-1-6684-9089-1.ch002

INTRODUCTION

Online education has become increasingly popular in the current era due to the widespread availability of the internet and advancements in technology. With online education, students attend classes from anywhere in the world, at any time that suits them, without the need to physically travel to a school or university. One of the major benefits of online education is its flexibility. Students access course materials and complete assignments at their own pace, and often schedule their classes around work, family, or other obligations. Online education is also often more affordable than traditional in-person education, as it eliminates the need for expensive campus facilities and infrastructure. However, online education also has its challenges. Students may struggle with motivation and staying engaged with their coursework without the structure and accountability of in-person classes. Additionally, online education may not be suitable for all types of learners, as some students may prefer a more hands-on, interactive learning experience Seale, J. K. (2013).

Need for the Online Education for Disabled Learners

Online education is a great option for disability learners as it provides them with a more accessible and flexible learning environment. Ensure that all learning materials are accessible to all students. This includes providing materials in different formats such as audio, video, and text, as well as ensuring that the materials are compatible with assistive technologies like screen readers. Incorporate inclusive design principles into your online education program. This means designing with accessibility in mind from the outset and considering the needs of all learners, including those with disabilities. Use a variety of teaching methods to accommodate different learning styles. For example, use videos, interactive quizzes, and discussion forums to provide different ways for learners to engage with the material. Provide learners with disabilities with additional support and accommodations as needed. This may include providing extra time for assignments or exams, offering assistive technology training, or providing additional resources and tutoring. Foster a supportive and inclusive environment by promoting respect and understanding among all learners. Encourage learners to ask questions and offer support to each other. By implementing these strategies, create an online education program that is accessible and inclusive for all learners, including those with disabilities.

Challenges in Online Learning

These challenges include inaccessible digital content, inadequate support for assistive technologies, and limited opportunities for social interaction and engagement. In general, individuals with disabilities face challenges when using non-technology components for studying. Some potential drawbacks may include:

1. Limited Accessibility: Some non-technology components, such as textbooks or printed materials, may not be accessible to learners with certain disabilities, such as visual impairments or dyslexia.
2. Physical Barriers: Non-technology components may require physical manipulation, such as turning pages, that may be difficult or impossible for learners with certain physical disabilities.
3. Lack of Customization: Non-technology components may not be easily customizable to meet the unique needs of individual learners with disabilities. For example, a textbook may not have adjustable font sizes or spacing.
4. Limited Interactivity: Non-technology components may not provide the same level of interactivity or engagement as technology-based resources. This makes it more difficult for some learners with disabilities to stay engaged and motivated.
5. Limited Feedback: Non-technology components may not provide immediate feedback, which is important for learners with certain disabilities, such as those who need to practice language skills or math problems.

It's important to note that many of these drawbacks are mitigated with accommodations or assistive technology. Additionally, not all learners with disabilities will face the same challenges when using non-technology components, as their individual needs and abilities may vary.

Objectives of This Study

The objectives of an online learning framework for disabled learners should focus on creating an accessible and inclusive learning environment that meets the unique needs of disabled learners. This chapter provides various online educational resources for learners with some disabilities and learning disability learners. Also proposes a framework that suggests a list of online educational resources for the above said learners.

Some key objectives could include:

1. Accessibility: Ensure that all online learning materials, platforms, and tools are fully accessible to disabled learners, including those with visual, hearing, and physical disabilities.

2. Flexibility: Provide flexible learning options that allow disabled learners to work at their own pace and on their own schedule, with the ability to adjust the pace and content to meet their individual needs.
3. Accommodation: Provide individualized accommodations and support services, such as closed captioning, sign language interpretation, assistive technology, and accessible course materials, to ensure that disabled learners have equal access to the same learning opportunities as their peers.
4. Engagement: Foster a supportive and inclusive learning community that actively engages and includes disabled learners, encouraging them to participate in class discussions, group projects, and other collaborative activities.
5. Assessment: Develop assessment strategies that take into account the unique needs of disabled learners, providing alternative ways of demonstrating knowledge and skills that do not rely solely on traditional testing methods.

The outline of the chapter are as follows: Following the introduction in Section 1, Background as related work is given in Section 2. Proposed framework is presented in Section 3 as Methodology. Finally the chapter is concluded with Section 4, Conclusion.

BACKGROUND

This section is shown in two subsections, one is the importance of online education for different learners with some disability and learning disability learner.

Importance of Online Education

Learners with Some Disability

1. Online learning provides disabled students with convenient access to educational materials from the comfort of their own homes. This setup eliminates the physical exhaustion faced by students with limited mobility, and it also simplifies the learning process for visually or hearing-impaired individuals.
2. As a result, they can explore the world in a more liberating environment, which reduces the obstacles to learning and encourages their active participation.

Learning Disability Learner

1. Online learning offers numerous advantages for learning disability students, including the absence of large crowds that can potentially cause discomfort.

2. Consequently, students are able to overcome shyness and experience increased motivation, as they are supported and encouraged by their loved ones who are present at home.

Literature Review

This section shows the related work done by the previous research in finding online resources for disabled learners. The following Figure 1, shows the various work done by researchers for providing digital tools amongst learners with disabilities. Online tools have revolutionized the way we learn, especially for disabled learners. These tools are designed to cater to the unique needs of disabled learners, making learning accessible and enjoyable.

Figure 1. Previous work on Online Learning for Disabled Learners

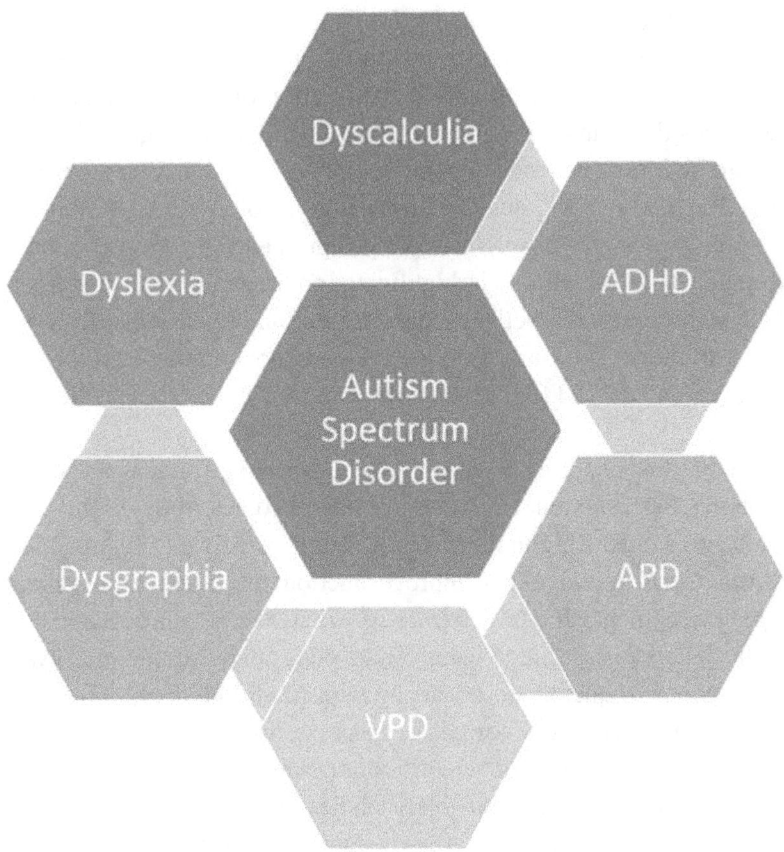

More and more students with disabilities are opting for online learning, and it's crucial to apply research-based interventions to keep them engaged in their education. Based on the literature on at-risk students with disabilities, we have identified five significant areas of impact that help improve student engagement. These areas are referred to as the "5 Cs," which include learner control, a flexible and challenging curriculum, a safe and supportive environment, a compassionate community, and connection to students' personal goals and aspirations. This article explores these 5 Cs in detail and provides examples of how they applied to online learning environments Cavanaugh, C et.al., 2013.

The aim of this qualitative research is to investigate the effectiveness of UDL in catering to the learning requirements of students with disabilities. The results indicate that the students' perspectives are in agreement with the principles of UDL, particularly the principle of multiple means of engagement. However, it was observed that addressing the learning needs of certain students may pose challenges for others. To address these limitations, UDL needs to focus on individual students' learning needs, not only through curricular and setting modifications but also through direct interventions Griful-Freixenet et al., 2017.

This article outlines how a university educator implemented the principles of Universal Instructional Design (UID) in the creation of three online courses. The article examines the course design process and demonstrates how UD principles were put into action to enhance "cognitive access" through various pedagogical techniques. The study showcases how instructors incorporate UID considerations throughout the five stages of instructional design: analysis, design, development, implementation, and evaluation. The highlighted courses utilized diverse assignment formats, adaptable options, a well-structured format, and consistent interactions with peers and instructors. The findings reveal that learners appreciated the structure, instructor engagement, and flexible options provided in the courses Rao, K et al., 2015.

The aim of this paper is to present an e-learning framework that aids in comprehending the functional restrictions linked to learning disabilities (LD). Additionally, it intends to highlight diverse learning principles and dimensions that help overcome these limitations and improve functional abilities within home, school, and college environments. It is important to note that the research concentrates solely on academic disorders associated with LD and does not delve into the physical and sensory limitations frequently co-occurring with LD Pirani et al., 2013.

A four-step process was followed for the systematic literature review, which included an exhaustive search of scientific literature databases, the application of exclusion and inclusion criteria to select studies, and the analysis and synthesis of literature. The main findings of the review are as follows: Firstly, there is a lack of e-learning studies addressing accessibility for individuals with cognitive impairments (N = 29), with the predominant focus being on design guidelines rather than assessing effectiveness. Secondly, accessibility standards were weakly included (N = 5). Thirdly, special education findings were weakly included (N = 3), with a focus on specific neuropsychological disorders or syndromes rather than on cognitive function impairments as promoted by the International Classification of Functioning. Fourthly, the review identified five families of accessibility functions (adaptive systems, game elements, accessible content, virtual agents, and accessible interfaces or environments) and their dependence on the activity-domain of learning Smith, S. J., & Basham, J. D. 2014 - Cinquin et la., 2019.

Despite the existence of numerous technical standards and guidelines for ensuring accessibility on e-learning platforms, there is a lack of emphasis on the pedagogical and didactic aspects of accessibility. This means that while disabled students may be able to access the e-learning platform, they may still face barriers in accessing the course content, resources, activities, collaboration tools, and interaction tools. In order for an e-learning course to truly be inclusive, it must address both the technical and pedagogical aspects of accessibility. To this end, this research aims to investigate various methods, tools, and practices and develop a reference model for designing accessible e-learning courses in the context of higher education Guglielman 2010. This research emphasizes the significance of incorporating student feedback into faculty disability awareness training in higher education. The study employs a mixed methods approach to assess the attitudes of instructors towards a Disability Awareness and Inclusive Teaching Video Training, which utilizes student voices as the principal instructional strategy Tarconish et al., 2023.

Using a questionnaire designed by the researchers to achieve the study's objectives, the research employed a descriptive-analytical method. The study involved 124 public-school teachers from Irbid's public schools. The survey results indicated that the teachers had a favorable attitude towards digital learning, with significant differences in responses based on gender and age Shater 2023.

Apart from the above related work, many researchers have done much work on digital resources for disabled learners. A survey of commonly used devices and software: This study conducted a survey of 157 university students with disabilities to determine their use of assistive technology. The results showed that the most commonly used devices and software included screen readers, voice recognition software, and electronic notetakers. Online tools such as screen readers, text-to-speech software, and image description tools have been found to be effective in supporting learners with visual impairments. Tools like captioning and transcription services have been found to be effective in supporting learners with hearing impairments. Digital resources like assistive technologies, learning management systems, and personalized learning platforms have been found to be effective in supporting learners with learning disabilities Smith et al., 2014 - Beyene et al. 2023.

Insights from the Study

1. From the above study, it is evident that the importance of online educational resources for learners.
2. And to add on this, some researchers have worked on developing new online resources for education purposes.
3. After this study, the chapter emphasizes that online resources are important not only for common learners but also for those who have learning disabilities and disabled learners in different physical forms.

METHODOLOGY

The framework is intended to be a practical guide for educators and administrators in designing and delivering online learning experiences that are equitable and inclusive for all students, including those with disabilities. It emphasizes the importance of a proactive, student-centered approach to accessibility and accommodation that is grounded in the principles of universal design. The framework also highlights the critical role of engagement in creating a positive and supportive learning environment for students with disabilities.

Prevalent Learning

1. Accessibility is a crucial component of the framework. Online learning materials, such as videos, audio recordings, and digital texts, should be designed to be accessible to all learners, including those with disabilities. This requires attention to the principles of universal design, such as providing multiple means of representation, action, and expression. Accessibility also involves providing support for assistive technologies, such as screen readers, speech-to-text software, and captioning tools.

Immersive reading refers to the experience of getting deeply engaged in a text, to the point where the reader becomes fully absorbed in the story or information presented. It involves the reader's complete attention and active participation, as they imagine and experience the scenes and events described in the text. Immersive reading is achieved through a variety of techniques, including vivid descriptions, well-developed characters, and an engaging storyline. This type of reading is both entertaining and educational, and has been shown to enhance empathy, creativity, and critical thinking skills. In recent years, advancements in technology have made it possible for readers to have even more immersive reading experiences through virtual and augmented reality applications.

2. Accommodation is the second component of the framework. Accommodations are individualized adjustments to online learning practices that enable students with disabilities to fully participate in the learning experience. Accommodations may include extended time for assignments or exams, alternative formats for materials, and support from disability services personnel. Accommodations should be designed to be flexible, responsive to student needs, and consistent with the principles of universal design.
3. The third component of the framework is engagement. Online learning is isolating, particularly for students with disabilities who may face additional barriers to social interaction and engagement. To address this, the framework emphasizes the importance of creating a sense of community and belonging in online learning environments. This is achieved through strategies such as peer support networks, synchronous online discussions, and opportunities for collaborative learning.

Discrete Learning

1. In this, the first component is Methodology where the approach is chosen according to the disability of the learners. Screen readers are software programs that read text aloud to people with visual impairments. They used to read digital books, web pages, and other digital content. Braille displays are electronic devices that convert digital text into braille, allowing people with visual impairments to read. Voice recognition software software allows people with mobility impairments to use their voice to control their computer, including typing and navigating. Augmentative and Alternative Communication (AAC) are devices that help people with speech impairments to communicate, such as text-to-speech software or communication boards. Captioning and transcripts tools provide written text versions of spoken content, which is helpful for people with hearing impairments. Assistive listening devices include devices like hearing aids and FM systems that amplify sound and make it easier for people with hearing impairments to hear. Tactile graphics are images that are designed to be felt, rather than seen, and helpful for people with visual impairments.
2. Evaluation is the second component that completes the learners evaluation through online mode depending on the disability. The evaluation process should start by identifying the learners' needs, such as the type of disability they have and the specific challenges they face in learning. This information will help

in selecting appropriate technology-based interventions. Once the learners' needs are identified, the next step is to select appropriate technology-based interventions that address their learning challenges. The interventions may include software, hardware, or assistive technologies. The technology-based interventions selected should be evaluated for their accessibility and usability for learners with disabilities. This involves testing the technology with the learners and assessing whether they use it effectively. The evaluation process should also assess the impact of the technology-based interventions on the learners' academic, social, and emotional development. This may involve using standardized tests, surveys, and observation to evaluate the learners' progress. Finally, the results of the evaluation should be analyzed to identify the strengths and weaknesses of the technology-based interventions. This will help in improving the interventions and making them more effective in supporting the learners' needs.

The following Figure 1 depicts the Online Learning System for Disabled Learners (OLSDL).

Figure 2. OLSDL

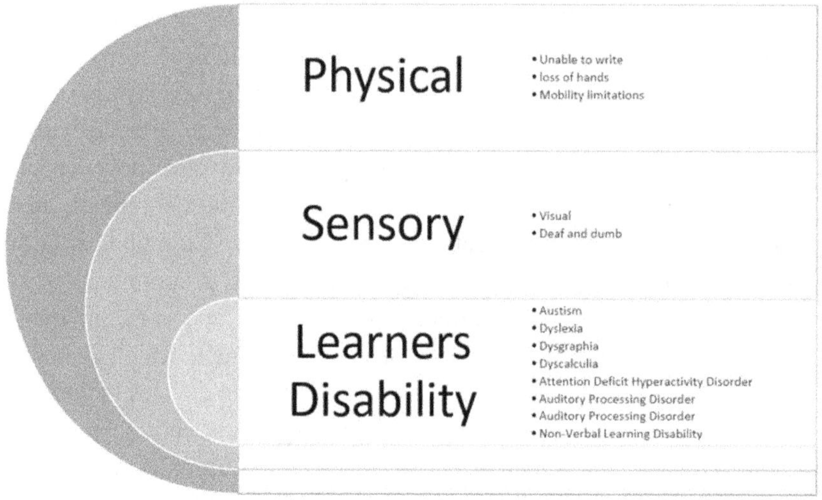

Different Forms of Learners

They are different forms of learners. Learners prefer visual aids such as charts, graphs, diagrams, and videos to comprehend information. Learners learn best by hearing information. They prefer listening to lectures, podcasts, and discussions. Learners learn best by experiencing and doing things themselves. They prefer hands-on activities and interactive learning experiences. Learners prefer reading and writing to learn new information. They enjoy taking notes, reading textbooks, and researching online. Learners learn best in group settings, through discussions, and collaboration with others. In this study, the following two forms of learners are considered.

1. Disabled Learners
2. Learning disabilities

Figure 3. Different forms of Learners

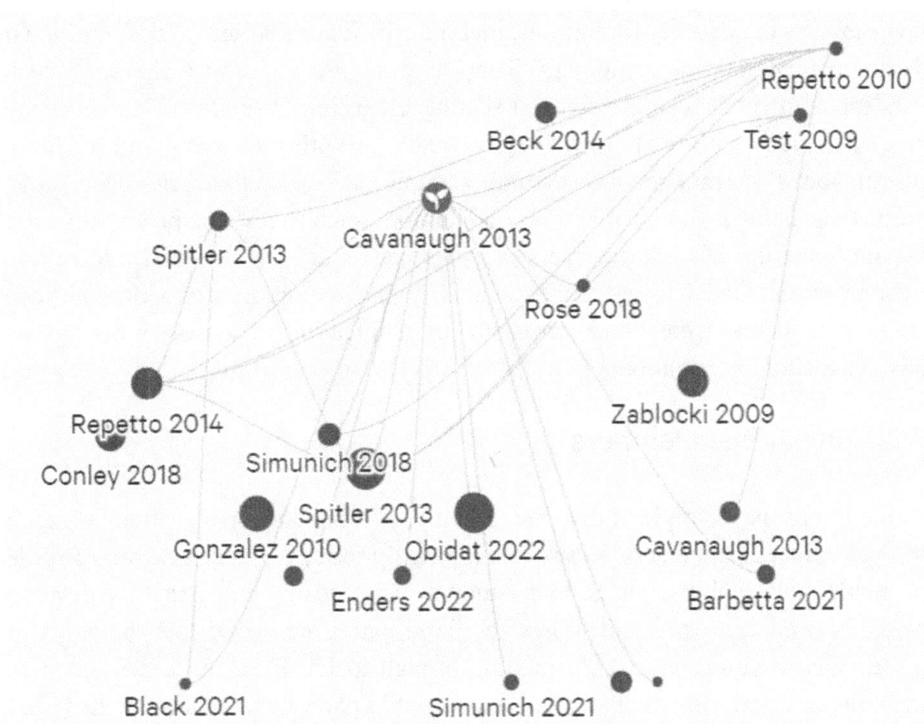

Disabled learners are individuals who experience a physical, sensory, or intellectual impairment that may impact their ability to learn or engage in educational activities. This includes individuals with physical disabilities, such as those who use wheelchairs or have mobility impairments, as well as individuals with sensory impairments, such as those who are deaf or hard of hearing, blind or have low vision. Additionally, disabled learners may also include those with intellectual or developmental disabilities, such as individuals with Down syndrome, autism spectrum disorder, or learning disabilities such as dyslexia or ADHD. These learners may require specialized educational interventions and accommodations to ensure that they have equal access to education and the opportunity to reach their full potential.

There is no one-size-fits-all approach to working with disabled learners as the needs and abilities of each individual vary greatly. Some different types of disabled learners are Visual Impairment learners and may have partial or total loss of vision. They may require access to materials in alternate formats such as braille or large print, or assistive technology such as screen readers or magnification software. Hearing Impairment learners may have partial or total loss of hearing. They may require access to materials in written or visual formats, or assistive technology such as captioning or sign language interpreters. Physical Disabilities learners may have mobility or dexterity limitations, making it difficult for them to use traditional classroom materials or equipment. They may require assistive technology such as adaptive keyboards or specialized seating. Learning Disabilities learners may have difficulties with reading, writing, or math, or with processing and retaining information. They may require accommodations such as extended time on exams, alternative testing formats, or assistive technology such as text-to-speech software. Autism Spectrum Disorder learners may have difficulties with social interaction, communication, and sensory processing. They may require accommodations such as a quiet workspace, clear and consistent communication, or sensory tools such as noise-cancelling headphones.

Visual Impairment learners

Braille is a prime example of discrete learning, as it breaks down written language into individual characters or letters that are easily read by touch. Learning Braille in small, manageable chunks help students with visual impairments develop the skills needed to read and write. Tactile graphics are raised-line drawings or models that used to convey information through touch. Breaking down complex diagrams or maps into smaller, more manageable parts help students with visual impairments better understand the information being presented. Audio books are broken down into shorter segments or chapters, making it easier for students with visual impairments to digest the information. Pausing after each chapter or section also provides an opportunity for students to reflect on and process the information. Multi-sensory activities that involve touch, sound, and movement help students with visual impairments learn in discrete steps. For example, a math lesson could involve using tactile manipulatives to explore the concept of addition, followed by an audio explanation of how to solve addition problems. Adaptive technology such as screen readers, magnifiers, and speech recognition software help students with visual impairments break down information into smaller, more manageable parts. For example, a screen reader reads a document aloud, allowing students to process the information at their own pace.

There are several online tools available for learners with visual impairments, including:

- JAWS (Job Access With Speech): JAWS is a screen reader that reads aloud the contents of the computer screen. It is designed to assist users with visual impairments in accessing and interacting with computer software and applications.
- NVDA (Non-Visual Desktop Access): NVDA is a free, open-source screen reader that provides access to the Windows operating system and popular applications. It reads web pages, documents, and other content, and allows users to navigate using keyboard shortcuts.
- ZoomText: ZoomText is a screen magnifier that enlarges and enhances everything on the computer screen. It is customized to suit the user's needs and preferences, and includes features such as text-to-speech, color and contrast adjustments, and more.
- Kurzweil 3000: Kurzweil 3000 is an assistive technology software that read electronic text aloud, highlights text as it is read, and provides definitions and translations of words. It also includes tools for note-taking, outlining, and organizing information.
- Learning Ally: Learning Ally is a digital library of audiobooks that provides access to a wide range of academic and popular titles for individuals with visual impairments or learning disabilities.
- Be My Eyes: Be My Eyes is a free app that connects visually impaired users with sighted volunteers through video chat. Volunteers assist with tasks such as reading labels or identifying objects, making it a useful tool for everyday tasks.
- Voice Dream Reader: Voice Dream Reader is a text-to-speech app that reads digital text aloud using natural-sounding voices. It also highlights text as it is read, and includes a variety of customization options.

- Braille Translator: Braille Translator is an online tool that converts digital text into Braille. It is useful for individuals who are learning Braille or who prefer to read in Braille.

Hearing Impairment learners

Sign language is broken down into individual signs, which are practiced and mastered one at a time. This allows students with hearing impairments to build their sign language skills in a structured and manageable way. Closed captioning is broken down into smaller segments, such as individual sentences or phrases, making it easier for students with hearing impairments to follow along and understand the information being presented. Visual aids such as diagrams, pictures, and videos broken down into smaller, more manageable parts, allowing students with hearing impairments to focus on one concept at a time. Assistive technology such as hearing aids and cochlear implants adjusted to provide discrete levels of sound amplification or sensory input, allowing students with hearing impairments to learn in a more controlled and structured way. Multi-sensory learning used to break down information into smaller, more manageable parts. For example, a science experiment could be broken down into discrete steps, with each step involving a different sensory experience, such as observing, touching, or smelling.

Here are some online tools that helpful for learners with hearing impairments:

1. Captioning Tools: These tools generate captions for audio or video content, allowing students with hearing impairments to read what is being said. Some examples include:
   - Google Meet captioning
   - Zoom captioning
   - Otter.ai
   - Rev.com
   - CaptionCreator.net
2. Sign Language Learning Platforms: These platforms teach American Sign Language (ASL) or other sign languages through video lessons, quizzes, and interactive exercises. Some popular platforms are:
   - SignSchool
   - Start ASL
   - ASLDeafined
   - Sign Language 101
3. Speech-to-Text Tools: These tools convert spoken words into written text in real-time, allowing students to follow along with lectures or conversations. Some examples include:

- Google Live Transcribe
- Ava
- TranscribeMe
- SpeechTexter

4. Hearing Aid Simulator: This tool allows students to experience what it's like to have different levels of hearing loss and how hearing aids help. Some examples include:
    - Phonak Hearing Aid Simulator
    - Widex Hearing Aid Simulator
5. Assistive Listening Devices: These devices used in conjunction with hearing aids or cochlear implants to amplify sound in specific settings, such as classrooms or lectures. Some examples include:
    - LooktoSpeak by Sarah Ezekiel & Google Creative Lab
    - Roger Pen by Phonak
    - FM Systems by Williams Sound
    - Soundfield Systems by Frontrow

Google Look to Speak is a communication app designed for people who have difficulty communicating verbally. The app uses eye-tracking technology to allow users to select pre-written phrases and words, which are then spoken out loud by the app. This is a helpful tool for people who may have difficulty communicating due to conditions such as ALS, cerebral palsy, or other speech impairments. To use the app, users need a compatible Android device, such as a smartphone or tablet, and an eye-tracking device, such as a Tobii Dynavox device. Once the app and eye-tracking device are set up, the user selects from a variety of phrases and words to communicate with others. Google Look to Speak is free to download and use, and it is a helpful tool for people who have difficulty communicating verbally.

Physical Disabilities

Universal Design for Learning (UDL) is a framework that provides multiple means of representation, expression, and engagement to help all learners access and engage with content. It is especially helpful for individuals with physical disabilities by providing different ways to interact with materials. Assistive technology are many different types of assistive technology available that support learners with physical disabilities, such as screen readers, speech-to-text software, and adaptive keyboards. These tools help individuals to access content and complete tasks that may be challenging otherwise. Multi-sensory learning involves engaging multiple senses (such as sight, sound, and touch) to reinforce learning. This is particularly helpful for individuals with physical disabilities, who may rely on different senses for information. Peer tutoring involves pairing a more experienced learner with a less experienced learner to provide support and guidance. This is an effective way to support individuals with physical disabilities, as it provides a more personalized learning experience. Project-based learning involves working on a real-world project or problem, which helps to engage learners and provide a more meaningful context for learning. This approach is especially helpful for individuals with physical disabilities, as it allows them to work at their own pace and focus on their strengths.

The following are the online tools and resources that help support discrete learning for learners with physical disabilities:

1. Assistive Technology: There are a variety of assistive technologies that help students with physical disabilities access online learning materials. Some examples include:
    - Voice recognition software such as Dragon NaturallySpeaking
    - On-screen keyboards and alternative input devices such as touchscreens, joysticks, and switches
    - Screen readers such as JAWS or NVDA
    - Braille displays and Braille translation software
    - Adaptive pointing devices such as trackballs, touchpads, and head pointers
2. Accessible Online Learning Platforms: Many online learning platforms offer accessibility features such as closed captions, transcripts, and alternative text descriptions for images. Some platforms that prioritize accessibility include:
    - Blackboard Ally
    - Canvas
    - Google Classroom
    - Microsoft Teams
    - Moodle
3. Adaptive Learning Tools: These tools are tailored to the specific needs of learners with physical disabilities, such as modifying the size of text or adjusting the speed of videos. Some examples include:
    - Read&Write by Texthelp
    - Snap&Read by Don Johnston
    - Kurzweil 3000 by Kurzweil Education
    - Learning Ally
    - Bookshare
4. Virtual Learning Experiences: Virtual field trips, simulations, and other online experiences provide engaging and interactive learning opportunities for students with physical disabilities. Some examples include:

- Google Expeditions
- Smithsonian Learning Lab
- Virtual Field Trips from Discovery Education
- The Ability Experience from Ability360

Learning Disabilities

Autism Spectrum Disorder

It is important to remember that each learner is unique and may require individualized accommodations and support to succeed in their learning environment. The term "learner disabilities" refers to a range of physical, cognitive, and emotional conditions that may impact an individual's ability to learn or engage in educational activities. There are several different types of learning disabilities that impact a person's ability to acquire and use specific skills such as reading, writing, and math. The following are the some of the learning disabilities shown in Figure:

Figure 4. Various Disabilities based on Autism

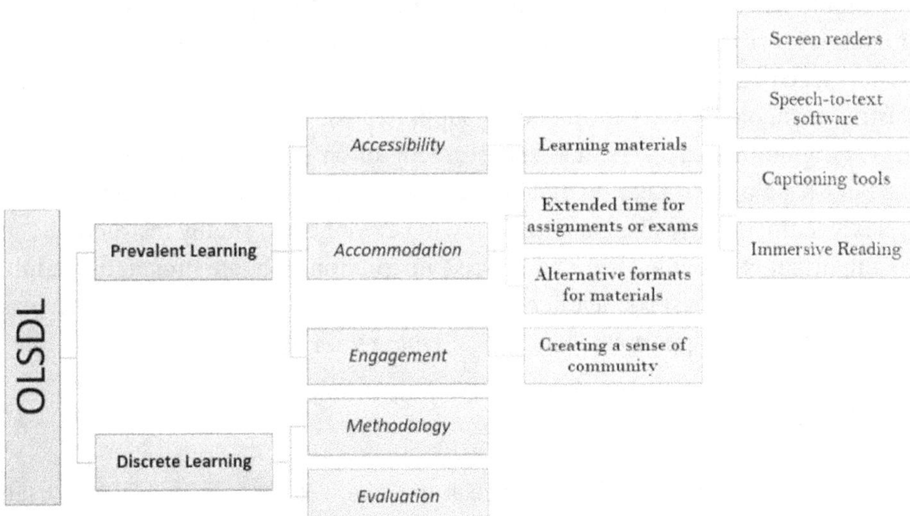

Dyslexia: A reading disability that makes it difficult to recognize words, spell accurately, and read fluently. Dysgraphia: A writing disability that makes it difficult to write legibly, organize thoughts on paper, and use proper grammar and punctuation. Dyscalculia: A math disability that makes it difficult to understand and work with numbers, perform calculations, and solve math problems. Attention Deficit Hyperactivity Disorder (ADHD): A condition that makes it difficult to focus, pay attention, and complete tasks. Auditory Processing Disorder (APD): A condition that makes it difficult to understand spoken language and follow directions. Visual Processing Disorder (VPD): A condition that makes it difficult to interpret visual information, recognize shapes, and distinguish between similar objects. Non-Verbal Learning Disability (NVLD): A condition that impacts social skills, spatial awareness, and understanding of non-verbal cues. It's important to note that learning disabilities vary greatly from person to person, and many individuals may have more than one type of learning disability. Early identification and appropriate support greatly improve the academic and personal success of individuals with learning disabilities.

Dyslexia

Online learning is a great resource for learners with dyslexia as it offers flexibility, customization, and accessibility. Dyslexic learners may find it easier to read text when a dyslexia-friendly font is used. Some examples of dyslexia-friendly fonts include OpenDyslexic and Dyslexie. Dyslexic learners often benefit from using multimedia resources such as videos, images, and interactive activities. These help to reinforce learning and make it more engaging. Text-to-speech software helpful for dyslexic learners as it reads out loud the text on the screen. Some examples of text-to-speech software include Read&Write and NaturalReader. There are many assistive technology tools available to support dyslexic learners, such as spell-checkers, grammar-checkers, and speech recognition software. Dyslexic learners may find it helpful to take regular breaks to rest their eyes and prevent fatigue. Dyslexic learners may struggle with following complex or lengthy instructions, so it is important to provide clear and concise instructions that are broken down into manageable steps. Dyslexic learners may benefit from additional support, such as a tutor or learning support assistant, to help them with their studies.

Online Tools

1. Text-to-Speech Tools: Text-to-speech tools help students with dyslexia listen to digital text instead of reading it, which improves reading comprehension and reduces reading fatigue. Some examples include:
    - NaturalReader
    - Read&Write by Texthelp
    - Snap&Read by Don Johnston
    - Microsoft Learning Tools

2. Dyslexia-Friendly Fonts: These fonts are designed to make reading easier for people with dyslexia. They have specific features such as increased spacing between letters and varying shapes for each letter to make them easier to distinguish. Some examples include:
   - OpenDyslexic
   - Dyslexie font
3. Lexend font: Audiobooks and EBooks: Audiobooks and ebooks help students with dyslexia engage with reading materials by providing an alternative way to access text. Some platforms that offer audiobooks and ebooks include:
   - Learning Ally
   - Bookshare
   - Audible
4. Multi-Sensory Learning Tools: These tools engage multiple senses to help students with dyslexia process information more effectively. Some examples include:
   - Nessy
   - Orton-Gillingham Online Academy
   - Reading Rockets
5. Writing and Spelling Tools: These tools help students with dyslexia improve their writing and spelling skills through features such as word prediction and speech-to-text. Some examples include:
   - Co: Writer Universal
   - Grammarly
   - Google Docs voice typing

Dysgraphia

Dysgraphic learners may find it easier to type than to write by hand. Allow them to complete assignments on a computer using word processing software. Speech-to-text software helpful for dysgraphic learners who struggle to write. They speak their ideas, which the software then transcribes into text. Consider offering alternative assignments that don't require a lot of writing, such as video presentations or verbal responses. Using digital tools such as note-taking apps or online whiteboards is helpful for dysgraphic learners as it allows them to organize their thoughts and ideas without struggling to write them down. Dysgraphic learners may need more time to complete written assignments. Consider giving them extra time or breaking assignments into smaller parts to make them more manageable. There are many assistive technology tools available to support dysgraphic learners, such as spell-checkers, grammar-checkers, and speech recognition software. Dysgraphic learners may find it helpful to take regular breaks to rest their hands and prevent fatigue.

Dysgraphic learners may benefit from additional support, such as a tutor or learning support assistant, to help them with their studies.

Online tools for Dysgraphia

1. Speech-to-Text Tools: Speech-to-text tools help students with dysgraphia convert spoken words into written text, which improve writing speed and accuracy. Some examples include:
    - Google Docs voice typing
    - Dragon NaturallySpeaking
    - Windows Speech Recognition
    - Apple Dictation
2. Word Prediction Software: Word prediction software helps students with dysgraphia improve their spelling and writing speed by suggesting words as they type. Some examples include:
    - Co: Writer Universal
    - WordQ
    - Ginger Software
3. Typing Programs: Typing programs help students with dysgraphia improve their typing speed and accuracy, which is helpful when taking notes or writing papers. Some examples include:
    - TypingClub
    - Typing.com
    - NitroType
4. Graphic Organizers: Graphic organizers help students with dysgraphia organize their thoughts before they begin writing, which makes the writing process less overwhelming. Some examples include:
    - Inspiration
    - Mindomo
    - Lucidchart
5. Audiobooks and E-Books: Audiobooks and e-books help students with dysgraphia engage with reading materials by providing an alternative way to access text. Some platforms that offer audiobooks and e-books include:
    - Learning Ally
    - Bookshare
    - Audible

Dyscalculia

Dyscalculia learners often benefit from visual aids such as diagrams, charts, and graphs. Online learning platforms often provide these aids, so be sure to take advantage of them. Dyscalculia learners may struggle with complex mathematical concepts. Online learning platforms often provide lessons that are broken down into smaller, more manageable segments. Make use of these resources to help you understand difficult topics. Dyscalculia learners may benefit from hands-on activities that allow them to physically manipulate objects. Online learning platforms often provide virtual manipulatives and games that are used to reinforce mathematical concepts. Dyscalculia learners may need to repeat concepts several times to fully understand them. Online learning platforms often provide practice problems and quizzes that are used to reinforce concepts. Dyscalculia learners may benefit from additional support such as tutoring or working with a learning specialist. Many online learning platforms offer these services, so be sure to take advantage of them if needed.

Online Tools for Dyscalculia:

1. Math Manipulatives: Online math manipulatives help students with dyscalculia visualize math concepts and make abstract ideas more concrete. Some examples include:
    - National Library of Virtual Manipulatives
    - Mathigon
    - Desmos
2. Dyscalculia-Friendly Math Fonts: These fonts are designed to make math equations and symbols easier to read for people with dyscalculia. Some examples include:
    - DyscalculiaFonts
    - OpenDyslexic Math
3. Math Games and Puzzles: Math games and puzzles help students with dyscalculia engage with math concepts in a fun and interactive way. Some examples include:
    - Math Playground
    - Hooda Math
    - Math Games by PBS Kids
4. Audio and Visual Math Resources: These resources help students with dyscalculia learn math concepts through audio and visual cues, which improve understanding and retention. Some examples include:
    - Math Antics
    - Khan Academy

  - Math is Fun
5. Math Assistive Technology: There are a variety of assistive technologies that help students with dyscalculia access online math materials. Some examples include:
  - Read&Write by Texthelp
  - MathType
  - MyScript Calculator

Attention Deficit Hyperactivity Disorder(ADHD)

People with ADHD benefit from a structured routine. Create a daily schedule that includes time for online learning, breaks, and other activities. Online learning full of distractions, so it's essential to minimize them. Turn off notifications, close unnecessary tabs, and use noise-cancelling headphones to block out background noise. Individuals with ADHD may have difficulty focusing for long periods. Breaking up lessons into smaller, more manageable chunks helps maintain focus and engagement. Visual aids, such as diagrams and videos, helpful for individuals with ADHD. They help to simplify complex concepts and make learning more engaging. Interactive activities, such as quizzes and games, help to keep learners engaged and motivated. Individuals with ADHD benefit from frequent feedback on their progress. Providing regular feedback help to keep learners on track and motivated. If online learning is not sufficient, consider online tutoring. An online tutor provide personalized attention and support to help individuals with ADHD succeed academically.

Online Tools for ADHD

1. Time Management Tools: Time management tools help students with ADHD stay organized and manage their time effectively. Some examples include:
  - Trello
  - Google Calendar
  - MyStudyLife
2. Focus Tools: Focus tools help students with ADHD minimize distractions and stay on task. Some examples include:
  - Focus@Will
  - Brain.fm
  - Forest App
3. Study and Note-Taking Tools: These tools help students with ADHD take effective notes and study more efficiently. Some examples include:
  - Evernote

- OneNote
- Quizlet

4. Mind Mapping Tools: Mind mapping tools help students with ADHD visualize and organize their thoughts, which improve their ability to brainstorm and plan. Some examples include:
   - MindMeister
   - XMind
   - Coggle
5. Study Groups and Online Communities: Joining a study group or online community provides support and accountability for students with ADHD. Some examples include:
   - ADHD Online Support
   - ADHD Reddit Community
   - ADDitude Magazine

Figure 5. Framework for Disabled Learners using Online Tools

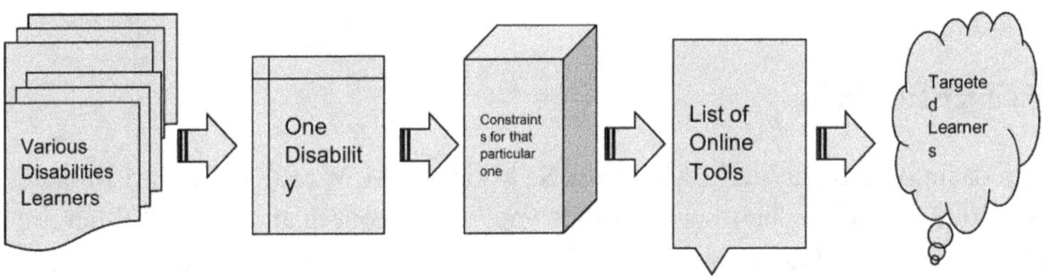

Procedure for the Proposed Framework

1. Load different disabilities learners in the database.
2. Identify the type of Disability- As mentioned in the above para, they are different forms of learners. First step in the framework is to identify the type of disability.
3. Build the Constraints - Depending upon the identified disability, a set of constraints are built. Constraints are the variables that are used to define what type of disability the learners are and their related digital tools suggestion.
4. Establish the Online Tools - Based upon the constraints the online learning tools are established and suggested for the respective learners
5. Provision to the Learners - Amongst all the online learning tools provided, the respective learners use the online tools depending on the interest.

The framework is given to suggest different tools for various disabled learners. For each disability, a list of digital resources are provided in this chapter.

CONCLUSION

This chapter concludes that the Equitable and Inclusive Online Learning framework provides a comprehensive approach for supporting students with disabilities in online learning environments. The framework emphasizes the importance of creating an inclusive culture, providing accessible digital content, using universal design principles, offering multiple means of engagement and representation, and leveraging assistive technology. This chapter provides various online educational resources with respect to the learners knowledge and their ability to learn the subjects. By implementing this framework, educators ensure that students with disabilities have equal access to educational opportunities and achieve success in online learning environments. It is crucial to prioritize equity and inclusivity in online learning, and this framework provides a valuable guide for achieving those goals.

REFERENCES

Aljedaani, W., Krasniqi, R., Aljedaani, S., Mkaouer, M. W., Ludi, S., & Al-Raddah, K. (2022). If online learning works for you, what about deaf students? Emerging challenges of online learning for deaf and hearing-impaired students during COVID-19: A literature review. *Universal Access in the Information Society*, •••, 1–20. PMID:35910240

Alnaim, N., & Alsarawi, A. (2023). Obstacles To Distance Education For Students With Learning Disabilities And Ways To Face Them: From The Point Of View Of Female Teachers. *Turkish Online Journal of Distance Education*, *24*(1), 200–219. doi:10.17718/tojde.1076073

Beyene, W. M., Mekonnen, A. T., & Giannoumis, G. A. (2023). Inclusion, access, and accessibility of educational resources in higher education institutions: Exploring the Ethiopian context. *International Journal of Inclusive Education*, *27*(1), 18–34. doi:10.1080/13603116.2020.1817580

Cavanaugh, C., Repetto, J., Wayer, N., & Spitler, C. (2013). Online learning for students with disabilities: A framework for success. *Journal of Special Education Technology*, *28*(1), 1–8. doi:10.1177/016264341302800101

Cinquin, P. A., Guitton, P., & Sauzéon, H. (2019). Online e-learning and cognitive disabilities: A systematic review. *Computers & Education*, *130*, 152–167. doi:10.1016/j.compedu.2018.12.004

Cooper, M. (2006). Making online learning accessible to disabled students: An institutional case study. *Research in Learning Technology*, *14*(1), 103–115. doi:10.3402/rlt.v14i1.10936

Crow, Kevin L. "Four types of disabilities: Their impact on online learning." TechTrends 52.1 (2008): 51.

Griful-Freixenet, J., Struyven, K., Verstichele, M., & Andries, C. (2017). Higher education students with disabilities speaking out: Perceived barriers and opportunities of the universal design for learning framework. *Disability & Society*, *32*(10), 1627–1649. doi:10.1080/09687599.2017.1365695

Guglielman, E. (2010, September). E-learning and Disability: Accessibility as a Contribute to Inclusion. In EC-TEL Doctoral Consortium (pp. 31-36).

Khribi, M. K., Othman, A., & Al-Sinani, A. (2022, July). Toward Closing the Training and Knowledge Gap in ICT Accessibility and Inclusive Design Harnessing Open Educational Resources. In *2022 International Conference on Advanced Learning Technologies (ICALT)* (pp. 289-291). IEEE. 10.1109/ICALT55010.2022.00093

Kim, J. Y., & Fienup, D. M. (2022). Increasing access to online learning for students with disabilities during the COVID-19 pandemic. *The Journal of Special Education*, *55*(4), 213–221. doi:10.1177/0022466921998067

Pirani, Z., Molvizadah, V., Sayyed, M. A., & Sasikumar, M. (2013). E-learning framework for learning disabled children. *International Journal of Computer Applications*, *63*(19), 38–42. doi:10.5120/10577-5703

Rao, K., Edelen-Smith, P., & Wailehua, C. U. (2015). Universal design for online courses: Applying principles to pedagogy. *Open Learning*, *30*(1), 35–52. doi:10.1080/02680513.2014.991300

Rice, M. F. (2022). Special education teachers' use of technologies during the COVID-19 era (Spring 2020—Fall 2021). *TechTrends*, *66*(2), 310–326. doi:10.100711528-022-00700-5 PMID:35098256

Roberts, J. B., Crittenden, L. A., & Crittenden, J. C. (2011). Students with disabilities and online learning: A cross-institutional study of perceived satisfaction with accessibility compliance and services. *The Internet and Higher Education*, *14*(4), 242–250. doi:10.1016/j.iheduc.2011.05.004

Seale, J. K. (2013). *E-learning and disability in higher education: accessibility research and practice*. Routledge. doi:10.4324/9780203095942

Shater, A., A. J. AlMahdawi, and M. A. S. Khasawneh. "The Digital Learning of Disabled Students: Perceptions of Teachers in Public Schools." (2023).

Shater, A., AlMahdawi, A. J., & Khasawneh, M. A. S. (2023). The Digital Learning of Disabled Students: Perceptions of Teachers in Public Schools.

Skaraki, E. (2023). Designing educational material to teach Braille to adult educators through the method of distance learning. *Advances in Mobile Learning Educational Research, 3*(1), 602–609. doi:10.25082/AMLER.2023.01.009

Smith, S. J., & Basham, J. D. (2014). Designing online learning opportunities for students with disabilities. *Teaching Exceptional Children, 46*(5), 127–137. doi:10.1177/0040059914530102

Smith, S. J., & Basham, J. D. (2014). Designing online learning opportunities for students with disabilities. *Teaching Exceptional Children, 46*(5), 127–137. doi:10.1177/0040059914530102

Tarconish, E., Lombardi, A., & Taconet, A. (2023). Disability Awareness & Inclusive Teaching Online Training Videos for College Instructors Featuring Students with Disabilities. *Journal of Postsecondary Education and Disability, 35*(4).

Tarconish, E., Lombardi, A., & Taconet, A. (2023). Disability Awareness & Inclusive Teaching Online Training Videos for College Instructors Featuring Students with Disabilities. *Journal of Postsecondary Education & Disability, 35*(4). Cooc, N. (2023). National Trends in Special Education and Academic Outcomes for English Learners With Disabilities. *The Journal of Special Education*, •••, 00224669221147272.

Thapliyal, M., Ahuja, N. J., Shankar, A., Cheng, X., & Kumar, M. (2022). A differentiated learning environment in domain model for learning disabled learners. *Journal of Computing in Higher Education, 34*(1), 60–82. doi:10.100712528-021-09278-y

Wald, M. (2009). E. A. Draffan, and Jane Seale. "Disabled learners' experiences of e-learning.". *Journal of Educational Multimedia and Hypermedia, 18*(3), 341–361.

Chapter 3
Youth Aspirations Towards Industry 4.0 Job Requirements:
The Example of the Serbian Labor Market

Mihajlo Dusan Djukic
https://orcid.org/0000-0001-5677-330X
Institute of Economic Sciences, Belgrade, Serbia

Dejana Pavlovic
https://orcid.org/0000-0001-5090-302X
Institute of Economic Sciences, Belgrade, Serbia

ABSTRACT

A push towards digitalization and a solid inflow of foreign investments has provided a basis to create new jobs and facilitate a smoother transition of youth into the Serbian labor market. Although still inordinately high when compared to its EU counterparts, the unemployment rate youth has decreased over the last decade following improved macroeconomic trends. Further positive outcomes necessitate targeted, labor market policies focused on strengthening employability - of interest to research are changes in the external environment (particularly Industry 4.0 requirements) and deficiencies caused by rigidity in formal education. Taking into consideration youth career aspirations and capacities, a Principal Component Factor Analysis was applied to data collected from the survey of youth (18-29 years old, n=125). The results help provide conclusions on youth attitudes related to learning preferences, additional skills required and the self-estimation of their digital-skill capacities. The analysis extracts five factors relevant for developing effective active labor market measures.

DOI: 10.4018/978-1-6684-9089-1.ch003

Due to the constant imperative to maintain and strengthen economic competitiveness, there is a need to adopt sustainable development policies whose effectiveness may be monitored, evaluated and continuously improved on. Systemic evidence-based support to youth is of great importance, as it serves as a foundation for future and sustained growth being very important in the context of optimization of human and labor resources in the following decades. Identifying areas in which there are good bodies of synthesized knowledge to inform policy and increasing investments aimed at obtaining evidence on youth employment interventions are necessary for developing systemic support to vulnerable youth (Apunyo et al. 2022). The value of research evidence is particularly important for dealing with youth facing multiple barriers (Thiem & Dasgupta, 2022) or within specific policy environment (e.g. COVID-19), (Ognjenovic & Djukic, 2023). Digitalization presents a challenge and opportunity at the same time, since dynamics of Industry 4.0 in the present and the future present a reality that makes changes imperative for Education 4.0 (Gonzalez-Perez & Ramirez-Montoya, 2022). Currently 73 million youth (15 to 24 years old) in 2022 either do not have a job or is looking for one but does not possess the appropriate skills or competencies to be fully employed (ILO, 2022). Digitalization of business processes provides various job opportunities for youth. However, not all of them possess knowledge required to adapt to the Industry 4.0 requirements and the jobs of "new generation". To ensure that as many youth are gainfully employed as possible, it is necessary to devote attention to vulnerable groups of youth who face disproportionate difficulties when acquiring digital skills and entering the labor market. Their digital literacy is prerequisite for both being competitive for the jobs created under Industry 4.0 revolution and further learning purposes. Therefore, the central idea of the paper is to provide insights into youth preferences related to acquiring adequate digital skills, their preferences and desirable capacity building mechanisms that could help them enhance their employability. This is particularly relevant for the youth with lower levels of education, ones residing in rural areas or facing other constraints with regards to participating in the labor market (i.e. youth with disabilities, carers, members of minority ethnic groups, etc.).

The overall global youth unemployment rate (for those 15 to 24 years of age) at the end of 2022 amounted to 15.6% (World Bank, 2023), with noticeable differences among regions. The worst situation could be detected in Middle East and North Africa which recorded youth unemployment rate of 25%, whereas Europe and Central Asia as well as East Asia and Pacific performed much better with rates of 15.3% and 11.7%, respectively (World Bank, 2023). Although the youth unemployment rate is determined by a range of factors, some of them are not directly correlated with economic development and other trends might indicate the effectiveness of youth labor policies under different socio-economic contexts. Developing work-school programs specific to each country and sector can allow the formation of human capital that is more in line with the demands in the labor market (Kang, 2021). As is the case with other countries in the Western Balkans, Serbia faces high and tenacious rates of youth unemployment (Pavlovic et al., 2019; Stanojevic & Pavlovic, 2022; Zubovic el al., 2015). According to the Statistical Office of the Republic of Serbia (SORS), the unemployment rate among youth 15-24 years of age in Serbia was 26.4% in 2021, while youth unemployment rate for youth age 15-29 was 20% for the same year (SORS, 2021). Youth unemployment rates in Serbia, similarly to the most of other European economies, are considerably lower if compared to a decade ago, benefiting from favorable macroeconomic trends. However, further decrease of youth unemployment could be hardly possible just resulting from macroeconomic stability even if that causal relationship exists. Youth unemployment rates have been on the rise since the 2008 global financial crisis (Tamesberger et al., 2014). In the Eastern Europe group of countries, the level of youth unemployment clearly follows macroeconomic situation showing decline from 26.1% to 15.4% over the period 2000-2008. Following the global economic crisis 2008-2009, the rates increased again climbing to 27% in 2013 and remaining stable afterwards. Improvements could be again noticed in the years prior to COVID-19 onset when rate halved to 12.7% in 2018 (Liotti, 2022). Furthermore, most Balkan countries are facing high unemployment rates among young people after 2008. Youth unemployment was hard hit originally due to the lingering effects of the financial crisis, which saw more than half of all youth becoming unemployed in Bosnia & Herzegovina, Northern Macedonia, Greece and Serbia. Despite of implementation of numerous strategies and policies to support them when entering in the labor market, young people in the Balkans continue to have difficulties (Figure 1), as evidenced by their unemployment rates which are still above the EU average (EUROSTAT, 2023; ILO, 2023).

It is questionable whether macroeconomic policy measures aimed to, for example to attract FDI, could be effective for reducing unemployment in the long term (Zdravkovic et al., 2017). More probably, that reform processes in the education system (Ivic & Pesikan, 2012), tax policy aimed at reducing inequality (Vladisavljevic et al., 2021) or other measures dealing with systemic barriers to labor market participation (Lebedinski & Vladisavljevic, 2022) could facilitate smooth transition from education to labor market. Concern with youth unemployment has been focusing on the group of youth who are not in education, employment, and training – NEET. There is an equally greater issue of NEET youth, those who are neither in employment nor in education or training from 15 to 29 years of age (the so-called "NEET rate"). In Serbia, NEET rate for people age 15-24 was 16.4% in 2021. In 2021, the NEET rate among 15-29 year-olds in Serbia was 18.8%, interrupting a long-term trend of decline due to 0.8% increase (year-on-year). The EU NEET rate for 2021 was 13.1% among youth within the same age group but not uniform among EU members. In 2021, the Netherlands and Sweden recorded the lowest NEET rates of 5.5 and 6%, respectively, while Italy and Romania recorded the highest, at 23.1 and 20.3%, respectively (Eurostat, 2021). Even when gaining employment, low-skilled or unskilled youth are at a significantly higher risk of becoming unemployed within their first year of employment (Pavlovic et al., 2017; Pavlovic et al., 2019). Moreover, in order to keep their first job, young people need to possess technical and core work skills such as problem solving, team-work, communication and learning to learn (Brewer, 2013). Rapid shifts in technological development have resulted in the necessity of possessing digital skills and competencies in the labor market. As low- and mid-skilled occupations are becoming precarious due to ongoing automation, the need for advanced analytical and technological skills is equally growing (OECD, 2015). Skill shortages result in difficulties obtaining reliable and long-term employment (Baird & Parayitam, 2019) as well as reduce employee performance (Akman & Turhan 2018). Accordingly, low-skilled youth are forced into poorly paid, precarious employment (OECD & ILO, 2014). Furthermore, due to increased automation, low-skilled youth coming from a vocational background are at a higher risk of becoming NEET as they are also forced to move from one precarious job to another as their skills grow outdated in relation to the job market (ILO, 2020).

Figure 1. Unemployment rate of youth in % (16-29 years old)
Source: EUROSTAT

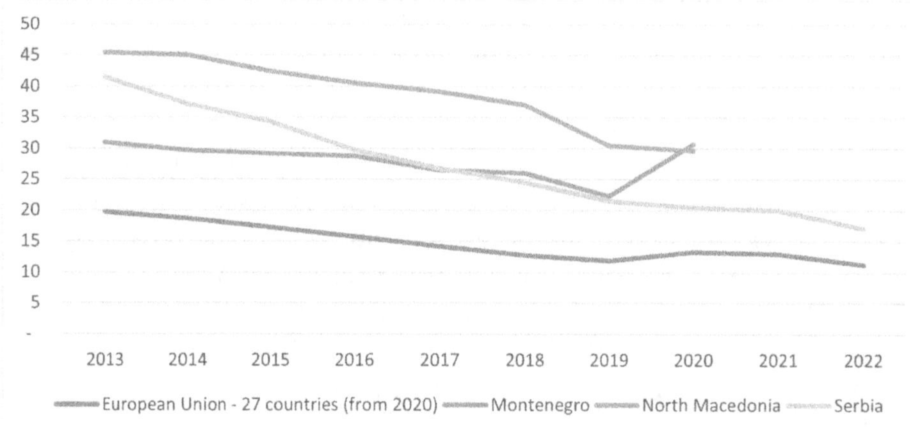

High unemployment rates among Serbian youth are commonly associated with misalignment between labor market demand and the supply of graduates in specific fields (Bijelovic Bosanac et. al., 2017), skill mismatch (Vasic, 2019) and a lack of relevant work experience (Pavlovic et al., 2019) as well as an absence of interest in gaining competencies beyond one's formal education (Jojkic & Mitrovic 2017).

Increased joblessness, in parallel with changing labor market prospects due to a greater prevalence of non-standard employment, has caused an interrupted school to work transition and has delayed youth's entrance into the labor market (Contini et al., 2019). Tamesberger & Bacher (2014) argue that conventional youth labor market indicators have limited significance as they fail to include individuals who are not in search of work or are detached from the labor market.

With regards to digital skills, Serbia performs well if global comparisons of digital literacy and skills are considered (Wiley, The Digital Skills Gap Index – DSGI, 2023). As per Wiley Digital Skills Gap Index (DSGI), with a score 4.8 out of 10, Serbia is ranked 34th among Europe and Central Asia countries, and 64th in overall. Singapore (7.8), United Arab Emirates (7.5) and Finland (7.5) hold leading position. However, the country is still lagging to the most of the EU peers and even some of the Western Balkan countries (Albania and Montenegro scored 5.2 and 5.0 respectively). Despite noticeable catch-up tendencies, participation in digital education in Serbia is still below EU average (Figure 2.). As digital skills are increasingly a basic need among all employers for all age groups and across economic sectors, this paper uses ICT skills as a determining factor in analyzing youth. The research focuses on the three important aspects of youth preparedness for the digital jobs of new generation and job market in general. Firstly, it has been assessed youth preferences on the specific capacity building areas that could improve their position in the labor market (i.e. software development, project management, digital security, etc.). The underlying idea is to assess what are the specific knowledge areas that youth consider relevant for strengthening their position in the labor market. Secondly, there have been analyzed youth attitudes towards general job requirements they lack and consider the most important for getting a job (i.e. high school/faculty degree, work experience, professional education and certificate, etc.). For this purpose, there have been listed job requirements that are the most common selection criteria listed in job posters. Finally, third aspect includes self-assessment of the current knowledge of basic digital skills (i.e. windows, MS office, internet search, etc.) necessary for conducting most of the jobs and for further learning purposes.

The paper is structured as follows. In the following chapter are provided state of the art insights in analyzing youth employment issues, particularly research papers applying factor analysis in assessing youth skills, attitudes and determinants of effective youth employment policies. The third chapter provides detailed elaboration of the research design, applied methodology and data collection methods. Research results and related discussion are presented in the fourth chapter. Finally, fifth section outlines the most relevant findings and provide recommendations for the policy makers and other stakeholders including business representatives, civil society organizations, and youth workers.

Figure 2. Individuals doing an online course (of any subject) in % (16-29 years old)
Source: EUROSTAT

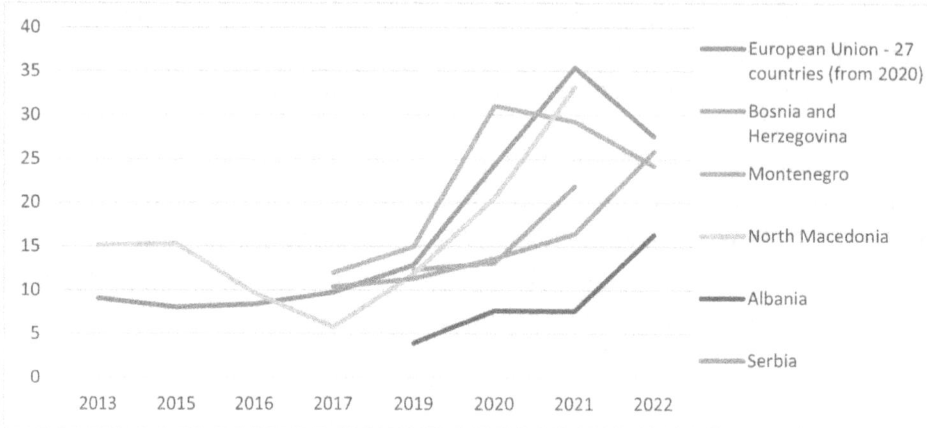

LITERATURE REVIEW

A plethora of studies worldwide have investigated employers' expectations of young professionals entering the labor market (Pang et al., 2019), many concluding there to be a deficiency or outright lack of skills in youth. Based on a literature review, Tsitskari et al. (2017) argue that, in addition to offering an academic degree, undergraduate programs often fail to equip students with the skills needed in the current job climate. This is a common perception among both employers and students in Serbia, who believe that educational institutions fall short of providing undergraduates with key ICT and soft skills (Matijevic & Šolaja, 2020) that are vital for the Industry 4.0. (Picatoste et al., 2018). Skill shortages can result in difficulties in obtaining employment (Baird & Parayitam, 2019) as well as poor employee performance (Akman & Turhan, 2018). Based on research surveying 361 graduate and under-graduate students, Islam (2022) reports that students are aware of the changing job market scenario, and they are trying to have those skills which will make them competent compared to the early years, but they are not prepared enough to accept the challenges faced in industry 4.0.

Overall, there is a significant demand for developed digital skills within the labor market as Picatoste et al. (2018) underscored that almost half of all new positions require them. Through analysis of the job advertisements aiming to identify skills and knowledge relevant for specific areas, Pejic-Bach et al. (2020) indicate that Industry 4.0 organizations are looking not for the traditionally educated industrial, software or electrical engineers, but for the experts with the multidisciplinary skills, which can help factories to create and improve low-end and high-end machine and device capabilities, using knowledge and skills in embedded and distributed system development. Further, according to Digital Opportunity traineeship, an EU initiative for developing digital competences and increasing the employment chances of youth, skills related to machine learning, quantum technology, big data and cybersecurity are highly in demand within the EU labor market. In addition, the initiative points to the significance of digital marketing competences, web design and software development, which are vital to run a successful business. Accordingly, a report from the association IAB Europe on human capital in the digital environment also highlights the importance of digital business skills in the contemporary labor market (IAB, 2019). Moreover, information and analytical skills together with competencies in the field of cross-media and social media are among the main training areas pertinent to recruiters, as found by a study conducted among employers and training specialists in the EU-28.

In the context of acquiring digital skills, diverse labor market preferences and constraints, as well as dissimilitude in competencies gained, the principal component technique of factor analysis has been often employed to help researchers in detecting the most important factors of youth professional aspirations as well as associations between specific determinants. Most studies conducted are within employee job satisfaction, employment expectations, digital competences, unemployment determinants and informal education programs. This technique has been employed in multiple studies such as Ozdamar-Keskin et al. (2015) who analyzed data gathered from 20,172 open and distant learners from Turkey. Aiming to group and classify the attitudes and statements of learners in their personal learning preferences, problem solving skills, project-work skills and abilities to use digital tools for learning purposes, their results indicate learners possess problem-solving and project-working skills which might help them overcome deficiencies resulting from poor formal education. However, low digital literacy limits their potential to utilize communication technologies for more demanding jobs and further learning purposes, with the authors going as far as to suggest that students prefer visual learning environments while being reluctant to collaborative online group work as found in modern business environments. They conclude that investing more effort into developing a collaborative approach to education is necessary to strengthen employability among students and life-long learners.

Digital competence is not merely important from the aspect of the Industry 4.0 requirements, but also as a general determinant of job confidence. Since belief in oneself, associated with courage and working independence, highly impacts job satisfaction (such as that job does not affect the private life) (Pavlovic et al., 2016) and future job and learning prospects (Clark, 2011; Lee, 2013). Todhunter (2015) analyzed the relationship between computing competence and confidence while focusing on tasks and circumstances under which this relationship might be observed.

The targeted population included first- and second-year nursing students between July 2008–April 2009, of whom a total of 375 students were examined (accounting for the 18.75% of the total student population). The study was conducted through a questionnaire based on a 4-point ordinal scale assessing their confidence in using specific personal computer applications (1 – Strongly disagree and 4 – Strongly agree). The results clearly showed a positive relationship between computing competences and confidence among nursing students, as users of ICT and computer-related resources, but later adopters of ICT technologies. The observable features included the ability to work independently and employ computing skills for further learning. Confirming the importance of rising ICT adoption in nurse education curricula, the authors indicated that frequent assessment of the learning outcomes is necessary to measure progress in adopting ICT skills and working familiarity in a computer environment.

León-Pérez et al. (2020) also shed particularly valuable light into how ICT skills may be achieved. Using a questionnaire method for self-assessment of the digital skills required for higher education purposes, the authors confirmed that students primarily use digital technology within the broader scope of academic projects. Digital technology skills have proved to be highly beneficial in terms of information management, critical thinking and managing mobile devices. However, research has shown that the use of ICT in teaching remains underutilized. Although students reportedly do not believe that ICT in the classroom could be helpful in terms of further developing digital skills, they consider academic projects as highly effective in gaining such skills. These results have also been presented by Oblinger (2003) who concluded there to be a gap between what students require from academic institutions and what their education offers them.

In terms of analyzing youth aspirations and expectations when conducting job searches, Pavlovic et al. (2016) argued that youth in Serbia prefer safe jobs characterized by low turnover. Relatively high-risk aversion in terms of job mobility might stem from the still present legacy of socialism and a job market characterized by the dominant role of the state as the employer (Gligorov et al., 2011), lack of labor mobility and underdeveloped entrepreneurial culture resulting from high corruption and tax evasion (Vladisavljevic et al., 2022) and similar issues. Earning an above-average salary is therefore a less important employment factor in Serbia for youth at the start of their career when compared to learning and training possibilities that could yield better opportunities in the future.

METHODOLOGY

The objective of this paper was to provide a detailed insight into youth' perspectives on adapting to the changing job market requirements under the Industry 4.0 context. Specifically, we aimed at providing response to the following research questions:

1. What are the youth preferences and perceptions in terms of capacity building and current knowledge with regards to adapting to Industry 4.0?
2. What are the attitudes of youth with regards to overall job requirements nowadays and what do they consider the most important factors of getting an adequate job?
3. What is the nature of relationship between youth capacities, aspirations and further learning potentials in the Industry 4.0 context?

To analyze the relationship between youth capacities, aspirations and further learning potential in conjunction with the basic digital skills necessary for adapting to Industry 4.0 requirements, this study applies a Principal Component Analysis (PCA) as a statistical data reduction technique falling under the factor analysis method. The rationale in applying this method is for the analysis to be self-reported data that may determine the strength of correlation between youth aspirations and competences, including preferences related to digital skills learning tools.

The research applied a questionnaire, as it is the most frequently used tool to measure attitudes and perceptions of the analyzed population of youth 18-29 years old. The online questionnaire was distributed to youth associations and civil society organizations working with the youth, particularly NEET youth or youth at risk of becoming NEET. The online questionnaire consisted of 23 questions divided into 4 separate sections:

1. Socio-demographic characteristics containing questions related to gender, age or area in which the respondent lives
2. Employment and employment opportunities consisting of 6 questions, the answers to which help define the respondent's current status in the labor market, the possible causes of their unemployment and their experience with informal training
3. Skills analyzing the degree to which skills have been developed, attitudes on improving their position in the labor market and career aspirations
4. Active training policies examining the motivation and opportunities to improve skills

Following a detailed review of the literature, the survey was developed using a five-point Likert-type scale ranging from strongly disagree (1) to strongly agree (5). Questions were developed respecting the core work-related skills as provided by the World Economic Forum (WEF, 2016) and consulting methodologies used in similar assessments of basic digital skills and learning preferences conducted by many authors over the last decade (Helsper et al., 2016; Van Deursen et al., 2014). The instrument was designed to be completed in 15 minutes to encourage higher participation with questions targeted to assess participants' self-assessment of youth preferences in specific capacity building areas that might be able to improve their position in the labor market, attitudes towards job requirements and self-assessment of their basic digital skills.

The research was conducted from October to December 2021. The total number of respondents was 125.

RESULTS

Sample Characteristics

Demographic analysis has been performed to assess the respondents' background including gender, age, settlement and the current working status. There have been also collected data on time preferences regarding eventual enrollment of vocational education programs, use of smartphones and computers, daily internet access and previous experience in vocational education programs. In overall, the sample is well balanced with regards to age (with 38.4% respondents are female) and age (35.2% falling under 19-24 age group). The majority of respondents live in urban areas. Around 72.8% and 10.4% live in cities and suburban areas respectively. In terms of current labor market status, about one third of the respondents are currently in NEET status (32.8%), meaning that they are neither employed or being in the process of education. Around half of the respondents are students or participating in informal education programs. All of them are interested in acquiring additional digital skills that could enhance their employability and increase their chances of getting adequate job in the future. However, their preferences with regards to the time they are willing to spend on further education significantly varies ranging from up to 5 hours (13.6%), 5-10 hours (28.8%), 10-20 hours (25.6%) and above 20 hours per week (21.6%). Almost all of them are well equipped, having daily internet access and smartphones (97.6%). More than three quarters of the participants own a personal computer (77.6%).

Data Analysis

Sampling adequacy could be assessed satisfactory with KMO Kaiser–Meyer–Olkin (KMO) measure at 0.81, which confirms sample reliability well exceeding the generally acceptable limit of 0.5 (Field, 2009). Cronbach's alpha for the final instrument was calculated at 0.87 which reflects a high degree of internal consistency. Additionally, Bartlett's test of sphericity shows p value is less than 0.05, which helped us rejecting the null hypothesis that the variables were not correlated (Table 1).

Table 1. Bartlett's test of sphericity, KMO and Cronbach Alpha

Bartlett test of sphericity	
Chi-square	1127.52
Degrees of freedom	136
p-value	0.000
H0: variables are not intercorrelated	
Kaiser-Meyer-Olkin Measure of Sampling Adequacy	
KMO	0.812
Test scale = mean(unstandardized items)	
Average interitem covariance:	0.510194
Number of items in the scale:	17
Scale reliability coefficient:	0.8695

Source: Authors' calculations

Following performed sample adequacy calculations, it could be concluded that Principal Component Analysis (PCA) is an appropriate method of analysis for the observed 125 items (greater than 100 items). After setting minimum Eigenvalues at 1.0 which is a common criterion (Morgan et al., 2011), there have been confirmed five significant components that accounted for 72.2% of the total variance in the original data. The vertical axis indicates the variance explained by each component on the horizontal axis (Figure 3).

PCA has been performed to assess the underlying structures of the analyzed 17 variables. In line with the above explained determination of the relevant constructs, five components were identified. After performing the Varimax rotation method there have been detected subcomponents/items with result loading above 0.4 for each of the selected components.

Figure 3. PCA application and minimum Eigenvalues per component

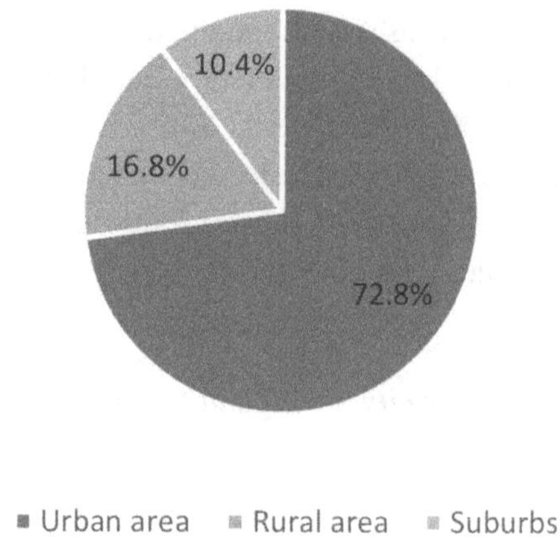

Component 1 – Basic Digital Skills

The subcomponents behind the Component 1 refer to self-assessment of the current competences of basic digital skills which represent precondition for performing almost any of the modern jobs nowadays. Identified skills include general knowledge of Windows operating system, internet browsing skills, e-mail account management and knowledge related to use of social networks for business purposes.

Component 2 – Training Requirements

The subcomponents referring to the Component 2 include perceived capacity building areas that could help strengthen employability of the respondents. The most important capacity building fields include software development, information technology security, telecommunications and internet of things. All of the perceived areas were stated as very important capacities relevant for the job demand requirements within Industry 4.0 processes (WEF, 2016).

Component 3 – Initial Working Experience

The subcomponent within Component 3 named "Initial working experience" refers to the skills acquired within the internship period and initial years of employment. It would be important to note that a letter of recommendation is considered to be particularly important since it could be a signal for future employers that the previous employer is satisfied with the performance of the specific employee. This could have important policy implications suggesting that policy makers should put more effort in promoting internships and supporting youth for acquiring first job experience. It should be noticed that the Government of the Republic of Serbia has already started the "My First Salary"[1] program for unemployed youth who have no working experience and have university or high school diploma. Program provided salary subsidies for the first employment and creating opportunities for training and gaining the first working experience (National Employment Agency of the Republic of Serbia, 2022). The first effects of the program are still to be evaluated.

Component 4 – Education

Component 4 named "Education" includes formal and informal education degrees and certificates. It signals that youth consider evidence of completed formal and informal education relevant determinants of the job market competitiveness. It shows the importance of adapting curricula of formal education programs to the requirements of the digitized economy. Additionally, knowledge acquired through informal programs is also important. The Serbian Government recently introduced a program of validation of non-formal and informal learning through the system of Publicly Recognized Activity Organizers (PRAO) established under the Law on Adult Education (RS Official Gazette, No 55/13,88/17–other law, 27/18–other law and 6/2020)[2] and the National Qualification Framework (RS Official Gazette, No 27/18, 6/20). It would be important to monitor the implementation of this program and provide additional support at the local level. The critical success factor would be involvement of the employers that could provide important inputs to the content and conditions for recognition of the informally acquired knowledge.

Component 5 – Project Management and Entrepreneurial Skills

Component 5 refers to the horizontal skills including project management and entrepreneurship skills which provide basis for self-employment and realization of the self-made working opportunities. Results show high loadings for both management and entrepreneurship skills reflecting the positive aspects of acquiring these skills through formal and informal education programs.

DISCUSSION

The results confirm a close relationship between digital competencies attained by youth and those sought within the labor market. Although continually exposed and immersed in new technologies throughout life, which should otherwise result in

Table 2. PCA component loadings with Varimax rotation (>0.4)

Variable	Comp1	Comp2	Comp3	Comp4	Comp5	Unexplained
sofdel		0.5280				0.2114
mngmt					0.6499	0.1772
projmngmt					0.6943	0.156
infsec		0.4902				0.2576
iot		0.4071				0.3415
telecom		0.4646				0.2781
formeduc				0.667		0.2661
profeduc				0.5668		0.2537
nonformeduc						0.4119
workexp			0.4956			0.3582
lettrecomm			0.6571			0.2467
genknowl						0.5537
win	0.4208					0.2327
micoffice						0.3314
internet	0.4865					0.1487
email	0.4515					0.1802
socialnet	0.4309					0.305

Source: Authors' calculations

semi-familiarity with the digital landscape, many Serbian youth fail to attain the most basic of digital competencies needed in the labor market (Kuzmanovic, 2017). The demand for digital skills stems directly from substantial labor-market changes caused by rapid technological development and disruption. As low- and mid-skilled occupations grow increasingly precarious due to technological disruption, there is an equal nascent demand for advanced analytical and technological skills to complement changes in the labor market. The World Economic Forum suggests that even though 85 million jobs might be replaced by technical innovation before 2025, another 97 million may emerge due to their potential transformation (WEF, 2020). In order to bridge between formal education, acquired competences and labor market needs, the 2017 Law on Dual Education introducing reforms in secondary vocational education was adopted in the Republic of Serbia (Law on Dual Education, Official Gazette of the Republic of Serbia, no. 101/2017 and 6/2020)[3], which was followed by Serbia's Law on the Dual Model of Studies in Higher Education adopted in 2019. The educational models introduced by this legislation allow for competences to be achieved through learning within the workplace. Such policy initiatives should assist in attuning educational practices to labor market needs and help in meeting employer expectations of the work-ready skills of recent graduates.

Eurostat data show that the digital competences of Serbian students are lower when compared to their peers in the EU, specifically in advanced digital competences, such as software usage or problem solving (Bradic Martinovic et al., 2019). An additional challenge is the lack of soft skills among Serbian youth, necessitating that employers invest more in training intended to tackle shortcomings from the education system rather than build on existing employee competencies (Cekic & Markovic, 2019). Analyzing the most frequently required skills and competencies of employers in Serbia, Ognjenovic (2018) concluded that technical and technological skills as well as ICT competencies are highly prized. Accordingly, the Regional Youth Employability Skills Portfolio included IT and computer skills as one of the key bases of learned knowledge needed to tackle high rates of youth unemployment in the Western Balkans (Belgrade Open School, 2017)[4].

Most literature concurs that digital skill development is a leading indicator of labor-market development. EU 2020 Strategy has launched seven leading initiatives, the most prominent of which focuses on digital skill development, which has been mirrored in the Digital Agenda for Europe directed toward the development of a unified digital market both for business and residential accompanied by the rapid growth of internet access. The Digital Agenda itself stresses the strong correlation between digital skills and competitiveness, resulting in both prospective and current EU members investing in primary and secondary education. Strangely, academia and more formal sectors in administration and economics continually adopt digital skills and tools to create a more productive environment. Thereby, electronic models have become key factors in the improvement of digital development itself, acting a self-perpetuating model to bring about total digital transformation.

The influence of digital skills on economic development has not only long been the subject of research but also an area of ready investment in order to improve individual economic standing. In 2000, for instance, the Government of Great Britain started a life-learning program to significantly increase the employment rate by educating on individual digital skills (Brown, 2000).

Digital literacy has also faced the prejudice of being unnecessary to insure life skills and one's standard of living. Challenging this notion, Titan, et al. (2014) conclude there to be an interactive connection between digital literacy and labor market as a result of inconsistencies between labor market needs and digital skills. In support, Pirzada & Khan (2013) show a positive correlation of established digital skills with employment with Bennett & Maton (2008) underscoring the importance of digital skills in education and long -term career advancement. Bennett & Maton (2008) concluded that those able to understand the uses of e-services have an advantage in educational progress and job development. Bradic Martinovic et al. (2019) also asserts that occupational digital knowledge and skills assist in advancement from lower to higher levels of employment. Frand (2000) as well as Djukic & Pavlovic (2020) go so far as to conclude digital skills ensure a basic level of keeping oneself informed, improving both one's life as well as ability to be gainfully employed.

Possessing even rudimentary digital skills helps empowering students to excel within life. Conversely, the digital divide between those possessing digital skills and those who do not also affect social and employment equality. In Serbia in 2012, a study on digital literacy evaluation was conducted in which 1077 respondents (68% employed youth and 32% students) (Lakic, et al., 2012) indicated that student digital literacy was significantly higher than that of those employed. To make matters worse, digital illiteracy has a lingering long-term effect on education and hence employment as it limits access to educational resources. According to Grip (2004) disparate digital education also forms similarly disparate groups capable of further education and acquiring skills. Partial evidence of the digital divide is that internet use has been rapidly growing even though IT expertise has remained relatively limited. Digital literacy among genders may also be unequal (Van Deursen and Van Dick, 2010) with males disproportionately possessing higher digital skills than females reflecting uneven socialization and education with digital technology over the last several decades (Helsper, 2016). Other authors (Eshet-Alkalai & Amichai-Hamburger, 2004 have attributed digital inequality between genders to have arisen due to stereotyped attitudes and perceptions regarding capability and self-assessment and skills/competencies. In real terms among genders, such skills are more or less equalized, with performance tests showing men and women differing little (van Deursen & van Dijk, 2011).

Van Deursen and Van Dick (2011), observed digital skill advancement from 2010 to 2013, concluding that a gain in operational and formal internet skills results in a better performance on information internet skills. Moreover, the highest level of education indicated the highest level of all observed internet skills. The skill gap between the higher levels of education and lower and middle levels increased while the gap between lower and middle levels of education decreased. Helsper (2016) found that even though there was no gender difference in computer usage among younger children (7–11 years old), gender differences expanded significantly by one's mid-teens (16–17 years old). By this time, males were found to use the Internet more frequently and performed a wider range of activities than girls.

CONCLUSION WITH RECOMMENDATIONS

This research has intended to investigate youth aspirations and perceptions of the factors which contributing successful labor market outcomes with a particular focus on digital skills. In line with specified research questions the following results could be extracted: (1) Youth capacities with regards to basic digital skills are well developed which represents a solid basis for further development of Industry 4.0-related skills. The most important capacity building fields include software development, information technology security, telecommunications and internet of things.; (2) Initial working experience has been detected critical factor of achieving favorable labor market outcome. Internships and initial working experiences are particularly important and should be incorporated in formal and non-formal education programs; (3) Since research confirmed relationship between different factors of successful labor market outcomes – skills, preferences and capacities, the focus of the capacity building activities should be on developing horizontal skills such as management and entrepreneurship which important link between individual capacities and Industry 4.0 requirements. Research results have helped detecting the five components critical for matching youth interests and labor market requirements identified through analyzing recent reports of the international organizations and authors performing empirical research in the field. The first group of skills refers to basic digital skills, owned by the most of participants and assessed necessary for performing any modern job nowadays. Participants' self-assessment results show high scores on basic digital skills, particularly use of internet search and e-mail communication with 4.4 and 4.3 points respectively (1 to 5 Likert scale, with 5 being the highest). On the other hand, use of social networks for business purposes seems to be still underexploited among Serbian youth, although considered very useful competence for conducting many jobs nowadays in the field of marketing and business communication. Use of social networks for business purposes could be incorporated in both formal and informal education curriculum in order to strengthen employability of incapacitated youth in Serbia. Second component encompasses potential areas for acquiring skills necessary to catch up with competences demanded for the jobs of new generation in the most propulsive ICT sectors such as internet of things, internet security, software development and telecommunications. It should be noted that these competences could be hardly learned just through informal education programs as acquiring would also depend on solid theoretical foundations. Therefore, the importance of the specific skills perceived by survey participants should be taken into account when developing formal education curricula and designing internship programs in collaboration with employers. This study also shows that learning aspirations and possibility to enroll in informal trainings are quite diverse reflecting available time and learning dynamics preferred. When designing learning programs, it would be important to take into account learning model preferences which would increase costs but probably improve learning outcomes. Further research is therefore necessary when designing specific learning programs including dynamics, time slots and duration. The third component refers to work experience and recommendation letter as evidence of initial career steps of job candidates often required by employers. Being perceived as very important characteristic when applying for the specific job improve chances to receive a job offer, providing links with employers through internships and volunteering should be further promoted within education system, starting from secondary school. As could be expected, formal and informal education completion under Group 4 – Education, are confirmed as important determinants of future labor market outcomes. In line with many other research that confirmed completion of formal education increases effectiveness of informal education and active labor market policy measures on labor market outcomes, dealing with NEET unemployment should at first point consider possibilities for reintegration to formal education system. Project management and entrepreneurship are perceived as significant horizontal skills, particularly important for strengthening self-employment mindset. As ICT sector has become one of the leading drivers of Serbian economic growth which recorded respectable growth rates over the last years, building capacities for entrepreneurship and developing horizontal skills is now a cornerstone of the economy based on knowledge. Development of start-up culture characterized by learning, flexible business approaches and supporting environment should be promoted from early education stages.

In line with similar research, this paper reveals importance of gaining basic digital skills among Serbian youth which are extremely important in terms of both strengthening employability of youth with different education levels and background, but also for the purpose of future learning. As youth and NEET are heterogenous group of people, support for completion of formal education should be accompanied with capacity building programs in the field of digital skills that should be considered horizontal as inevitable for the most of required job profiles nowadays. Apart from the labor market, owning basic digital skills is extremely important in other aspects of life satisfaction including socialization of youth and marginalized groups suffering from social exclusion. The practical value of the research refers to use of the obtained results for designing training and formal education programmes as well as specific labour policy measures aimed at enhancing youth competitiveness at the labour market. Research results are therefore primarily relevant to institutions in the field of education including state authorities, primary and secondary schools, academia and civil sector.

This research has several limitations. For example, it is based on single data sources – self-reported data on youth population skills and perceptions on factors that could contribute to employment. Use of other data sources would enhance reliability of the obtained results. Additionally, although sample size of 125 participants is justified for using the PCA as statistical data reduction technique, it is not sufficient for providing general conclusions, especially for the subgroups analysis – gender, education level, settlement, etc. Hence, future studies in this domain coupled with secondary data could focus on comparisons between different subgroups facing various obstacles such as females, youth in NEET status or rural population. It would be also interesting to draw on interregional differences and to provide comparisons between aspirations and skills among youth from other countries of the Central and Eastern Europe. Finally, practical use of the results obtained within this research and other similar studies should not neglect the supply side of the digital transformation processes – teachers and education providers. Knowledge of teachers as well as attitudes and capacities of organizations providing learning support should be important elements of designing effective upskilling and reskilling programs.

ACKNOWLEDGMENT

The research presented in this paper was funded by the Ministry of Science, Technological Development, and Innovation of the Republic of Serbia.

REFERENCES

Akman, I., & Turhan, C. (2018). Investigation of employers' performance expectations for new IT graduates in individual and team work settings for software development. *Information Technology & People*, *31*(1), 199–214. doi:10.1108/ITP-01-2017-0020

Apunyo, R., White, H., Otike, C., Katairo, T., Puerto, S., Gardiner, D., Kinengyere, A. A., Eyers, J., Saran, A., & Obuku, E. A. (2022). Interventions to increase youth employment: An evidence and gap map. *Campbell Systematic Reviews*, *18*(1), e1216. doi:10.1002/cl2.1216 PMID:36913191

Baird, A. M., & Parayitam, S. (2019). Employers' ratings of importance of skills and competencies college graduates need to get hired: Evidence from the New England region of USA. *Education + Training*, *61*(5), 622–634. doi:10.1108/ET-12-2018-0250

Belgrade Open School. (2017). *The Regional Youth Employability Skills Portfolio*, Belgrade, Serbia. https://bos.rs/rs/uploaded/Portfolio%2025_7.pdf

Bennett, S., Maton, K., & Kervin, L. (2008). The digital natives. *British Journal of Educational Technology*, *39*(5), 775–786. doi:10.1111/j.1467-8535.2007.00793.x

Bijelovic Bosanac, B., Pavlovic, Lj., & Martinic, M. (2017). *Stanje ducausegve politike zapošljavanja mladih u Republici Srbiji*, Beograd: Udruženje građana: Inicijativa za razvoj i saradnju. http://www.poslodavci.rs/wp-content/uploads/2017/09/stanje-i-perspektive-politike-zapoljavanja-mladih.pdf

Bradic-Martinovic, A., Pavlovic, D., & Zdravkovic, A. (2019). *Digitalne veštine studenata u Srbiji. Naucni skup "Trendovi razvoja: Kvalitet visokog obrazovan"a"*, Fakultet tehničkih nauka, Novi Sad, Srbija, 210-213. http://www.trend.uns.ac.rs/stskup/trend_2019/radovi/T4.1/T4.1-1.pdf

Brewer, L. (2013). Enhancing youth employability: what? Why? and how? Guide to core work skills, published by International Labor Organization-ILO, Geneva.

Brown, T. (2000). Lifelong learning: An organizing principle for reform. *Australian Journal of Adult Learning*, *58*(3), 312. https://www.researchgate.net/publication/329574090_Lifelong_Learning_An_organising_principle_for_reform#fullTextFileContent

Cekic, J., & Markovic, J. (2019). *Work-based learning in Serbia. Turin: European Training Foundation. (Report)*, ETF Europe. https://www.etf.europa.eu/sites/default/files/2019-07/Work-based%20learning%20Serbia.pdf

Clark, A. (2011). Embedding transferable skills and enhancing student learning in a political science research methods module. Evidence from the United Kingdom. *PS, Political Science & Politics*, *44*(1), 135–139. doi:10.1017/S1049096510002039

Contini, D., Filandri, M., & Pacelli, L. (2019). Persistency in the NEET state: A longitudinal analysis. *Journal of Youth Studies*, *22*(7), 1–22. doi:10.1080/13676261.2018.1562161

Djukic, M., & Pavlovic, D. (2020). *Istrazivanje o potrebama anketara za sprovodjenje integrisanih usluga za zaposljivost NEET mladih sa fokusom na 4.0 industrijsku revoluciju*, Youth Work Centre, Serbia. https://socijalnoukljucivanje.gov.rs/wp-content/uploads/2021/12/Istrazivanje_o_potrebama_aktera_za_sprovodjenje_integrisanih_usluga_za_zaposljivost_NEET_mladih.pdf

Eshet-Alkalai, Y., & Amichai-Hamburger, Y. (2004). Experiments in digital literacy. *Cyberpsychology & Behavior*, *7*(4), 421–429. doi:10.1089/cpb.2004.7.421 PMID:15331029

Europe, I. A. B. (2019). *Human Capital in the Digital Environment. Report*. Retrieved from https://iabeurope.eu/wp-content/uploads/2019/09/HCitDE-Report-2019-FINAL-1.pdf

European Commission. (2019). *Digital Opportunity traineeships: boosting digital skills on the job*. https://ec.europa.eu/digital-single-market/en/digital-opportunity-traineeships-boosting-digital-skills-job

European Commission. (2020). *Digital Education Action Plan (2021-2027); Resetting education and training for the digital age*. https://ec.europa.eu/education/education-in-the-eu/digital-education-action-plan_en

European Foundation for the Improvement of Living and Working Conditions. (2013). *Working conditions of young entrants to the labor market*. ETF Report. Retrieved from https://www.eurofound.europa.eu/sites/default/files/ef_files/docs/ewco/tn1306013s/tn1306013s.pdf

Eurostat (2021). *Database Labor Market Survey*, Retrieved from https://ec.europa.educauseat/data/database

Eurostat. (2021). *Youth unemployment rate by sex, age and country of birth*. Retrieved from: https://appsso.eurostat.ec.europa.eu/nui/show.do?dataset=yth_empl_100&lang=en

Field, A. (2009). Discovering Statistics Using SPSS (and Sex and Drugs and Roc' 'n' Roll), Los Angeles [i.e. Thousand Oaks, Calif.], SAGE Publications.

Frand, J. (2000). The information-age mindset. *International Journal of Higher Education*, *35*(2), 14–24.

Gligorov, V., Ognjenovic, K., & Vidovic, H. (2011). *Assessment of the labour market in Serbia* (No. 371). WIIW Research Report.

González-Pérez, L. I., & Ramírez-Montoya, M. S. (2022). Components of Education 4.0 in 21st century skills frameworks: Systematic review. *Sustainability (Basel)*, *14*(3), 1493. doi:10.3390u14031493

Helsper, E. J. (2016). *Slipping through the net report: Are disadvantaged young people being further left behind in the digital era?* in Socio-digital Skills and Wellbeing of Disadvantaged Young People. Retrieved from https://www.lse.ac.uk/media@lse/research/DiSTO/DiSTO-NEETs.aspx

International Labor Organization. (2020). *Global Employment Trends for Youth 2020: Technology and the future of jobs*. Retrieved from: https://www.ilo.org/wcmsp5/groups/public/---dgreports/---dcomm/---publ/documents/publication/wcms_737648.pdf

International Labor Organization. (2023). Youth Unemployment Database. Available at: https://www.ilo.org/shinyapps/bulkexplorer2/?lang=en&id=UNE_3UNE_SEX_AGE_EDU_NB_A, accessed at July, 22nd, 2023

Islam, M. A. (2022). Industry 4.0: Skill set for employability. *Social Sciences & Humanities Open*, *6*(1), 100280. doi:10.1016/j.ssaho.2022.100280

Ivic, I., & Pesikan, A. (2012). Education system reforms in an unstable political situation: The case of Serbia in the first decade of the [21]st century. *CEPS Journal, 2*(2), 31–53. doi:10.26529/cepsj.383

Jojkic, I., & Mitrovic, P. (2017). *Da znanje ostane u Srbiji: Studija o položaju mladih osoba na tržištu rada u Republici Srbiji*. Beograd: Friedrich-Ebert-Stiftung. https://library.fes.de/pdf-files/bueros/belgrad/13784.pdf

Kang, Y. D. (2021). Determinants of youth unemployment: Empirical analysis of OECD and EU member countries in 2000-2017. *Journal of Economic Development, 46*(3), 109–131.

Kuzmanovic, D. (2017). *Empirical validation of digital literacy construct and analysis of predictors of achievement* (doctoral dissertation). National Repository of Dissertations in Serbia (1234567899324).

Lakic, D., Parojcic, J., Kovacevic, N., & Antonijevic, M. (2012). Digital literacy of employers and students in Serbia. Conference Employer Engagement in a Digital Age. University of Greenwich, UK.

Lebedinski, L., & Vladisavljević, M. (2022). Parenthood and labor market outcomes in Serbia. *Ekonomski anali*, 67(232), 7-47.

Lee, J. (2013). Employment and earnings in high tech ethnic niches. *Social Forces, 91*(3), 747–784. doi:10.1093fos199

Liotti, G. (2022). Labour market regulation and youth unemployment in the EU-28. *Italian Economic Journal, 8*(1), 77–103. doi:10.100740797-021-00154-3

Matijevic, M., & Solaja, M. (2020). *ICT in Serbi– - At a Glance*, 2020. Novi Sad: Vojvodina ICT Cluster & Deutsche Gesellschaft für Internationale Zusammenarbeit (GIZ) GmbH. https://www.ict-cs.org/fileadmin/user_upload/ICT-in-Serbia-At-a-Glance-2020.pdf

Ministry of Education, Science and Technological Development of the Republic of Serbia. (2021). *The Law on the Dual Model of Studies in Higher Education*. Laws. https://www.mpn.gov.rs/wp-content/uploads/2020/01/ZAKON.pdf

Morgan, G. A., Barrett, K. C., & Leech, N. L. (2011). IBM SPSS For Intermediate Statistics: Use and Interpretation, 4 ed. Routledge, Taylor & Francis group, New York.

Oblinger, D. (2003). *Boomers, Gen-xers, and Millennials: Understanding th' 'new students'*. Educause, 4:38-47. https://er.educause.edu/-/media/files/article-downloads/erm0342.pdf

OECD. (2015). *OECD Skills Outlook 201– - Youth, Skills and Employability*. https://www.oecd-ilibrary.org/education/oecd-skills-outlook-2015_9789264234178-en

Ognjenovic, K. (2018). Training Intentions and Skills Needs in the Private Sector Companies in Serbia. *Journal of Wom'n's Entrepreneurship and Education, 1-2*(1-2), 68–85. doi:10.28934/jwee18.12.pp68-85

Ognjenovic, K., & Djukic, M. (2023). *Response of the Economic System to COVID-19: A Review of Evaluation Methods Applied in Assessing Economic Policies' Effects in Serbia. Reimagining Systems Thinking in a Post-Pandemic World.* IGI Global. doi:10.4018/978-1-6684-7285-9.ch006

Organization for Economic Co-operation and Development. (2015). *OECD Skills Outlook 201– - Youth, Skills and Employability*. https://www.oecd-ilibrary.org/education/oecd-skills-outlook-2015_9789264234178-en

Organization for Economic Cooperation and Development & International Labor Organization-OECD & ILO. (2014). *Promoting better labor market outcomes for youth. Report on youth employment and apprenticeships prepared for the G20 Labor and Employment Ministerial Meeting*. https://www.oecd.org/g20/topics/employment-and-social-policy/OECD-ILO-Youth-Apprenticeships-G20.pdf

Ozdamar-Keskin, N., Ozata, F. Z., Banar, K., & Royle, K. (2015). Examining digital literacy competences and learning habits of open and distance learners. *Contemporary Educational Technology*, *6*(1), 74–90. doi:10.30935/cedtech/6140

Pang, E., Wong, M., Leung, C. H., & Coombes, J. (2019). Competencies for fresh graduates' success at work: Perspectives of employers. *Industry and Higher Education*, *33*(1), 55–65. doi:10.1177/0950422218792333

Pavlovic, D., Bjelica, D., & Domazet, I. (2019). What Characteristics in the Youth Labor Market of Serbia Are Likely to Result in Employment? *Stanovnistvo*, *57*(2), 35–47. doi:10.2298/STNV190823006P

Pavlovic, D., Zubovic, J., & Obradovic, V. (2017). Relationship of the Youth Unemployment and Determinants of the Labour Market in the Balkan Countries. *Industry*, *45*(4), 153–166. doi:10.5937/industrija45-15155

Pavlovic, D., Zubovic, J., & Zdravkovic, A. (2016). Youth expectations in job search in Serbia. *Industry*, 44(4). https://www.researchgate.net/publication/314027512_Youth_expectations_in_job_search_in_Serbia#fullTextFileContent

Pejic-Bach, M., Bertoncel, T., Meško, M., & Krstić, Ž. (2020). Text mining of industry 4.0 job advertisements. *International Journal of Information Management*, *50*, 416–431. doi:10.1016/j.ijinfomgt.2019.07.014

Picatoste, J., Pérez-Ortiz, L., & Ruesga-Benito, S. M. (2018). A new educational pattern in response to new technologies and sustainable development. Enlightening ICT skills for youth employability in the European Union. *Telematics and Informatics*, *35*(4), 1031–1038. doi:10.1016/j.tele.2017.09.014

Pirzada, K., & Khan, F. N. (2013). Measuring Relationship between Digital Skills and Employability. *European Journal of Business and Management*, *5*(24), •••. https://www.researchgate.net/publication/258205764_Measuring_Relationship_between_Digital_Skills_and_Employability

Stanojevic, D., & Pavlovic, D. (2022). A Profile of Youth not in Employment, Education or Training (NEET) in Serbia. Monography "Human Capital and Welfare", published by University of Antwerp, Antwerp, Belgium.

Statistical Office of the Republic of Serbia. (2021). *Labor Force Survey in The Republic of Serbia, 2020*. Bulletin. Retrieved from https://publikacije.stat.gov.rs/G2021/PdfE/G20215671.pdf

Tamesberger, D., & Bacher, J. (2014). NEET youth in Austria: A typology including socio-demography, labormarket behaviour and permanence. *Journal of Youth Studies*, *17*(9), 1239–1259. doi:10.1080/13676261.2014.901492

Tamesberger, D., Leitgöb, H., & Bacher, J. (2014). How to combat NEET? Evidence from Austria. *Inter Economics*, *49*(4), 221–227. doi:10.100710272-014-0503-1

Thiem, K. C., & Dasgupta, N. (2022). From precollege to career: Barriers facing historically marginalized students and evidence-based solutions. *Social Issues and Policy Review*, *16*(1), 212–251. doi:10.1111ipr.12085

Todhunter, F. (2015). Using principal components analysis to explore competence and confidence in student nurses as users of information and communication technologies. *Nursing Open*, *2*(2), 72–84. doi:10.1002/nop2.19 PMID:27708803

Tsitskari, E., Goudas, M., Tsalouchou, E., & Michalopoulou, M. (2017). Employers' expectations of the employability skills needed in the sport and recreation environment. *Journal of Hospitality, Leisure, Sport and Tourism Education*, *20*, 1–9. doi:10.1016/j.jhlste.2016.11.002

Van Deursen, A. J. A. M., Helsper, E. J., & Eynon, R. (2014). *Measuring Digital Skills. From Digital Skills to Tangible Outcomes project report*. www.oii.ox.ac.uk/research/projects/?id=112

Van Deursen, A. J. A. M., & van Dijk, J. A. G. M. (2009). Using the Internet: Skill related problems in users' online behavior. *Interacting with Computers, 21*(5-6), 393–402. doi:10.1016/j.intcom.2009.06.005

Van Deursen, A. J. A. M., & van Dijk, J. A. G. M. (2010). Measuring Internet skills. *International Journal of Human-Computer Interaction, 26*(10), 891–916. doi:10.1080/10447318.2010.496338

Vasic, V. (2019). *Skills Mismatch Measurement in Serbia*. Turin: European Training Foundation. (Report). https://www.etf.europa.eu/sites/default/files/2019-07/Skills%20mismatch%20measurement_Serbia_0.pdf

Vladisavljevic, M., Zubovic, J., Djukic, M., & Jovanovic, O. (2021). Inequality-Reducing Effects of Tobacco Tax Increase: Accounting for Behavioral Response of Low-, Middle-, and High-Income Households in Serbia. *International Journal of Environmental Research and Public Health, 18*(18), 9494. doi:10.3390/ijerph18189494 PMID:34574420

Vladisavljevic, M., Zubovic, J., Jovanovic, O., Djukic, M., Najdovska, N. T., Pula, E., ... Gjika, A. (2022). Tobacco tax evasion in Western Balkan countries: Tax evasion prevalence and evasion determinants. *Tobacco Control, 31*(Suppl 2), s80–s87. doi:10.1136/tobaccocontrol-2021-056879 PMID:35022328

Wiley (2023). Digital Skills Gap Index 2021. Available at: https://dsgi.wiley.com/, accessed at June 21st, 2023

World Bank. (2023). Database on youth unemployment, retrieved from the International Labour Organization. "Labour Force Statistics database (LFS)" ILOSTAT. ilostat.ilo.org/data. Available at: https://data.worldbank.org/indicator/SL.UEM.1524.NE.ZS, accessed at June 21st, 2023

World Economic Forum. (2016). The Future of Jobs Report 2016, Retrieved from https://www3.weforum.org/docs/WEF_Future_of_Jobs.pdf

World Economic Forum. (2020). The Future of Jobs Report 2020. Retrieved from https://www.weforum.org/reports/the-future-of-jobs-report-2020

Zdravkovic, A., Djukic, M., & Bradic-Martinovic, A. (2017). Impact of FDI on unemployment in transition countries: Panel cointegration approach. *Industrija, 45*(1), 161–174. doi:10.5937/industrija45-13548

Zubovic, J., Zdravkovic, A., & Pavlovic, D. (2015). Effects of regulation on youth unemployment: Evidence from European countries. *Industrija, 43*(2), 129–144. doi:10.5937/industrija43-8483

ADDITIONAL READING

Bartodziej, C. J. (2017). The concept industry 4.0. In *The Concept Industry 4.0* (pp. 27–50). Springer Gabler. doi:10.1007/978-3-658-16502-4_3

Eynon, R., & Geniets, A. (2015). The digital skills paradox: How do digitally excluded youth develop skills to use the internet? *Learning, Media and Technology, 41*(3), 463–479. doi:10.1080/17439884.2014.1002845

Ministry of Education, Science and Technological Development of the Republic of Serbia. (2021). *The Law on the Dual Model of Studies in Higher Education*. Laws. https://www.mpn.gov.rs/wp-content/uploads/2020/01/ZAKON.pdf

Neagu, G., Berigel, M., & Lendzhova, V. (2021). How Digital Inclusion Increase Opportunities for Young People: Case of NEETs from Bulgaria, Romania and Turkey. *Sustainability (Basel), 13*(14), 7894. doi:10.3390u13147894

Puriwat, W., & Tripopsakul, S. (2020). Preparing for Industry 4.0 — Will Youths Have Enough Essential Skills?: An Evidence from Thailand. *International Journal of Instruction, 13*(3), 89–10. doi:10.29333/iji.2020.1337a

Rojko, A. (2017). Industry 4.0 concept: Background and overview. [iJIM]. *International Journal of Interactive Mobile Technologies, 11*(5), 77–90. doi:10.3991/ijim.v11i5.7072

The Government of the Republic of Serbia. (2021). *Dual education - New knowledge for the new age. Announcement*. Retrieved from https://www.srbija.gov.rs/tekst/en/129780/dual-education.php

World Economic Forum. (2020). *The Future of Jobs Report 2020*. Retrieved from https://www.weforum.org/reports/the-future-of-jobs-report-2020

KEY TERMS AND DEFINITIONS

Industry 4.0: – an industrial revolution which refers to use of automation and data exchange in manufacturing processes. It is often associated with the use of advanced technologies such as Internet of Things (IoT), Artificial intelligence (AI), Big data, Robotics and many others.

Youth policy: – a set of policy instruments such as strategies, programs, and other public policy measures aimed at providing better opportunities for youth population.

Employment policy: – a set of policies aimed at supporting individuals to find the best possible job with respect to their education background, skills and job aspirations. It also includes measures supposed to help creating better working environment including equal opportunities, protection of the workers' rights, improvement of the capacities through education and trainings, etc.

Factor analysis: – refers to statistical method used to describe variability among correlated variables. Its logic relies on reducing a large set of variables to smaller number of factors that have some common characteristics and are easily understandable.

Principal component analysis: – one of the techniques used within factor analysis approach. It reduces the number of variables making them easier for interpretation and visualization. As not requiring strong assumptions with regards to data, it could be useful tool for data exploration and analysis.

NEET: – youth in NEET status are youth who are not employed and not receiving an education or vocational training support. They belong to vulnerable youth population often facing difficulties and barriers to participate in labour market.

ICT skills: – refer to the skills and abilities to use information and technology tools for different purposes – communication, data analysis, learning, etc. They are very important for job career prospects and therefore considered vital for conducting most jobs nowadays.

ENDNOTES

1 https://oecd-opsi.org/innovations/my-first-salary/

2 https://www.ilo.org/dyn/natlex/natlex4.detail?p_lang=en&p_isn=95227&p_country=SRB&p_count=176

3 https://www.paragraf.rs/propisi/zakon-o-dualnom-obrazovanju.html

4 https://bos.rs/rs/uploaded/Portfolio%2025_7.pdf

Chapter 4
Profiling Industry-Relevant Employability Skills:
A Study on the Healthcare Industry in India

Swati Sisodia
GD Goenka University, India

Sumaira Jan
Department of Humanities, National Institute of Technology, Srinagar, India

ABSTRACT

The demand for skilled and qualified manpower in India is likely to remain elevated and robust in the coming years, both nationally and internationally. To encourage, continuous development of employees' regular competency mapping of skills with changing job roles is essential to create a productive and expert employee. Competitiveness is closely related to the concept of employee performance, which has a direct relationship with training. The professional quality and the perceived satisfaction vary from levels to levels in a healthcare organization. Hence, conducting job profiling is important to identify skills across the hierarchy. This research article aims to explore the Employability Skills required by the professionals working in the healthcare sector at Medical Service Vertical, further, the utility of each skill is identified and then their relative importance is studied at each hierarchical level. Digital technology and sustainable manpower can boost the creation of new jobs.

DOI: 10.4018/978-1-6684-9089-1.ch004

INTRODUCTION

Healthcare Industry is fast growing both in terms of revenue & employment generation. Labor demand in India is substantial and robust in the coming years, both nationally and internationally. But that would require skilled and qualified labor. As a result, each hospital should strive to increase the quality of its human capital. (Baroudi & Ginzberg, 1986). Training is essential to augment an employee's performance and reconstructing people's behaviour (Marimuthu et al., 2009). Currently, training is regarded as a continuous activity and one of the most basic perpendiculars of HRM (Human Resource Management) activities. (Al Karim 2019).

Everywhere in the world, it is increasingly realised that economic well-being and productive efficiency are a function of man's intellectual and professional abilities. It cannot be denied that an effective human resource base is crucially significant in today's highly competitive environment. Training/learning is a method of assisting employees in gaining knowledge, altering attitudes, and up-skilling. (Obisi 2001& 2011). The magnitude of a customized training programme can only be evaluated with a clear comprehension of its direct effect on employee's performance & satisfaction level (Giousmpasoglou & Ikonen, 2019). No organization can function effectively pipelining money, machines, method and minutes, human resource is an indispensable element of success (Agarwal & Ahuja, 2014).

Role of frontline workers changes with change in situations. Like many other countries in the world India is also facing a gigantic burden from the COVID-19 pandemic. The physiological, psychological, & emotional well-being of frontline health-care workers is being jeopardised as a result of this. They are, without a doubt, constantly under pressure as a result of extended work hours, limited medical equipment (such as PPE kits), & the fear of becoming infected. (The Lancet, 2020).

India's Healthcare industry has evolved alongside emerging disruptive technologies. With the increase in the competition, healthcare providers are recruiting and retaining employees having soaring competency skills to handle the functions and complexity of advance technologies.

The private sector contributes for over 80% of total healthcare spending and is a significant player in the healthcare industry. (Talib et al., 2013). There is a global shift in the healthcare policy, therefore healthcare providers need to adopt strategies for 'patient-cantered' care (Lawn et al., 2009). Manipal Group, Medanta Mediclinic, Max Healthcare, Apollo Hospital, & Fortis Healthcare are among of India's most well-known corporate hospital networks.

Employability skill is a combination of technical skills, also known as hard skills and generic skills. However, in today's workplace, these hard skills are not adequate to keep individuals employed when organizations are planning for restructuring (Mora et al., 2011). As a result, Employability Skills are referred as the most viable bases of competitive edge in service industry (Vij et al., 2014). Employability Skills (ES) should be industry and job-specific. It will lead to the effective acquisition of generic skills by enhancing the competency level of employees. There is a demand for highly educated/skilled, talented, experienced, & trained individuals in such a high-risk profession where healthcare providers are striving hard to establish a zero-negligence environment. (Mora et al., 2011).

One of the most difficult components of the selection process is determining/ classifying talent & acquiring manpower in accordance with the KSA set to do a given task. Fundamental ES expand on & upsurge competency levels gained via primary education, such as reading & writing, as well as the hard skills required to execute specialised tasks. Individuals must consistently acquire and convey new information and abilities obtained at work because of the demand for such fundamental job skills. These skills are not only significant for the professional working at Medical Service Vertical (MSV) but also critical in ameliorating lifelong learning process. Competitive employees have certain personal attributes such as enhanced level of job satisfaction, reliable, punctual and loyal.

As a result, before developing any training programmes, it is critical to first classify the skill gaps in the workforce. This approach of skill improvement will create a blueprint of skill deficiencies. As a result, it become easy for human resource practitioners to design customized training programme for their employees (Colombo & Stanca, 2008). Employability skills (ES) make it simpler to grasp work prospects, to stay in them for a longer period of time, and to embrace the evolving needs of the labour market, as evidenced by recent study. Thus, healthcare providers are putting huge investments in the training and development of its employees (Blickstein, 1996; Politt, 2001).

This research paper aims to explore the ES required in the Healthcare Industry (HI). This study focus on Medical Service Vertical, the utility of each skill is identified and then their relative importance is studied at each hierarchical level. This is the only industry that ensures that soft skills and hard skills must be adequately developed for better performance and greater outcome. This high-end industry is growing exponentially in terms of employment and six sigma practices, thus it is essential to select and retain the best of the talent, to keep it error-free.

Contemplating the importance of the healthcare industry and the scope of its direct customer interaction, it becomes imperative; to identify the necessary Employability Skills required to perform the specific job. The effectiveness of any Training Program lies in the fact that how effectively and efficiently the Training Program fills the gaps between actual performance and set benchmarks. It is pivotal to recognize the areas of Proficiency/ Sufficiency/ Deficiency before developing any training program.

The motivation behind conducting a study on profiling industry-relevant employability skills in the healthcare industry in India stems from the critical need to bridge the gap between academia and the job market. In recent years, the healthcare sector has experienced significant growth and transformation, presenting numerous employment opportunities. However, there is a persistent challenge of finding skilled and industry-ready professionals to meet the industry's demands.

By understanding the specific employability skills required in the healthcare industry, this study aims to enhance the preparedness of aspiring healthcare professionals and improve their chances of employability. Profiling industry-relevant employability skills involves identifying the key competencies, knowledge, and attitudes that are sought after by healthcare employers.

Additionally, the study seeks to address the unique context of the healthcare industry in India. India's healthcare sector is diverse, encompassing hospitals, clinics, pharmaceutical companies, medical research institutions, and more. Each sub-sector may have distinct requirements and expectations from potential employees. Therefore, a comprehensive analysis of the employability skills within the Indian healthcare industry is essential to cater to the specific needs of the industry.

The findings of this study will provide valuable insights for educators, policymakers, and training institutions to align their curriculum and programs with industry expectations. It will aid in the development of targeted training initiatives that equip students and professionals with the necessary skills and competencies required to excel in the healthcare industry.

Ultimately, by conducting a study in the healthcare industry in India, we aim to foster a stronger connection between educational institutions and the job market. This will empower individuals pursuing careers in healthcare to enhance their employability, contribute effectively to the industry, and address the ever-growing healthcare challenges in the country.

LITERATURE REVIEW

The structure of Employability Skills (ESs) has altered as a result of rapid workplace transformation and an extremely competitive labour market. Earlier ESs are labelled as the minimum skills prerequisite by an individual to get employed but now the essence of transferable skills has changed to not only employment-related but also vital for doing well on the job. Employability Skills are not a one-size-fits-all set of abilities; rather, they vary between industries and employment levels, from the lowest-level labourer to the highest-level executive. In today's world, the most difficult challenge for professionals is finding the appropriate individual for the right job. After performing an extensive literature review researchers have identified that there are limited studies describing all the skill typologies needed in the Healthcare Sector. The majority of the work is around teamwork, communication, & problem-solving skills/abilities. First, the basic employable skills prerequisite in the MSV were established through a literature review. viz. *Conceptual & Analytical, Planning & Organizing, Resource & Self-management, Communication, Teamwork, Interpersonal, Work Psychology, Information Communication and Technology, Critical Thinking & Problem Solving* (Sisodia & Agarwal, 2017). Some of the skills namely Teamwork, Critical Thinking & Problem Solving & work psychology, are necessarily required to cope and survive in the crises. In such situations, the demand for skilled manpower surge manifold. Furthermore, newly hired ICU (intensive care unit) employees often lack the psychological/emotional training to cope with demanding business environments, which can put them in extremely tough situations when it comes to balancing work and personal life.

The majority of work is oriented towards identifying/analysing prowess in the HI, have identified prowess either holistically or in verticals such as Medical Education & Medical Research (Colliver, 2000; Pedersen, 2009). Most of the studies conducted in past focused on graduate attributes. Limited studies focus on developing the existing workforce to make them more adaptable and decreasing the gaps, created by the inclusion of the new workforce. There are few studies on employability skills in the healthcare industry that have been undertaken in India. Sisodia & Agarwal (2019), revealed the model of "Whirl of Prowess" which embraces nine fundamental ESs (communication, ICT, interpersonal skills, teamwork, and others) prerequisite for professionals working at MSV (Table 1.).

Further, these nine skills were subdivided into twenty-two skill-typologies representing each core skill (Table 2.). Employees can be rated on these 22 employability traits, with certain benchmarks showing Proficiency/Sufficiency/Deficiency levels and customized training programmes. The peripheral abilities were the subsets of each core skill that represented distinct areas of expertise.

Research Gaps

Despite the growing importance of employability skills in the healthcare sector of Delhi/NCR, there is a significant research gap in understanding and addressing the specific needs and requirements at diverse hierarchical levels. Previous studies have predominantly focused on general employability skills, lacking a comprehensive examination of the skills specifically demanded within the medical services field.

Table 1. Employability Skills

Employability Skills
Communication
Information Communication & Technology
Work Psychology
Team work
Interpersonal
Critical Thinking & Problem Solving
Self-Management
Planning & Organizing
Conceptual & Analytical

The first research objective aims to identify the skills required at different hierarchical levels within the healthcare sector. While some studies have explored the skills necessary for entry-level positions or specific job roles, there is a lack of research that encompasses the entire hierarchical structure, from frontline healthcare workers to management and leadership positions. This research gap hampers the development of targeted training programs and career advancement opportunities tailored to each level within the healthcare sector.

While some existing frameworks exist for employability skills in general, there is a lack of a comprehensive and specific chart that covers the unique requirements of the healthcare industry. This research gap inhibits the development of a standardized framework that can guide educational institutions, policymakers, and training providers in designing curriculum, training programs, and recruitment processes that align with the industry's demands.

While there have been efforts to identify key employability skills in healthcare, there is limited research that assesses the significance and relevance of each skill segment. Understanding the relative importance of different skills will enable a more targeted approach to training and development initiatives, ensuring that resources are allocated efficiently and effectively.

Table 2. Encyclopaedic list of Employability Skill Set

Skill Set	Factor
Communication	Firebrand
	People-person
Information Communication & Technology	Fireball
	Auditor
Work Psychology	Autopilot
	Equanimity
Teamwork	Trouper
	Initiator
	Dove
Interpersonal	Buoyant
	Amenable
	Magnetic
Critical thinking & Problem solving	Decision-maker
	Value- added seller
Self management	Administrator
	Assiduous
Planning & Organizing	Stimulator
	Mettlesome
	Band- Aid
Conceptual & Analytical	Starry-eyed
	Coal digger
	Cogent

Although some studies have explored training needs in healthcare, there is a research gap in utilizing employee profiling to tailor training programs to individual employees' skill gaps and career aspirations. A comprehensive understanding of specific employee training needs will enable organizations to design personalized training plans, leading to improved employee performance, job satisfaction, and retention.

Addressing these research gaps will contribute to the advancement of knowledge and practices in the healthcare sector, leading to improved employability outcomes, career progression opportunities, and enhanced overall performance within the industry.

RESEARCH METHODOLOGY

Research Objectives

The research focuses on making the Training Programs more effectual. The accomplishment of any Training Programs lies in Training Need Analysis (TNA) of the employees. For doing TNA, it is important to have an in depth knowledge of the skills needed in the Healthcare Industry. Thus, the major emphasis of the learning is:

1. To identify skills required at diverse hierarchical levels in the Healthcare sector of Delhi/NCR.
2. To construct an "Employability Skills Chart", encompassing all aspects of skills required in medical services. 3. To identify the importance and utility of every skill segment on which training needs to be imparted.
3. Employee profiling to aid in the construction of a "Training Chart" for understanding specific employee Training needs in medical services.

A research instrument with twenty-two closed-ended questions was used to conduct the conclusive study. Data was gathered from 100 respondents with minimum 10 years of experience operating at MSV using a simple random sample technique. The respondents are the professional employed in the Medical Service Vertical at various multi-specialty hospitals of Delhi/NCR region. Delhi/NCR was chosen for its world class health care infrastructure and maximum number of multi-specialty hospitals.

Firstly, the essential ESs prerequisite in the MSV were recognised through a literature review. (Sisodia & Agarwal, 2017). Furthermore, using the Empirical Formula and K-means cluster analysis, the utility and relative importance of these skill typologies were assessed. The close-ended questionnaire having twenty-two skills was administered on the healthcare professionals employed in the MSV with minimum 10 years of working experience (namely; CEO, supervisors, executives, and managers). They have provided weight to each of these twenty-two skills. Further, they were requested to assign weights to each prowess on the scale of 0-1 (all positive values are included).

- *Demographic Outline*

Table 6 exhibits the demographic features of the respondents. The scope is confined to the Delhi/NCR region due to a geographical restriction. Private multispecialty hospitals with at least 500 beds and 200 employees were included in this research. The rationale behind selecting such hospitals & employees were, a) private hospitals are chosen as most of them have well defined hierarchies and often conduct training programme at different hierarchical levels as compared to public hospitals. b) Moreover, the researcher has considered hospitals with huge infrastructure and revenue as training programs requires investment and proper infrastructure ensures division of hierarchy as per different levels of management.

The following private multi-speciality hospital having at-least 200 beds were selected for the study (Table 3). The hospitals were located in different part of Delhi-NCR such as Delhi, Noida and Ghaziabad. The list of the hospitals are as given below.

Table 3. List of Targeted Hospitals

Region	Name of Hospitals
Delhi	Manipal Group of Hospitals **(H1)**, Maharaja Agrasen Hospital **(H2)**, Jaipur Golden Hospital **(H3)**, Max Shalimar Bagh **(H4),** Fortis Shalimar Bagh **(H5)**
Noida	Jaypee Hospital **(H6)**, Vinayak Hospital **(H7)**
Ghaziabad	Columbia Asia Hospital **(H8)**
Gurugram	Medanta - The Medicity **(H9)**, Fortis Memorial Hospital **(H10)**

The distribution of sample across the three hierarchical levels viz. Top-level, Middle level and Lower level are represented in the Table 4

Table 4. Data Collection from Hospital across Organizational Hierarchy

ORG. HIERARCHY	**HOSPITALS**										**Total**
Levels	**H1**	**H2**	**H3**	**H4**	**H5**	**H6**	**H7**	**H8**	**H9**	**H10**	
Lower-Level	4	6	5	5	3	3	4	3	4	3	35
Middle-Level	4	4	5	5	4	5	4	5	4	3	43
Top-level	3	2	3	3	2	2	2	3	1	1	22
Total	11	9	12	12	9	10	10	11	9	7	**100**

The distribution of the sample across the three verticals of the Medical Service is represented below in the Table 5. Line Service vertical comprises of Emergency, Out-patient, In-patient, ICU, and Operations. Support Services vertical comprises of Supportive Services, Central Sterile Supply, Radiology, Nursing, and Diet Services. Auxiliary Services consist of Registration, Records, Stores, Mortuary and Security.

Table 5. Data Collection from Hospitals across Medical Service Verticals

MEDICAL SERVICES	HOSPITALS										Total
Verticals	H1	H2	H3	H4	H5	H6	H7	H8	H9	H10	
Line Services	3	3	6	3	5	3	3	3	3	3	35
Support Services	3	4	6	5	4	3	5	5	4	4	43
Auxiliary Services	2	3	3	3	3	2	2	2	1	1	22
Total	8	10	15	11	12	8	10	10	8	8	**100**

Table 6. Demographic Outline of the Respondents from Hospitals in Delhi/NCR

Characteristic		Number	Percentage (%)
Gender			
Male (M)		52	52
Female (F)		48	48
	Total	100	
Age (years)			
22-25		22	22
25-28		34	34
28-32		39	39
32-35		05	05
	Total	100	
Highest Degree Earned			
Post-Graduation		12	12
Graduation		49	49
Others (diploma holders etc)		39	39
	Total	100	
Hierarchical Level			
Middle Level		43	43
Lower Level		35	35
Higher Level		22	22
	Total	100	

ANALYSIS

Analysis of Research Instrument

Result of Empirical Formula-I

The focus group interview was conducted on 100 respondents using a structured questionnaire. The interview aimed to identify the utility of each skill set. Respondents were asked to rate the gravity of each skill on the scale of 0-1 as per the utility across three hierarchical levels. Thus, the relative importance of each construct is obtained by summing up the average score (weight) of each factor multiplied by the total score (weight) of items (responses were gathered from focus group interview).

Level of Relative Importance of Factors (Individual Construct) = $_{n=1}\sum^{500}$ [Average Weight of Individual Factor provided by Focus Group* Total weight of Items of that Factor]

The result attained from Empirical Formula I (Agrawal & Ahuja, 2014) represents the importance of nine constructs for healthcare professionals.

Table 7. Employees Skill Set by their Relative Importance & Utility along with Weight

Employability Prowess/ Skills	Factor:1	Factor:2	Factor:3	Total (Weight)
Communication	People person (wt.=18.53)	Firebrand (wt.= 10.98)	NA	(wt. 29.51)
Information Communication & Technology	Fireball (wt.= 10.37)	Auditor (wt.= 10.78)	NA	(wt. 21.15)
Work Psychology	Autopilot (wt.= 17.95)	Equanimity (wt.= 9.08)	NA	(wt. 35.86)
Teamwork	Initiator (wt.= 8.83)	Trouper (wt.= 8.93)	Dove (wt.= 16.77)	(wt. 34.53)
Interpersonal	Amenable (wt.= 15.31)	Buoyant (wt.= 15.78)	Magnetic (wt.= 11.38)	(wt. 42.47)
Critical Thinking & Problem Solving	Value-added Seller (wt.= 13.82)	Decision maker (wt.= 7.342)	NA	(wt. 28.96)
Self Management	Assiduous (wt.= 20.29)	Administrator (wt.= 13.29)	NA	(wt. 33.58)
Planning & Organizing	Mettle some (wt.= 10.75)	Stimulator (wt.= 12.86)	Band-aid (wt.= 11.33)	(wt. 34.94)
Conceptual & Analytical	Starry-eyed (wt.= 13.96)	Coal-digger (wt.= 10.07)	Cogent (wt.= 7.77)	(wt. 31.80)

Result Obtained from Empirical Formula-II

The researcher has calculated the relative importance & utility of every employability skill set to know the relative significance of each skill at the time of work and in the workplace.

Relative Importance of Each Employability Skill= [Total weight of Factor 1 + Total Weight of Factor 2 + Total Weight of Factor 3]

The relative importance of nine constructs is calculated by summing up the weight (wt) assigned by 100 respondents to all the factors. The result obtained from equation 2 is notable in table 8.

Table 8. The sequence of Employability Skill Set by Relative Need at Workplace

Employability Prowess	Weights (Wt.)
Interpersonal	**42.47**
Work Psychology	**35.86**
Planning & Organizing	**34.94**
Teamwork	**34.53**
Self-management	**33.58**
Conceptual & Analytical	**31.80**
Communication	**29.51**
Critical thinking & Problem Solving	**28.96**
Information Communication & Technology (ICT)	**21.15**

Table 8 shows the final result, which is in decreasing order, with the most important skill allocated first place in contrast to the other skills. The employability skill which is highly needed in the Healthcare Industry for Medical Service Professionals is Interpersonal skills which include counselling, mentoring, leadership, adaptability & sensitivity. Further, Work Psychology is important in Healthcare Industry which includes willingness, courtesy, punctuality, Integrity and Initiative is few aspects. Least required Employability Skills is ICT skill as it is a cross-functional prowess and can be handled by a separate department.

Employers look for manpower with high interpersonal skills, work psychological skill and great at planning & organizing the task in critical situations in the best possible way; according to managers, people must be self-motivated and excellent in time-management.

Creation of Training Chart

The changing entail of the society and the workplace has a strong influence on the organizational choice for human capital development programs. In order, to ameliorate and intensify job performance and employee satisfaction, the skills & knowledge and behaviour learned during the training program have to transfer to the workplace and require maintenance over time.

The research aims to describe the significance of a tailored training program. Healthcare Industry is a high-risk industry in which professionals have to practice zero negligence. In such a risk driven sector, the role of training & development has assumed paramount consequence. The research framework used in this study was created by ***Dan S. Chiaburu & Amanuel G. Tekleab*** (Chiaburu & Tekleab, 2005). He has investigated the individual & environmental determinants of learning transfer, training generalisation, learning, & training maintenance in the workplace.

This model depicts all three phases of training viz., pre-training, training & evaluation/follow-up, which are pivotal for any training activity to forge the training program's effectiveness. The "Training Chart" is similar to the "Health Chart" which portrays a clear picture of an employee's current skill sets, the job-specific sine qua non, organizational goals, and skill gaps. Training Chart is a building block of any training program. For this study, a training chart is developed after conducting a pragmatic analysis of the Employability Skill Set required at the medical service vertical (Table 1). This assist in creating a cluster profiling (Table 7) and comprehensive profiling of employee skill across the organizational hierarchy (Figure 4).

Employees Cluster Profiling based on Prowess Typology across the levels

The 22 factors (the result of factor analysis) were created around nine Employability Skills after carrying-out K-means cluster analysis using SPSS 23.0, on the sample size of 100 professionals. These 22 skill typologies are divided into three clusters. The K-means cluster analysis is shown in Table 9 below, for all the Employability Skills needed by the Healthcare professionals.

The nine Employability Skill sets are divided into three core clusters, were named as Specialized Skills, Gaffer Skills and Hegemony Skills all are mutually exclusive from each other's & cluster contain the skills needed in that domain (Figure 3.).

Table 10, shows a detailed view of the cluster profile of 22 skill typologies. This cluster profile of an employee will be a gauge available with the hiring authorities at the time of recruitment and designing of the compensation package.

Table 9. Result of K-means Cluster Analysis

Initial Cluster Centers

	Cluster		
	Specialized Skills	Gaffer Skills	Hegemony Skills
Firebrand	3	3	5
People-person	4	3	5
Fireball	5	4	4
Auditor	5	3	4
Autopilot	2	5	4
Equanimity	3	4	5
Initiator	2	4	5
Trouper	2	5	3
Dove	2	5	2
Amenable	2	4	5
Buoyant	2	2	5
Magnetic	2	4	5
Value-added seller	2	5	4
Decision-maker	2	5	3
Assiduous	5	3	4
Administrator	3	5	4
Mettlesome	5	2	3
Stimulator	4	5	3
Band-aid	2	5	4
Starry-eyed	4	5	3
Coal-digger	5	3	2
Cogent	4	5	3

Comprehensive Outlining of Individual Prowess across the hierarchy

The 22 Skill Typologies of the employees', has varied significantly across the hierarchical levels. The healthcare industry is divided into three levels: lower, middle, and top. The skills requisite at each level ramify based on the job description & job specification. Research Instrument-II was administrated over 100 respondents having at least 10 years of experience. These participants were questioned in groups, either in person or through communicative media. They were asked to rate the skill typologies according to their importance and utility across three organizational levels on the scale of 0-1 (including decimal values). Once the responses were gathered, using Empirical Formula-III [$\sum_{i=1}^{100}\text{Skillatlevel}\left(\textbf{Top / Middle / Lower}\right)$ **per construct**] (Agarwal & Ahuja, 2014) result was evaluated. Later, as per the values obtained the skills were segregated across the levels as shown in Figure 4.

Table 10. Cluster Profiling of Employees Build on Skill Typologies

Cluster	Skills-Typologies	Description
Specialized Skills	Fireball, Auditor, Assiduous, Mettlesome, Coal-digger	This cluster contains technical & subject-specific skills like auditors, fireball, coal-digger, etc. This cluster highlights the technical abilities that medical service providers require.
Gaffer Skills	Autopilot, Trouper, Dove, Decision-maker, Administrator, Band-aid, Starry-eyed, Cogent, Stimulator, Value added seller	These are motivational skills. These are those leadership skills needed at work which helps in taking agile and effective decisions.
Hegemony Skills	Amenable, Buoyant, Magnetic, Firebrand, People-person, Equanimity, Initiator	These are life skills. These are those minimum managerial skills required for a smooth transaction of instructions and operating various events.

CONCLUSION & LIMITATIONS

This study was undertaken to curtail the gaps between the existing skills and the skills needed to perform a specific position in the Healthcare Industry. As stated in the previous studies, Healthcare Industry is a risk-based industry wherein, the nature & quality of services rendered need specialized skills and competencies. To be more effective and efficient constant training of professionals is required. But, as human resources are limited and have immense pressure to perform multi-tasking, time for taking training is compromised. Competency based/Skill based hiring is a key to recruitment and retention of manpower in the organization.

This study is helpful to both employee and Human Resource (HR) practitioners in the Healthcare Industry. Employees can identify and work on their shortcomings. Whereas, this will help the HR practitioners in determining the trainability of the personnel prior designing a training programme. This research paper would assist the employers of the organization at various phases such as, performing TNA (Training Need Analysis) & developing a decisive compensation package. Detailed job analysis will aid in identifying an explicit roles & correlated skills required to accomplish a task. The skill typologies highlighted in this paper will aid an HR manager in the recruitment process by allowing them to create a talent pool based on the job position's demand. Management may discern and recruit the finest of the talent by using the chart of diverse skill typologies indispensable, minimizing the cost and rate of failure.

The study only examines one aspect of the healthcare industry. Medical Education, Training, & Research are three additional verticals with a lot of potential for training & development. The scope of this study is restricted to private multispecialty hospitals in Delhi/NCR.

REFERENCES

Agarwal, N., & Ahuja, V. (2014). Creation of the Training-Chart: A Step Forward to Make the Training More Effective. [IJSSMET]. *International Journal of Service Science, Management, Engineering, and Technology*, *5*(2), 1–18. doi:10.4018/ijssmet.2014040101

Al Karim, R. (2019). Impact of different training and development programs on employee performance in Bangladesh's perspective. *International Journal of Entrepreneurial Research*, *2*(1), 8–14. doi:10.31580/ijer.v1i2.497

Blickstein, S. (1996). Does training pay off? *Across the Board*, *33*(6), 16–20.

Chiaburu, D. S., & Marinova, S. V. (2005). What predicts skill transfer? An exploratory study of goal orientation, training self-efficacy and organizational supports. *International Journal of Training and Development*, *9*(2), 110–123. doi:10.1111/j.1468-2419.2005.00225.x

Colliver, J. A. (2000). Effectiveness of problem-based learning curricula: Research and theory. *Academic Medicine*, *75*(3), 259–266. doi:10.1097/00001888-200003000-00017 PMID:10724315

ColomboE.StancaL. (2008). The impact of training on productivity: evidence from a large panel of firms. *Available at* SSRN.

Giousmpasoglou, C., & Ikonen, E. (2019). The importance of genuine service and human interaction for millennials' guest experience: the case of hotel reception and front line employees.

Lawn, S., Battersby, M., Lindner, H., Mathews, R., Morris, S., Wells, L., & Reed, R. (2009). What skills do primary health care professionals need to provide effective self-management support? Seeking consumer perspectives. *Australian Journal of Primary Health, 15*(1), 37–44. doi:10.1071/PY08053

Marimuthu, M., Arokiasamy, L., & Ismail, M. (2009). Human capital development and its impact on firm performance: Evidence from developmental economics. *The journal of international social research, 2* (8), 265-272.

Mora, M., O'Connor, R., Raisinghani, M., Macías-Luévano, J., & Gelman, O. (2011). An IT service engineering and management framework (ITS-EMF). *International Journal of Service Science, Management, Engineering, and Technology, 2*(2), 1–15. doi:10.4018/jssmet.2011040101

Obisi, C. (2001). Employee development, Issues and dimensions. *Unical Journal of public Administrator Sept, 1*.

Obisi, C. (2011). Employee Training and Development in Nigerian Organizations: Some Observations and Agenda for Research. *Australian Journal of Business and Management Research, 1*(9), 82–91. doi:10.52283/NSWRCA.AJBMR.20110109A09

Pedersen, R. (2009). Empirical research on empathy in medicine—A critical review. *Patient Education and Counseling, 76*(3), 307–322. doi:10.1016/j.pec.2009.06.012 PMID:19631488

Pollitt, D. (2001). A flexible workforce for the modern economy: The role of education and training. *Education + Training, 43*(7), 345–376. doi:10.1108/EUM0000000005857

Sisodia, S., & Agarwal, N. (2017). Employability Skills Essential for Healthcare Industry. *Procedia Computer Science, 122*, 431–438. doi:10.1016/j.procs.2017.11.390

Sisodia, S., & Agarwal, N. (2019). Examining Employability Skills of Healthcare Services in India: A Descriptive Literature Review. *The International Journal of Social Science, Management Engineering and Technology (IJSSMET), 10* (3).

Talib, F., & Rahman, Z. (2013). Current health of Indian healthcare and hospitality industries: a demographic study.

The Lancet (2020) COVID-19: Protecting healthcare workers. The Lancet 395(10228): 922.

Vij, P., & Sharma, M. (2014). The effect of human capital management practices on employee positive behavioural outcomes. *International Journal of Management Practice*, *7*(3), 204–221. doi:10.1504/IJMP.2014.063594

Werner, A., & Schneider, J. M. (1974). Teaching medical students interactional skills: A research-based course in the doctor-patient relationship. *The New England Journal of Medicine*, *290*(22), 1232–1237. doi:10.1056/NEJM197405302902206 PMID:4825853

Chapter 5
Learning Innovations for Developing Teachers' Competencies in Thai Communication in the 21st Century

Rungarun Rojrattanadamrong Chaisri
https://orcid.org/0000-0002-3926-6880
Srinakharinwirot University, Thailand

Suppawan Satjapiboon
Srinakharinwirot University, Thailand

Adisorn Rueangkitchanan
Srinakharinwirot University, Thailand

ABSTRACT

This research aims to evaluate current and desired conditions, develop innovative learning methods, and measure their effectiveness in enhancing Thai communication competencies among teachers in the 21st century. The core competencies include: Principles of the Thai language, listening, speaking, reading, and writing; the use of the language; communication etiquette; Thai language aesthetics; moral and communication ethics. The results showed that teachers' knowledge and skill competencies were at a moderate level, but their desire for improvement was high. Data analysis evidenced the effectiveness of the Thai language learning innovation; indeed, the posttest score of the teachers' Thai communication competency assessment was statistically significantly higher (70%) than the benchmark (p < .05).

DOI: 10.4018/978-1-6684-9089-1.ch005

INTRODUCTION

Thailand's National Education Plan 2017–2036 focuses on improving the quality and standard of education within the socio-economic context of the country and the world, driving innovation and creativity as well as the dynamics of intellectual society (Office of the Education Council, Ministry of Education, 2017). The educational process must encourage students to develop naturally and reach their full potential. In this regard, personnel's competencies are paramount; they are related to performance and consist of knowledge, skill, and ability, among other characteristics. Competency is a behavioural attribute that allows personnel in the organization to perform their work more prominently than others (Hay Group, 2003; Narongwit Saenthong, 2007; Office of the Education Council, Education Council, 2018; Office of the National Education Commission, Office of the Prime Minister, 2002; Royal Academy, 2013; Tatchapong Sethabutr, 2012).

For a person to be successful, it is more important to have competencies than intelligence quotient (Hay Group, 2003). For example, Finland promotes the development of transversal competencies in learners using innovative multidisciplinary learning modules (Phongsathorn Mahawichit, 2017). Singapore provides education for the competency development of people in the 21st century, including information and communication skills, to prepare for the changing world (Ministry of Education, Singapore, 2010; UNESCO, 2010). In Thailand, this is also a priority. Student outcomes must include core subjects, learning and innovation skills, information media and technology skills, life and career skills, as well as internal and external factors contributing to learning (Pridi Pluemsamrankit, 2017; Wijarn Panich, 2013). Therefore, competency classification can be classified as a core competency, job competency or other aspects as appropriate (Narongwit Saenthong, 2007; Office of the Education Council, Ministry of Education, 2019; Sukanya Rassameethammachot, 2008).

Innovation, information, and communication technologies play a role in the learning process of Thai communities, and society is changing amid the advancement of science and technology through step-by-step development into educational innovation (Boerrcher, 2010; Kirati Yodyingyong, 2009; Krirk Thuaamklang, 2012). Especially in tertiary education, it is advisable to encourage the pursuit of learning habits by being self-taught and continuously seeking knowledge. Being prepared for the world of work is a crucial skill for the 21st century (Facione, 2011; Office of the Education Council, Ministry of Education, 2009; Partnership for 21st Century Skills, 2011; Trilling & Fadel, 2009). Podcasts, in particular, are a form of learning innovation that is easily accessible to learners, who can watch whenever they want to and repeat as many times as possible. Connecting to the Internet is optional. Therefore, podcasts are a very interesting medium to apply to the development of language skills for communication (Boerrcher, 2021; Witthaya Wichit, 2009). Podcasts can be used successfully to mitigate some challenges inherent in developing learners' communication skills (Suvarnaphaet & Suvarnaphaet, 2023).

A teacher's language communication competency is what impresses students and influences their behavior. Reverse communication allows teachers to listen to students' feedback or comments, which enables them to communicate with a variety of objectives and perform their duties effectively, especially if online learning resources are promoted and made available to teachers at all times, allowing them to better apply their competencies in their daily lives (Borsheim et al., 2008; Cope & Kalantzis, 2000; Office of the Education Council, Education Council, 2018; Prozesky, 2014; Salas Pilco, 2013; Thuan Thongkaew, 2007; Trilling & Fadel, 2009; UNESCO, 2011). Thus, podcast technology has great potential for further development, especially in the context of the actualization of distance learning (Basenko & Baskakova, 2021).

According to the preceding, competency is a fundamental characteristic of an individual motivating a person to work. In particular, the competencies of 21st century teachers are specifically important and necessary for teachers' professional practice, as well as the creation of innovative podcasts, which can be accessed any time and from any location to develop quality people for the global society. This has led researchers to study and develop learning innovations to enhance teachers' competencies in Thai communication in the 21st century. Language communication competency is an important feature that will enable teachers to transfer knowledge, skills, and attitudes, in order to develop students into citizens who will drive the country and be an important force for the nation in the future.

RESEARCH OBJECTIVES

The authors' research objectives were as follows:

1. To assess current and desirable conditions to enhance teachers' competencies in Thai communication in the 21st century.
2. To develop learning innovations that enhance teachers' competencies in Thai communication in the 21st century.
3. To assess the effectiveness of learning innovations that enhance teachers' competencies in Thai communication in the 21st century.

TERMINOLOGY

"Teachers' competencies in Thai communication in the 21st century" refers to a person's ability to organize ideas by systematically composing words. This ability consists of three aspects, namely knowledge, skills, and attributes. In addition, this ability implies using knowledge and skills in the main field of the Thai language: Listening, speaking, reading, and writing, use of Thai language, including communication etiquette, moral and communication ethics, as well as appreciating beauty in Thai language aesthetics.

BENEFITS

1. The findings of the research resulted in new knowledge about the current and desirable conditions for enhancing teachers' competencies in Thai communication in the 21st century.
2. The findings of the research resulted in a learning innovation to enhance teachers' competencies in Thai communication in the 21st century.

RESEARCH FRAMEWORK

The authors developed a framework on the research and development of learning innovations to enhance teachers' competencies in Thai communication in the 21st century, drawing on UNESCO's (2010) core competency concept (Office of the National Economic and Social Development Council, 2020). Under Thailand's Strategy 12 of the Master Plan, learning development is achieved by enhancing the country's capacity and strengths to provide Thai people with more quality education. This goal meets international standards and the skills needed in the 21st century world to solve problems, adapt, communicate, and collaborate with others more effectively, learn throughout life, and adhere to the competency concept of the Office of the Education Council, Ministry of Education (2018). The researchers were able to synthesize nine core competencies in Thai communication (Figure 1).

The researchers used the Thai communication subcompetency framework, which consists of knowledge, skill, and attributes, as a guideline for enhancing teachers' competencies in Thai communication in the 21st century. The Thai communication subcompetency framework is classified according to core and job competencies to determine the current and desirable conditions related to the teachers' Thai communication. It can also be used as a guideline for the development of learning innovations that enhance teachers' competencies in Thai communication in the 21st century.

Figure 1. Core Competencies in Thai Communication

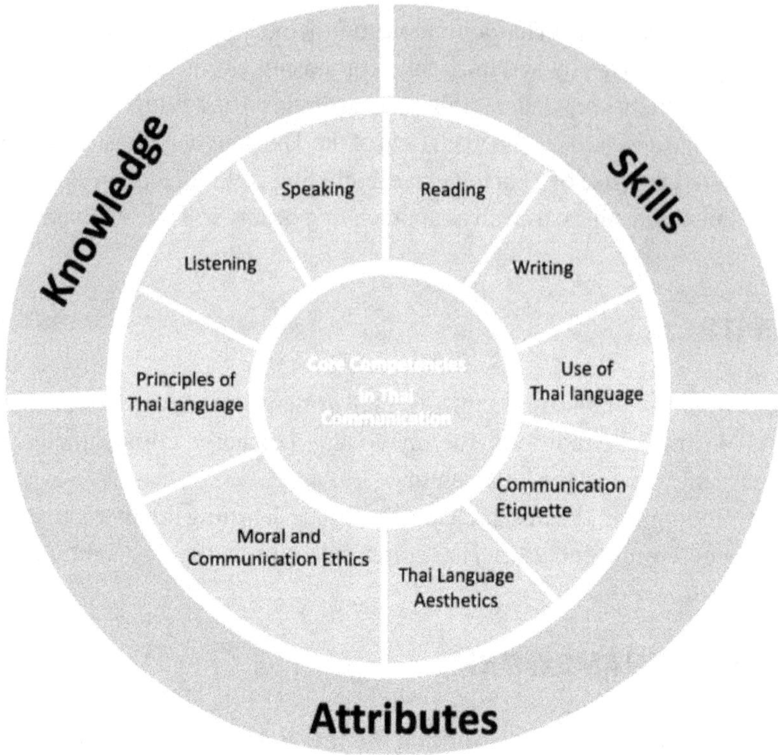

Figure 2. Competencies Framework for Teachers' Thai Language Communication in the 21st Century

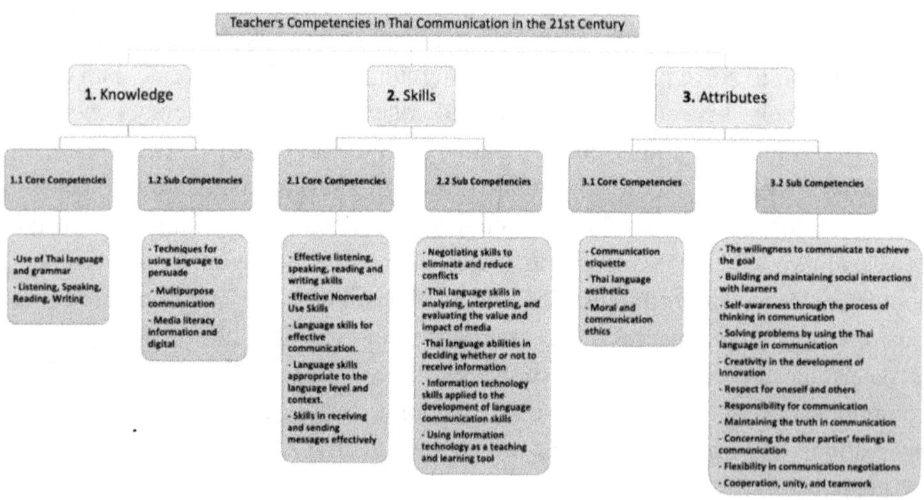

According to the Thai language communication core competency framework, nine core competencies have been classified into subcompetencies in knowledge, skills, and characteristics as a guideline for developing teachers' competencies in Thai communication in the 21st century. It is divided into a core competency and job competency. The researchers analysed indicators and behavioural factors related to teachers' competencies in Thai communication to use them as a guideline for creating an assessment of teachers' competencies in Thai communication.

RESEARCH METHODS

The authors organized this study into the following four phases:

Phase One: Examining current and desirable conditions to enhance teachers' competencies in Thai communication in the 21st century.

Phase Two: Developing learning innovations to enhance teachers' competencies in Thai communication in the 21st century.

Phase Three: Examining the effectiveness of learning innovations to enhance teachers' Thai communication competencies in the 21st century.

Phase Four: Evaluating learning innovations to enhance teachers' competencies in Thai communication in the 21st century.

The subsections below provide a detailed description of each phase.

Phase One: Examining Current and Desirable Conditions to Enhance Teachers' Competencies in Thai Communication in the 21st Century

The first phase of this research included the following steps:

1. Exploring data from relevant documents and research.
2. Developing a framework for teachers' competencies in Thai communication in the 21st century by classifying nine core competencies: Principles of Thai language, listening, speaking, reading, writing, use of Thai language, communication etiquette, Thai language aesthetics, and moral and communication ethics.
3. Examining current and desirable conditions. The researchers invited three experts (i.e., a curriculum and teaching expert, an assessment and evaluation expert, and a Thai language expert) to assess the questionnaire based on its consistency and objectives. The experts found that all the questions had an Indian Oil Corporation value of 1, which indicated a selection of questions that met the appropriate selection criteria and the applicable consistency.

The samples in the study were school managers and teachers. The researchers drew the sample from a multistage sampling of 24 people, 12 of whom were school managers and 12 of whom were teachers. More in detail:

a. The researchers randomly selected a province for each region. The authors used the simple random sampling method, with one. They elected one province for each region; thus, the number of provinces in the sample was four.
b. The researchers randomly selected schools from provinces obtained from the samples in point a above. The authors used schools as random units in the simple random sampling method, thus one school from each province. The sample included 4 schools and 3 school managers for each of them, for a total of 12 people.
c. The researchers randomly selected teachers from the schools they had obtained from the samples in point b above. The authors selected three teachers from each school by using the simple random sampling method. They used the teachers as a random unit. The total number of teachers in this sample was 12.

Phase Two: Developing Learning Innovations to Enhance Teachers' Competencies in Thai Communication in the 21st Century

The second phase of this research included the following steps:

1. The researchers formulated synthesized documents and survey results of current and desirable conditions to use them as a guideline for developing Thai language learning and Thai communication competency assessments.
2. The researchers created an innovative media podcast with content about the development of Thai communication competency and consisted of six topics: The art of listening, how to say it nicely, how to read well-rounded, principles of writing effectively, communication skills in the teaching profession, and digital media literacy-podcasts.
3. Five experts examined Thai language learning to determine the appropriateness of relevant innovations. They consisted of experts in educational technology, assessment and evaluation, and the Thai language. The assessment results validated Thai language learning innovations. In addition, three experts incurriculum and teaching, assessment and evaluation, and Thai language assessed communication competencies. The results of the assessment determined that Thai communication competencies were appropriate, as was the consistency between the indicators, the behavioural list, and the Thai communication competency with an Indian Oil Corporation value of 1.

4. The researchers conducted a pilot study with teachers studying in a master's degree program. Srinakharinwirot University, which was not a sample, selected 11 people for purposive sampling to evaluate the rater agreement index (RAI) between assessors, where the RAI value ranges from 0-1. If the value approaches 1, it shows that the three assessors are very consistent. These data will show the quality of the 10 aspects of the Thai communication competency assessment. The RAI was as follows: 1) Principles of Thai language: 0.99; 2) listening: 0.994; 3) speaking: 0.984; 4) reading: 0.990; 5) writing: 0.986; 6) use of Thai language: 0.9844; 7) communication etiquette: 0.98 9) Thai language aesthetics: 1.000; 10) moral and communication ethics: 0.988. Then, the authors conducted experiments on the target audience using innovative Thai language learning and Thai communication competency assessment.

Phase Three: Examining the Effectiveness of Learning Innovations to Enhance Teachers' Thai Communication Competencies in the 21st Century

The third phase of this research included the following steps:

1. In this research, the target group was teachers studying in a master's degree program at Srinakharinwirot University, purposive sampling of six people.
2. The authors conducted this research over four weeks, three hours per week, for a total of 12 hours.

Phase Four: Evaluating Learning Innovations to Enhance Teachers' Competencies in Thai Communication in the 21st Century

The fourth phase of this research included the following steps:

1. The authors assessed the effectiveness of Thai language learning innovations and Thai communication competency. They validated the results for analysis. Three assessors evaluated the RAI.
2. Data analysis aimed to determine the effectiveness of innovative Thai language learning by analysing the results from the experiments using a computer program and comparing the threshold by 70%.

FINDINGS

The data analysis allowed the authors to identify current and desirable conditions to enhance teachers' competencies in Thai communication in the 21st century.

Results on Current and Desirable Conditions to Enhance Teachers' Competencies in Thai Communication in the 21st Century

The authors divided the analysis into two parts. In the first part, the researchers dealt with general information about the participants. The authors divided the participants into two groups, namely school managers and teachers, with basic information (Table 1).

Table 1. Overall Participant Status

Status		Total (N=24)	
		Number (pp)	Percentage
Participants	School Director	1	4.17
	Deputy Director of the School	3	12.50
	Head of Learning Group	8	33.33
	School Teachers	12	50.00
Gender	Male	6	25.00
	Female	18	75.00
Age	21-30 years old	5	20.83
	31-40 years old	6	25.00
	41-50 years old	8	33.33
	51 years old and above	5	20.83
Education Level	Bachelor's degree	10	41.67
	Master's degree	13	54.17
	Doctorate	1	4.17
Region	North	6	25.00
	Central	10	41.67
	Northeast	6	25.00
	South	2	8.33

Table 1 shows that the overall status of the participants included 1 school director (4.17%), 3 deputy school directors (12.50%), 8 heads of learning groups (33.33%), and 12 school teachers (50%). Most of them graduated with a master's degree (54.17%).

In the second part of data analysis, the authors classified the general data according to current and desirable conditions to enhance teachers' competencies in Thai communication in the 21st century and to the types of competencies in each area (Table 2).

In general, teachers' competencies in Thai communication were currently at a moderate level ($\bar{x}$ = 3.97, SD = 0.75). However, the overall desirable condition was very high ($\bar{x}$ = 4.70, SD = 0.49). This showed that teachers had a greater desire to improve their Thai communication skills under the current conditions.

The teachers had an overall moderate level of knowledge in the core competency ($\bar{x}$ = 3.69, SD = 0.78) in terms of Thai language principles, listening, reading, writing, and the use of Thai. However, the results showed an overall high level of core competency ($\bar{x}$ = 4.59, SD = 0.61) for the teachers' desirable Thai communication competency. When the researchers considered individual aspects, they found a high level of demand for the development of knowledge in all areas.

The teachers had an overall moderate level of knowledge for job competency ($\bar{x}$ = 3.82, SD = 0.70). However, the results showed an overall high level of job competency ($\bar{x}$ = 4.49, SD = 0.63) for teachers' desirable Thai communication competencies. Teachers had a need for self-development in terms of knowledge, understanding of media literacy information and digital technologies for their teaching profession, Thai language strategies, and multipurpose communication, respectively.

The teachers had an overall moderate level of skills for the core competency ($\bar{x}$ = 3.86, SD = 0.62). However, the results showed an overall high level of core competency ($\bar{x}$ = 4.72, SD = 0.51) for teachers' desirable Thai communication competencies. When the researchers considered individual aspects, they found that the average of all aspects was very high. According to the data, the teachers had a much higher level of effective reading skills than in other areas. However, the teachers also had the highest reading development needs, which showed that they paid much attention to reading skills.

The teachers had an overall moderate level of skills in the job competency ($\bar{x}$ = 3.58, SD = 0.74). When the researchers considered individual aspects, they discovered that the average of all aspects was moderate, such as Thai language abilities in deciding whether or not to receive information, Thai language skills in evaluating and selecting information and communication technology, researching for teaching and learning, and for teachers' professional development, and Thai language skills in analysing, interpreting, and evaluating the value and impact of media. However, the results showed an overall high level of skill competency ($\bar{x}$ = 4.65, SD = 0.50) for teachers' desirable Thai communication competencies.

Table 2. Current and Desirable Conditions to Enhance Teachers' Competencies in Thai Communication in the 21st Century in General

Teachers' competencies in Thai communication in the 21st century	Current conditions		Interpretation	Sequence	Desirable conditions		Interpretation	Sequence	PNI modified
	$\bar{x}$	(SD)			$\bar{x}$	(SD)			$\bar{x}$
Knowledge	3.75	0.74	Moderate	3	4.54	0.62	High	3	3.54
Skill	3.85	0.68	Moderate	2	4.69	0.50	HIgh	2	3.69
Attributes	4.29	0.82	High	1	4.87	0.35	High	1	3.87
Total	3.97	0.75	Moderate	-	4.70	0.49	High	-	3.70

Table 3. Current and Desirable Conditions for Enhancing Teacher's Competencies in Thai Communication in the 21st Century, Listed Under Knowledge in the Core Competency Category

Teacher's competencies in Thai communication in the 21st century listed under knowledge in the core competency category	Current conditions		Interpretation	Sequence	Desirable conditions		Interpretation	Sequence	PNI modified
	$\bar{x}$	(SD)			$\bar{x}$	(SD)			$\bar{x}$
Knowledge of Thai language principles	3.46	0.78	Moderate	6	4.54	0.66	High	5	3.54
Knowledge of listening principles	3.75	0.79	Moderate	2	4.58	0.58	High	4	3.58
Knowledge of speaking principles	3.67	0.82	Moderate	5	4.67	0.56	High	1	3.67
Knowledge of reading principles	3.79	0.78	Moderate	1	4.63	0.58	High	2	3.63
Knowledge of writing principles	3.75	0.74	Moderate	3	4.62	0.65	High	3	3.62
Knowledge of the use of Thai language principles	3.71	0.75	Moderate	4	4.50	0.66	High	6	3.50
Total average	3.69	0.78	Moderate	-	4.59	0.61	High	-	3.59

Table 4. Current and Desired Conditions for Enhancing Teachers' Competencies in Thai Language Communication in the 21st Century, Listed Under Knowledge in the Job Competency Category

Teachers' competencies in Thai communication in the 21st century listed under knowledge in the job competency category	Current conditions		Interpretation	Sequence	Desirable conditions		Interpretation	Sequence	PNI modified
	$\bar{x}$	(SD)			$\bar{x}$	(SD)			$\bar{x}$
Knowledge of Thai language strategies for the teaching profession	3.79	0.78	Moderate	2	4.46	0.72	High	2	3.46
Knowledge of multipurpose communication	3.79	0.72	Moderate	3	4.42	0.66	High	3	3.42
Knowledge and understanding of media literacy information and digital technologies for the teaching profession	3.88	0.61	Moderate	1	4.58	0.50	High	1	3.58
Total Average	3.82	0.70	Moderate	-	4.49	0.63	High		3.49

Table 5. Current and Desired Conditions for Enhancing Teachers' Competencies in Thai Language Communication in the 21st Century, Listed Under Skills in the Core Competency Category

Teachers' competencies in Thai communication in the 21st century as listed under skills in the core competency category	Current conditions		Interpretation	Sequence	Desirable conditions		Interpretation	Sequence	PNI modified
	$\bar{x}$	(SD)			$\bar{x}$	(SD)			$\bar{x}$
Effective listening skills	3.83	0.70	Moderate	4	4.75	0.53	High	2	3.75
Effective speaking skills	3.71	0.62	Moderate	6	4.67	0.48	High	5	3.67
Effective reading skills	4.00	0.51	High	1	4.83	0.48	High	1	3.83
Effective writing skills	3.88	0.61	Moderate	3	4.71	0.46	High	4	3.71
Communication skills in both verbal and nonverbal languages	3.79	0.72	Moderate	5	4.63	0.65	High	6	3.63
Language skills for effective communication	3.96	0.55	Moderate	2	4.75	0.44	High	3	3.75
Total average	3.86	0.62	Moderate	-	4.72	0.51	High	-	3.72

Table 6. Current and Desired Conditions for Enhancing Teachers' Competencies in Thai Language Communication in the 21st Century, Listed Under Skills in the Job Competency Category

Teachers' competencies in Thai communication in the 21st century as listed under skills in the job competency category	Current conditions		Interpretation	Sequence	Desirable conditions		Interpretation	Sequence	PNI modified
	$\bar{x}$	(SD)			$\bar{x}$	(SD)			$\bar{x}$
Negotiating skills to eliminate and reduce conflicts	3.67	0.82	Moderate	6	4.67	0.56	High	2	3.67
Thai language skills in analyzing and interpreting, and evaluating the value and impact of media	3.92	0.72	Moderate	3	4.63	0.49	High	5	3.63
Thai language abilities in deciding whether or not to receive information	3.96	0.81	Moderate	1	4.71	0.46	High	1	3.71
Thai language skills in evaluating and selecting information and communication technology, researching for teaching and learning, and for teachers' professional development	3.96	0.75	Moderate	2	4.67	0.48	High	3	3.67
Information technology skills applied to the development of language communication skills	3.83	0.76	Moderate	4	4.67	0.48	High	4	3.67
Thai language skills in using information technology as a teaching and learning tool in a variety of settings	3.75	0.61	Moderate	5	4.58	0.51	High	6	3.58
Total average	3.85	0.74	Moderate	-	4.65	0.50	High	-	3.65

Table 7. Current and Desired Conditions for Enhancing Teachers' Competencies in Thai Language Communication in the 21st Century, Listed Under Attributes in the Core Competency Category

Teachers' competencies in Thai communication in the 21st century as listed under attributes in the core competency category	Current conditions		Interpretation	Sequence	Desirable conditions		Interpretation	Sequence	PNI modified
	$\bar{x}$	(SD)			$\bar{x}$	(SD)			$\bar{x}$
Communication etiquette	4.29	0.81	High	2	4.92	0.28	High		4.29
Thai language aesthetics	4.00	0.78	High	3	4.88	0.34	High		4.00
Moral and communication ethics	4.46	0.59	High	1	4.88	0.34	High		4.46
Total average	4.25	1.09	High	-	4.89	0.32	High		4.25

The teachers had an overall high level of attributes for core competency ($\bar{x}$ = 4.25, SD = 1.09); when the researchers considered individual aspects, they found that the average of all aspects was very high. Moreover, the results showed a high level of attributes ($\bar{x}$ = 4.89, SD = 0.32) for the teachers' desirable Thai communication competencies. However, when the researchers considered individual aspects, the teachers showed a growing desire to develop their competencies in all the aspects of their attributes, which included communication etiquette, Thai language aesthetics, morals, and communication ethics.

The teachers had an overall high level of attributes in the job competency ($\bar{x}$ = 4.33, SD = 0.55); when the researchers considered individual aspects, they found that most attributes were at a high level, such as respect for oneself and others in communication, the willingness to communicate in achieve the goal, maintaining the truth in communication with other parties, and cooperation, unity, and teamwork through communication. Only creativity in innovation development and application in communication was at a moderate level ($\bar{x}$ = 3.88, SD = 0.61). However, the results showed a high level of attributes in all aspects for teachers' desirable Thai communication in the job competencies, especially in terms of respect for oneself and others in communication, responsibility for other parties in communication, maintaining the truth in communication with other parties, concerning the other parties' feelings in communication, and flexibility in communication negotiations.

The foregoing evidences that teachers in the 21st century acknowledge the importance of Thai communication competency. According to the research findings, desirable conditions were higher than the current conditions in all aspects in terms of knowledge, skills, and attributes, which were necessary for the teachers. Therefore, the researchers used the results of this study as a guideline for research and development of learning innovations and to investigate the effectiveness of learning innovations to enhance the Thai communication competency of teachers in the 21st century.

Results on the Effectiveness of Learning Innovations that Enhance Teachers' Competencies in Thai Communication in the 21st Century

The authors divided the analysis into two parts. In the first part, the researchers analyzed the Assessor's conformity index (RAI) to determine the effectiveness of Thai language learning innovations and Thai communication competency assessments. The researchers considered their validated results to analyze the RAI of three assessors from the assessment of the nine aspects of Thai communication competency (Table 9).

Table 8. Current and Desired Conditions for Enhancing Teachers' Competencies in Thai Language Communication in the 21st Century, Listed Under Attributes in the Job Competency Category

Teachers' competencies in Thai communication in the 21st century listed under attributes in the job competency category	Current conditions		Interpretation	Sequence	Desirable conditions		Interpretation	Sequence	PNI modified
	$\bar{x}$	(SD)			$\bar{x}$	(SD)			$\bar{x}$
The willingness to communicate to achieve the goal	4.54	0.59	High	2	4.83	0.38	High		4.54
Building and maintaining social interactions with learners in communication	4.25	0.53	High	8	4.79	0.51	High		4.25
Self-awareness through the process of thinking in communication	4.08	0.58	High	9	4.75	0.53	High		4.08
Solving problems by using the Thai language in communication	4.08	0.65	High	10	4.79	0.51	High		4.08
Creativity in innovation development and application in communication	3.88	0.61	Moderate	11	4.79	0.41	High		3.88
Respect for oneself and others in communication	4.63	0.49	High	1	4.92	0.28	High		4.63
Responsibility for other parties in communication	4.46	0.51	High	5	4.92	0.28	High		4.46
Maintaining the truth in communication with other parties	4.50	0.51	High	3	4.92	0.28	High		4.50
Concerning the other parties' feelings in communication	4.33	0.56	High	7	4.88	0.34	High		4.33
Flexibility in communication negotiations	4.42	0.50	High	6	4.88	0.34	High		4.42
Cooperation, unity, and teamwork through communication	4.50	0.51	High	4	4.88	0.34	High		4.50
Total average	4.33	0.55	High	-	4.85	0.38	High		4.33

Table 9. Assessors' RAI Values Based on Components of the Assessment of Thai Communication Competency

Components of the assessment of Thai communication competency	Assessor'ss' RAI values
1. Principles of the Thai language	0.839
2. Listening	0.906
3. Speaking	0.839
4. Reading	0.958
5. Writing	0.894
6. Use of Thai language	0.889
7. Communication etiquette	1.000
8. Thai language aesthetics	1.000
9. Moral and communication ethics	0.822

The researchers found that the assessors' RAI values were consistent with the nine factors in the assessment of Thai communication competency, ranging from the highest to the lowest degree of consistency, respectively: Communication etiquette (RAI = 1.000), Thai language aesthetics (RAI = 1.000), reading (RAI = 0.958), listening (RAI = 0.906), writing (RAI = 0.894), use of Thai language (RAI = 0.889), principles of Thai language (RAI = 0.839), speaking (RAI = 0.839), and moral and communication ethics (RAI = 0.822).

In the second part of the analysis, the researchers assessed the effectiveness of the Thai language learning innovation by comparing the benchmark of 70% (equal to 28%) with the hypothetical score of the posttest of Thai communication competency above the threshold of 70%.

The researchers used t-test one sample statistics to compare the difference between posttest scores and benchmark scores. The purpose of this study was to compare the mean score of the Thai communication competency assessment with benchmarks in a single sample group with unknown variance at the statistical significance level of. The hypotheses were as follows:

Hypothesis Zero: The grade point average of the posttest was 28% or 70%
Hypothesis One: The grade point average of the posttest was greater than 34.1% or more than 70%.

Table 10 reports the results of this analysis. It shows the mean, standard deviation (SD), t-test statistical value, and the statistical significance level, which the researchers determined by comparing the benchmark of 70% with the teachers' posttest scores

Table 10. The Statistical Significance Level, Which Was Determined by Comparing the Benchmark of 70% with the Teachers' Posttest Scores

Group	N	Full score	Mean	SD	% of mean	t
Posttest	18	40	34.1	4	85.25	6.46

Note. * Statistically significant at .05 (t-table .05 = 6.46, df = 17)

Table 10 shows that the average score of teachers on the posttest was 34.1, representing 85.25%, and when the researchers compared the benchmark and the teachers' post-Thai communication competency assessment scores, they found the teachers' scores were statistically significantly higher than the .05 benchmark. Therefore, they concluded that the Thai communication competency scores for the posttest were above the 70% threshold.

DISCUSSION AND CONCLUSION

Current and Desirable Conditions to Enhance Teachers' Competencies in Thai Communication in the 21st Century

The authors conducted a study of current and desirable conditions to enhance teachers' competencies in Thai communication in the 21st century. They grounded their investigation on their synthesis of nine core competencies in Thai communication, which consisted of skills, knowledge, and attributes as subcompetencies to evaluate the effectiveness of Thai language learning innovations to enhance Thai communication competency. The authors looked at current and desirable conditions by surveying managers and teachers who differed in gender, age, education levels, and schools in each region. Overall, the authors found that, in general, the teachers' current Thai communication competency was at a moderate level, and that teachers in the 21st century had a great desire to improve their Thai communication competency. In addition, when the researchers considered each type of competency, they found that, at the time of the study, the teachers had moderate Thai communication competencies in terms of knowledge and skills, but they had a great desire to develop high competencies in these areas. Furthermore, the teachers' attributes included a high level of Thai communication competency and an aim to develop such competencies to a large extent as well. This demonstrated the importance of improving Thai communication skills for teachers, regardless of whether they teach Thai or another subject. The teachers had had a greater desire to improve their Thai communication skills than they did at the time of the study in every way, especially the following skills, which must be developed in teachers' Thai communication competencies: Reading skills, listening skills, use of language skills for communication, writing skills, speaking skills, and verbal and nonverbal communication skills. This provided a clear view of competency development based on the teachers' experience of what it was and what teachers expected, which was to collect information from teachers who taught Thai and other subjects by prioritizing them to get the opinions that best meet teachers' needs.However, the authors found that the teachers lacked speaking skills compared to other competencies; hence, developing principles of language alone was not sufficient. Also, it was important to develop speaking skills to help teachers gain confidence in teaching and know how to use language to communicate aesthetically. Teachers' development needs should be addressed in terms of both teaching style and communication skills, as well as respect for diversity. This ranking depended on the teacher's language proficiency. Competency was therefore a fundamental feature of an individual's ability to perform tasks to achieve the desired result and reflect Thainess, thus forming a life competency in the daily routine that the teacher intends to add to the learner's daily life (Chalermlarp Thongart, 2018; Hay Group, 2003; Office of the Education Council, Education Council, 2018). In addition, competencies were relevant in both linguistics and communication for practical approaches that the authors associated with competence and proficiency, specifically the ability to use language for communication and application. It was necessary to have components of both linguistic ability (verbal and nonverbal) and communication ability. This included adapting and having a flexible mindset that met the needs of executives or employers who wanted people in the organization to be able to adapt to changing situations and environments (Cohen et al., 2017; Kivunja, 2015; Llurda, 2000; Louviere et al., 2015; Trilling & Fadel, 2009). Further, some scholars developed a framework for evaluating competencies in learning management, such as Upawan Namhiran (2016), who found that knowledge and understanding from the study of specific subjects, knowledge of teaching methods, and competencies in learning management were important pieces of information for organizing guidelines on teaching profession experience to improve the quality of the development of both knowledge and professional skills for learners. Furthermore, Kunphat Pongpanich (2018) studied the components and indicators of teaching and learning competencies of primary school teachers who used Thai as a second language. Kunphat Pongpanich found that teachers' teaching and learning competencies should take learning management, classroom management, a curriculum and learning management plan, assessment and evaluation, and teachership into account. However, social support factors affected motivation for success, attitudes, performance, and the impact on teachers' competencies (Chamnong Chamchantharawong et al., 2017). Therefore, enhancing Thai language competency required establishing a language identity, such as through speaking skills that promoted negotiation and coordination, and writing skills that strengthened summary report writing and academic writing (Pouangplub et al., 2016). Therefore, the teacher's language communication competency was what would impress the student, influence their behavior, and reverse communication (Charoywan Prasertphon et al., 2013; Nattachai Tatongjai et al., 2015). The analysis of current and desirable conditions to enhance teachers' competencies in Thai communication in the 21st century played an important role in the development of innovative learning materials that would enable teachers to develop their Thai communication competencies and be able to properly organize the learning process.

The Effectiveness of Learning Innovations That Enhance Teachers' Competencies in Thai Communication in the 21st Century

The authors investigated the effectiveness of learning innovations that enhanced teachers' competencies in Thai communication in the 21st century in two ways, namely by the RAI of Thai language learning innovations to enhance Thai communication competency and by a posttest.

The RAI of Thai Language Learning Innovations to Enhance Thai Communication Competency

The analysis of the RAI of Thai language learning innovations to enhance Thai communication competency aimed to find out the effectiveness of Thai language learning innovations and Thai communication competency assessments by using a podcast, which was an innovative way to use this medium. It covered information on how to develop Thai communication competencies (i.e., speaking, reading, writing, essential communication for teachers, and media literacy). This was an important concept that teachers should apply to improve their communication competencies. Then, the researchers invited the teachers to design the activities during the class and produce their podcast materials. The teachers could consider the content on their own to integrate Thai communication competencies, including the posttest that presented the learning management approach. The results of the analysis can be summarised with the conformity of the RAI according to the elements in the assessment of Thai communication competency, in the following order: Communication etiquette, Thai language aesthetics, reading, listening, writing, use of Thai language, speaking, and moral and communication ethics. The development of innovative media to empower learners in the 21st century required critical thinking, solutions, verbal communication, coordination with others, information technology, and media literacy to help learners analyze and select the media appropriately. Whether it was the Internet or YouTube, it was all information technology media that could be used to facilitate learners' motivation to learn. It entertained and helped learners improve their language learning on their own (Bahadorfar & Omidvar, 2014; Nantaka Warinin, 2014; Tomlinson, 2012). Innovation, therefore, contributed to solving educational problems by connecting materials with advanced human creativity (Dodgson, 2021; National Education Association, 2010; National Innovation Agency, 2010; Schilling, 2008; Thitsana Khammanee, 2014), as well as developing innovative media to cover the skills needed in the 21st century. This approach could encourage learners to develop positive thinking processes for learning and language use, and develop innovative materials that were challenging and reflected the learners' needs. Learners were also encouraged to design innovative media on their own to build leadership and enable them to conduct learning activities that focused on the learners, so that they could interact together and learn how to use creativity, develop innovations, and use the media, information, and digital literacy involved in working effectively (Andrade, 2016; Bundsgaard & Hansen, 2011; Howard, 2018; Office of the Education Council, Education Council, 2018).

A Posttest on Thai Language Learning Innovations to Enhance Thai Communication Competency

The authors also used a posttest to carry out their analysis of data on the effectiveness of Thai language learning innovations; the posttest determined the effectiveness of these innovations. The researchers carried out the posttest after they had presented the learning management approach to teachers. The authors used learning activities to encourage teachers to connect their learning, use language competencies, and apply different learning management methods to creatively solve classroom problems and allow them to offer a variety of activities on their own. In particular, the researchers proposed a practical learning activity that used creative media and concrete assessment and evaluation. When the authors compared the posttest results to the benchmark, the results revealed that the teachers' post-Thai communication competency assessment score was statistically significantly higher than the benchmark of .05. Thus, the conclusion is that the posttest score of the assessment of Thai communication competency was above the 70% benchmark. Therefore, teachers should obtain the necessary development in both core competency and job competency. Teachers' basic or prior knowledge of a language plays an important role in integrating communication skills and in enabling them to select innovative media according to their creativity. Indeed, as a result, teachers can design a learning management system, learners can have appropriate access to information, and both teachers and learners can create a favorable teaching and learning environment, smooth communication, and negotiation to avoid conflicts, which also help to create and maintain social interactions for learners. This learning atmosphere makes learners feel at ease, helping them develop better language skills and communication etiquette that corresponds to the Thai cultural context. Learning and teaching management designed with specific competencies based on learner needs will benefit students in terms of learning management, speaking skills, and communication skills. Besides, students would appreciate the beauty of language and respect cultural differences. For example, in activities that encourage learners to communicate effectively with others, have communication etiquette in both listening and speaking, and respect themselves and others in society, language ability is not less important than the contents, although effective content is used in the teaching and learning process. Furthermore, if teachers choose technology to complement the content, teaching and learning will be more efficient and effective, including assessments that focus on practicum (Eyibe, 2009; Mishra & Koehler, 2009; Office of the Education Council, Education Council, 2018; Okoli, 2017). Moreover, it reflects leadership in collaborative work; work ethics are created both inside and outside the classroom because communication competency plays a vital role in the challenges of work and reflects professional success as well as preparedness for all sorts of communication in everyday life (Aliyu, 2017; Crosling & Ward, 2002; Fasawang Pattanapichet, 2009; Riemer, 2007; Thitsana Khammanee, 2014; Tran et al., 2016; Trilling & Fadel, 2009). Education in the 21st century should emphasize the development of learners' competencies in terms of knowledge, skills, and attributes, including proficiency in language communication skills, problem-solving, creativity and innovation, the use of information technology, and media literacy. These are essential for helping learners compete on a global scale amid a diverse and ever-changing society (Ledward & Hirata, 2011; Tan et al., 2017; Menggo et al., 2019). It is in line with the researchers who were interested in developing innovations to promote competency and communication. For example, Pornthip Chaiyaso et al. (2013) studied the development of teaching and learning management innovations to enhance the competencies in learning assessment of student teachers; they found that more than 50% of student teachers had a very good level of learning assessment competencies, and more than 90% of them passed the 60% criteria. In addition, Muhamad Salaebing and Kanyanee Chareonchang Nuchmee (2017) investigated the outcomes of creating a podcast to promote Malay communication skills. They designed a podcast that took into account the concept of language learning and focused on the sequence of language development while also providing great satisfaction to the learners. Also, Evans (2008) studied the effectiveness of mobile learning in the form of podcasts for reviewing lectures in higher education and found learners were able to use podcasts for reviewing lessons after classes effectively. Furthermore, Bolliger et al. (2010) found podcasts could stimulate learners' interest in learning and result in higher achievement as well.

SOLUTIONS AND RECOMMENDATIONS

The following recommendations could be made based on the discussion and conclusion of the results:

1. Thai language teachers are actively encouraged to use innovations to enhance teachers competencies in Thai communication.
2. It is recommended that teachers be trained in the production and design of innovations and that they be used in the educational process.

FUTURE RESEARCH DIRECTIONS

The following recommendations for future research are based on the research and development of learning innovations to enhance teachers' competencies in Thai communication in the 21st century:

1. Research should be conducted on competencies in other areas that are essential to the professional development of teachers in the 21st century.
2. Research and development of innovations focused on teacher competency should be encouraged.

REFERENCES

Aliyu, M. M. (2017). Developing oral communication skills through collaborative learning: Implications for Nigerian teachers. *International Journal of English Literature and Social Sciences*, *2*(5), 239225. doi:10.24001/ijels.2.5.15

Andrade, M. S. (2016). Curricular elements for learner success: 21st century skills. *Journal of Education and Training Studies*, *4*(8), 143–149. doi:10.11114/jets.v4i8.1743

Bahadorfar, M., & Omidvar, R. (2014). Technology in teaching speaking skill. *Acme International Journal of Multidisciplinary Research*, *2*(4), 9–13.

Basenko, G., & Baskakova, V. (2021). Podcasts in the teaching media space. In *E3S Web of Conferences* (pp. 1–8). EDP Sciences., doi:10.1051/e3sconf/202127312122

Boettcher, J. V., & Conrad, R. M. (2021). *The online teaching survival guide: Simple and practical pedagogical tips*. John Wiley & Sons.

Bolliger, D. U., Supanakorn, S., & Boggs, C. (2010). Impact of podcasting on student motivation in the online learning environment. *Computers & Education*, *55*(2), 714–722. doi:10.1016/j.compedu.2010.03.004

Borsheim, C., Merritt, K., & Reed, D. (2008). Beyond technology for technology's sake: Advancing multiliteracies in the twenty-first century. *The Clearing House: A Journal of Educational Strategies, Issues and Ideas*, *82*(2), 87–90. doi:10.3200/TCHS.82.2.87-90

Bundsgaard, J., & Hansen, T. I. (2011). Evaluation of learning materials: A holistic framework. *Journal of Learning Design*, *4*(4), 31–44. doi:10.5204/jld.v4i4.87

Chaiyaso, P., Piyaphimolsit, C., Ekwarangkun, P., Sritanyu, E., Chantra, C., Wanichwasin, P., Taksino, P., & Gamket, W. (2013). *Management innovation development. Teaching and learning to enhance the learning assessment competencies of student teachers*. Research and Development Institute of Kasetsart University.

Chamchantharawong, C., Chanjarain, I., Suwanjinda, C., & Siririrth, W. (2017). Factors affecting the functional competency of teacher of graduate diploma students in teaching profession at Chiangrai College. Veridian E-Journal. *Silpakorn University*, *10*(1), 269–280.

Cohen, L., Manion, L., & Morrison, K. (2017). *Research methods in education*. Routledge. doi:10.4324/9781315456539

Cope, B., & Kalantzis, M. (Eds.). (2000). *Multiliteracies: Literacy learning and the design of social futures*. Psychology Press.

Crosling, G., & Ward, I. (2002). Oral communication: The workplace needs and uses of business graduate employees. *English for Specific Purposes*, *21*(1), 41–57. doi:10.1016/S0889-4906(00)00031-4

Dodgson, M. (2021). The strategic management of technology and innovation. In Oxford Research Encyclopedia of Business and Management. doi:10.1093/acrefore/9780190224851.013.145

Evans, C. (2008). The effectiveness of m-learning in the form of podcast revision lectures in higher education. *Computers & Education*, *50*(2), 491–498. doi:10.1016/j.compedu.2007.09.016

Eyibe. (2009). *Modern curriculum development* (second edition). Adson Educational Publisher.

Facione, P. A. (2011). *Measured reasons and critical thinking*. The California Academic Press.

Hay Group. (2003). *Using competencies to identify high performers: An overview of the basics*. Hay Group, Inc.

Howard, Y. (2018). *Developing 3R8C skills to become innovative leaders for Thailand 4.0*. Tribune.

Khammanee, T. (2014). *Teaching science: Body of knowledge for an effective learning process*. Publishing House of Chulalongkorn University.

Kivunja, C. (2015). Teaching students to learn and to work well with 21st century skills: Unpacking the career and life skills domain of the new learning paradigm. *International Journal of Higher Education*, *4*(1), 1–11. doi:10.5430/ijhe.v4n1p1

Ledward, B. C., & Hirata, D. (2011). *An overview of 21st century skills: Summary of 21st century skills for students and teachers, by Pacific Policy Research Center.* Kamehameha Schools–Research & Evaluation.

Llurda, E. (2000). On competence, proficiency, and communicative language ability. *International Journal of Applied Linguistics*, *10*(1), 85–96. doi:10.1111/j.1473-4192.2000.tb00141.x

Louviere, J. J., Flynn, T. N., & Marley, A. A. J. (2015). *Best-worst scaling: Theory, methods, and applications*. Cambridge University Press. doi:10.1017/CBO9781107337855

Mahawichit, P. (2017). Innovation in learning from Finland. *IPST Magazine*, *46*(209), 40–45.

Menggo, S., Suastra, I., Budiarsa, M., & Padmadewi, N. N. (2019). Needs analysis of academic-English speaking material in promoting 21st century skills. *International Journal of Instruction*, *12*(2), 739–754. doi:10.29333/iji.2019.12247a

Ministry of Education. Singapore. (2010). *Learning of 21st-century competency.* https://www.moe.gov.sg/media/press/2010/03/moe-to-enhance-learning-of-21s.php

Mishra, P., & Koehler, M. (2009). Too cool for school? No way! Using the TPACK framework: You can have your hot tools and teach with them, too. *Learning and Leading with Technology*, *36*(7), 14–18.

Namhiran, U. (2016). A development of framework of competencies assessment for Thai student teachers' learning management and specific content knowledge. *Rommayasan: Journal of Humanities and Social Sciences*, 14(2), 207-—214.

National Education Association. (2010). *Preparing 21st century students for a global society: An educator's guide to the "four Cs."* https://www.nea.org/assets/docs/A-Guideto-Four-Cs.pdf

National Innovation Agency. (2010). *Innovation management for executives (IMEs)* (3rd ed.). Innovation Culture Promotion Department National Innovation Agency.

Office of the Education Council, Education Council. (2018). *Report on the results of research and development of the competency framework for students at the lower elementary school for basic education courses.* NESDB.

Office of the Education Council, Ministry of Education. (2009). *Educational reform proposals in the second decade (2009-2018).* Office of the Council of Education Secretariat.

Office of the Education Council, Ministry of Education. (2017). *National Education Plan 2017 –2036.* Office. Secretary of the Education Council.

Office of the Education Council, Ministry of Education. (2019). *Framework for learner competency at the basic education level and Lower primary education (Grade 1-Grade 3).* NESDB.

Office of the National Economic and Social Development Council. (2020). *National strategy 2018–2037.* Office of the Prime Minister.

Office of the National Education Commission, Office of the Prime Minister. (2002). National Education Act 1999, amended (2nd edition) 2002. Office of the Education Commission, National Office of the Prime Minister.

Okoli, A. C. (2017). Relating communication competence to teaching effectiveness: Implication for teacher education. *Journal of Education and Practice*, *8*(3), 150–154.

Panich, W. (2013). *Enjoy learning in the 21st century*. Sodsri-Saritwong Foundation.

Partnership for 21st Century Skills-P21. (2011). *Framework for 21st century learning.* http://www.p21.org/storage/documents/1._ p21_framework_2-pager.pdf

Pattanapichet, F. (2009). *A development of the competency-based English oral communication course for undergraduate public relations students* [Unpublished doctoral dissertation]. Chulalongkorn University Intellectual Repository (CUIR).

Pluemsamrankit, P. (2017). Factors affecting 21st century skills for learners. *Valayalongkorn Review Journal Humanities and Social Sciences*, *7*(3), 141–158.

Pongpanich, K. (2018). Analysis of factors and competency indicators of the teaching and learning of kindergarten teachers using Thai as a second language. *Journal of Yala Rajabhat University*, *13*(1), 1–12.

Pouangplub, S., Suriyamegha, O., Ngamnikorn, T., Chatdee, P., & Khumsat, T. (2016). The enhancement of Thai language usage competencies in accordance with identity of study fields in Panyapiwat Institute of Management. 동남아연구, *26*(1), 89—107.

Prasertphon, C., Kornpuang, A., Jansila, V., & Chatruprachewin, C. (2013). A model for developing learning management competency of new teachers at general education private schools. *Journal of Education Naresuan University*, *15*(5), 43–53.

Prozesky, D. R. (2014). Role of communication in teaching. Teaching Eye Health. University of Pretoria.

Rassameethammachot, S. (2008). *Guidelines for human potential development by competency-based learning* (6th ed.). Knowledge Compilation Division, Research Division, National Productivity Institute.

Riemer, M. J. (2007). Communication skills for the 21st century engineer. *Global Journal of Engineering Education*, *11*(1), 89–100.

Royal Academy. (2013). *Royal Institute Dictionary 2011*. Royal Academy.

Saenthong, N. (2007). *Let's get to know competency*. HR Center.

Salaebing, M., & Nuchmee, K. C. (2017). Result of a podcast learning to promote Malay language communication skills for undergraduate students. *Al-Hikmah Journal. Fatoni University*, *7*(13), 37–44.

Salas Pilco, S. Z. (2013). Evolution of the framework for 21st century competencies. *Knowledge Management & E-Learning*, *5*(1), 10–24.

Schilling, M. A. (2008). *Strategic management of technological innovation* (2nd ed.). McGraw-Hill Education.

Sattabut, T. (2012). A study of competency needs of human resource staff for industry sector. *Veridian E-Journal*, *5*(2), 426–448.

Sun, C. (2015). The cultivation of cross-cultural communication competence in oral English teaching practice. *English Language Teaching*, *8*(12), 7–10. doi:10.5539/elt.v8n12p7

Suvarnaphaet, K. M., & Suvarnaphaet, P. (2023). Use of podcasts to develop students' speaking skills. Humanities. *Arts and Social Sciences Studies*, *23*(1), 1–8. doi:10.14456/hasss.2023.1

Tan, J. P. L., Choo, S. S., Kang, T., & Liem, G. A. D. (2017). Educating for twenty-first century competencies and future-ready learners: Research perspectives from Singapore. *Asia Pacific Journal of Education*, *37*(4), 425–436. doi:10.1080/02188791.2017.1405475

Tatongjai, N., In Chanjarain, Weeraphan Siririth. (2015). Teacher's Competency in Vocational Institutions in the Upper Northern Thailand after Completing the Graduate Diploma Program in Teaching Profession. Journal of Education. *Mahasarakham University*, *9*, 138–152.

Thongart, C. (2018). Thai language learning management using 21st century competencies-based. *Journal of Education Studies*, *46*(1), 171–184.

Thongkaew, T. (2007). *Competency: Principles and practices.* Mahawatiyaraya Suan Dusit Rajabhat University.

Thuaamklang, K. (2012). *Development of educational media/innovations to promote academic standing*. Sathaporn Books.

Tomlinson, B. (2012). Materials development for language learning and teaching. *Language Teaching, 45*(2), 143-—179. doi:10.1017/S0261444811000528

Tran, L. T. B., Ho, N. T., & Hurle, R. J. (2016). Teaching for creativity development: Lessons learned from a preliminary study of Vietnamese and international upper (high) secondary school teachers' perceptions and lesson plans. *Creative Education*, *7*(7), 1024–1043. doi:10.4236/ce.2016.77107

Trilling, B., & Fadel, C. (2009). *21st century skills: Learning for life in our times.* John Wiley & Sons.

UNESCO. (2010). *World data on education 7th edition 2010/11.* http://www.ibe.unesco.org

UNESCO. (2011). *UNESCO ICT competency framework for teachers.* https://unesdoc.unesco.org/images/0021/002134/213475e.pdf

Warinin, N. (2014). *The model of teachers' competency development in learning management for primary schools in KamphaengPhet primary educational service area office 2*. [Unpublished doctoral dissertation]. Nakhon Sawan Rajabhat University.

Wichit, W. (2009). *Podcast with teaching and learning*. http://www.wittaya.net

Yodyingyong, K. (2009). *Innovative organizations: Concepts and processes*. Chulalongkorn University Book Center.

Chapter 6
Digital Skills and Competencies for a Changing Labor Market in Indonesia

Hani Gita Ayuningtias
https://orcid.org/0000-0003-2299-7406
Telkom University, Indonesia

ABSTRACT

Digital skills and competencies are becoming increasingly important in today's rapidly changing labor market in Indonesia. The adoption of new technologies and the increasing digitalization of work processes have created a significant demand for individuals who possess the necessary digital skills and competencies to succeed in the modern workplace. Many industries in Indonesia are rapidly adopting digital technologies to improve productivity and efficiency. This means that workers lacking digital skills and competencies are disadvantaged in the job market and may struggle to find employment. As more tasks become automated, workers who possess digital skills and competencies will be better positioned to adapt to new job requirements and remain employable. The growth of the digital economy in Indonesia is creating new job opportunities in fields such as e-commerce, digital marketing, and data analytics. Workers with digital skills and competencies will be better positioned to take advantage of these opportunities and succeed in these industries.

DOI: 10.4018/978-1-6684-9089-1.ch006

INTRODUCTION

Automation technology offers substantial opportunities for governments and businesses to boost production, efficiency, and growth. Automation can boost productivity and GDP growth in Indonesia for workers while also expanding market prospects for businesses (Das et al, 2019). Through the CIPG (Center for Innovation Policy and Governance), the Indonesian government issued legislation in 2018 for the growth of big data, artificial intelligence, blockchains, and financial technology (fintech), which became the beginning point for the rapid adoption of technology (Ramadhan and Putri, 2018). Furthermore, according to Das et al. (2019), Indonesia will transform and face automation and the future of work in 2030. In response to the situation, Indonesia developed a program called Vision Indonesia Gold 2045 in 2020. This vision is supported by four major pillars, which are as follows: human development and science and technology, economic equality, and strengthening resilience and national governance (Bappenas, 2020). Human capital is an intangible asset that must be cultivated extensively. Therefore, companies must invest in the development of their human capital, which has a significant impact on the performance of the company (Ayuningtias and Anggadwita, 2015). The Indonesian government is now focusing on developing excellent human resources and adapting to rapid technological changes. Competence in human resource management is one of the best ways for succeeding in competing worldwide (Ayuningtias, Anggadwita and Prasetia, 2018). The role of digital transformation in increasing productivity, effective service delivery, and job creation is highlighted in Indonesia's national strategic development plan. Indonesia's Ministry of Communication and Informatics (MCI) is developing a "Digital Nation Roadmap" to boost the country's digital readiness and competitiveness. The MCI commissioned this study to assist in the development of this "Digital Roadmap" and to give an assessment of the digital skills landscape in Indonesia, which will then lead to the formulation of actual actions and policies (Diningrat et al., 2022)

Indonesia has made significant progress in developing digital competencies and skills in recent years. The government has recognized the importance of digital technology in driving economic growth and has implemented various initiatives to promote digital literacy and skills development (Ameliah, Negara and Rahmawati, 2022). Following are some of the important developments in this field:

1. *Growth of Internet use*: Internet use in Indonesia is expanding from year to year, with around 196.7 million active Internet users in Indonesia in 2020 (APJII, 2020). This has enabled the development of the digital industry, including e-commerce, fintech, and games.

2. *Increased investment in the technology sector:* Indonesia has become an attractive place for technology investment, especially in terms of startup funding. In 2020, technology startups in Indonesia managed to raise more than $2.8 billion in investment funds, an increase of 15% from the previous year (Bappenas, 2020).
3. *Increasing number of the digital workforce*: More and more Indonesian companies need employees with digital skills (Ramadhan and Putri, 2018). This has increased the demand for a workforce with programming, data analysis, and digital marketing skills.
4. *Increasing digital awareness*: There is increasing awareness of the importance of digital competencies and skills in Indonesia (Diningrat et al., 2022). The Indonesian government has introduced digital training programs to help increase digital skills among Indonesians.
5. *Development of digital education*: Many universities in Indonesia have introduced digital and technology programs and professional training to help produce a digitally skilled workforce (Ministry of Education, Culture, Research, and Technology, 2021)

One of the key initiatives is the National Movement of One Million Digital Talent (NEMDT) program launched in 2018 by the Indonesian government, which aims to develop one million digital talents in Indonesia by 2023. The program provides various training and certification programs to develop skills in areas such as coding, data analytics, and digital marketing. According to IMD World Competitiveness Center (2022), Indonesia is ranked 51st in the Digital Competitiveness Index, which considers a variety of factors, including knowledge, technology, future readiness, expertise in the digital sector, and the number of digital talents in a country. The need for digital talent in Indonesia continues to increase every year (Wahyuningtyas et al., 2021). Based on research by IMD World Competitiveness Center (2022) The rankings from 2020 through 2022 The overall performance grade for Indonesia has improved. In 2020, it ranked 56th, followed by 53rd in 2021 and 51st in 2022. This ranking is founded, for instance, on knowledge, technology, and readiness for the future. The knowledge aspect has increased from the previous year, and the availability aspect has contributed to this increase. Talent in addition to an increase in training and education. Additionally, the technological aspect has grown, which is supported by the regulatory framework and capital. The sub-aspects such as adaptable attitudes, business agility, and IT integration that support the third factor, namely future readiness, have declined. To close the gap in developing digitally competent people, governments must prioritize workers' access to ICT education and suitable training to increase their digital abilities. The goal, which was adopted from the RPJMN, Indonesia's Short to Medium Term Development Plan, sets a target of "50% of workers having intermediate and advanced digital skills by 2024." In Indonesia, basic to intermediate digital skills account for 50% of the labor force, whereas advanced digital skills account for less than 1%. It will be impossible for the goal to meet its 2024 target unless the nation's IT skills uptake improves. Despite these efforts, Indonesia's digital workforce still has a significant skills gap. Many individuals and businesses, particularly those in rural areas, still lack the necessary digital literacy and skills to fully participate in the digital economy. However, the government and private sector continue to invest in digital skills development to bridge this gap and ensure that Indonesia is well-positioned to take advantage of the opportunities the digital age offers.

DIGITAL LITERACY

Digital literacy refers to the ability to use digital tools and technologies effectively. This includes basic computer skills, knowledge of software applications, and the ability to navigate the Internet and social media (Ministry of Communication and Information, 2022). The e-commerce sector supports much of Indonesia's digital economic growth around 34%, or equivalent to IDR 1,900 trillion in 2030. In addition, the business-to-business (B2B) sales model in the digital sector is projected to increase by 13%, or 763 trillion IDR, between 2021 and 2030. The health technology sector is also estimated to reach IDR 471.6 trillion or grow up to 8% (Ameliah, Negara and Rahmawati, 2022). There have been several recent studies on the development of digital literacy in Indonesia. Overall, the findings suggest that, while digital literacy is growing in the country, there are still significant gaps that need to be addressed. The government, through the Ministry of Communication and Information, has programs to improve digital competence, namely the National Digital Literacy Movement program. The ongoing program is expected to target the widest possible community and enhance their respective digital capabilities. According to UNESCO (2018), the government's offer of simple access to digital social services will assist in reducing poverty. In addition, the public will be able to obtain information and benefit from digital media via an enormous variety of data. Access restrictions and obstacles to obtaining a quality education can be fixed through the capacity of the community to use digital information (UN, 2022).

One study conducted in 2020 by the Indonesian Internet Service Providers Association (APJII) found that digital literacy levels in the country had improved significantly in recent years, with 56% of Indonesians having basic digital literacy skills, up from just 7% in 2014. However, the study also found that there are still significant gaps in digital literacy between different regions, age groups, and socioeconomic backgrounds. For example, while 72% of people aged 15-19 had basic digital literacy skills, only 20% of those aged 60 and over had the same level of skill. Another study conducted in 2021 by the Indonesian Ministry of Education and Culture found that, while many students in the country were proficient in using social media and messaging apps, their skills in using digital tools for learning and problem-solving were lacking. The study recommended that the education system in the country focus more on developing digital literacy skills, particularly in areas such as critical thinking, problem-solving, and collaboration. The Indonesia Digital Literacy Survey, 2020, demonstrated that digital literacy in Indonesia is mostly low, particularly in terms of the skills required to manage digital information and data. Indonesia's low levels of digital literacy are primarily due to a lack of critical thinking when obtaining information online. The Ministry of Communication and Informatics annually conducts research on the status of digital literacy in Indonesia referring to the Indonesian government's digital literacy roadmap for 2020-2024.

Figure 1. Indonesia Digital Literacy Road Map 2020- 2024
Source: (The Ministry of Communication and Informatics, 2020)

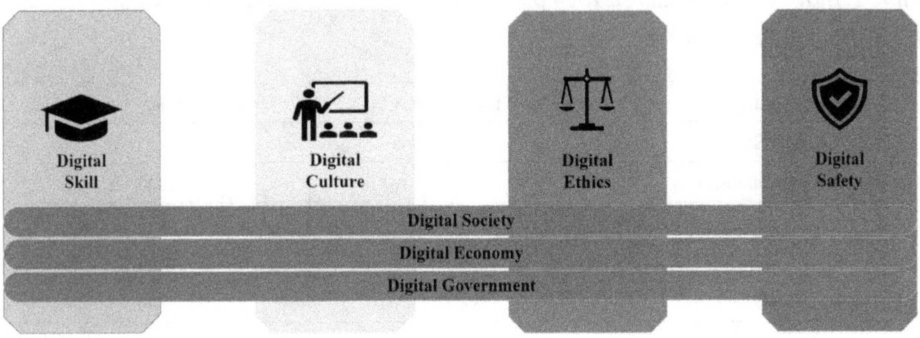

Based on Figure 1, there are four pillars that are part of the curriculum development framework for digital literacy, namely Digital Skills, Digital Ethics, Digital Safety, and Digital Culture (Ameliah, Negara and Rahmawati, 2022).

1. *Digital Skills*: the individual ability to know, understand, and use ICT hardware and software and systems digital operations in everyday life.
2. Digital Culture: the individual ability to read, describe, familiarize, examine, and build national insights, the values of Pancasila and Bhinneka Tunggal Ika in everyday life, and the digitalization of culture using ICT. Pancasila and Bhineka Tunggal Ika are two important principles in the Republic of Indonesia, reflecting the country's national identity and guiding principles (Indonesia

Information Portal, 2023). Pancasila is the philosophical foundation of the Indonesian state. It consists of five principles that serve as the core values for the nation. Bhineka Tunggal Ika is an Indonesian phrase that means "Unity in Diversity." It reflects the multicultural and multi-religious nature of Indonesia. The value of Bhineka Tunggal Ika lies in its recognition and acceptance of the diverse cultural, ethnic, and religious identities that coexist within the nation

3. *Digital Ethics*: individual's ability to realize, exemplify, adapt, rationalize, consider, and develop digital ethical governance (netiquette) in everyday life.
4. *Digital Safety*: user capabilities in recognizing, patterning, applying, analyzing, weighing, and increasing awareness of personal data protection and deep digital security in everyday life.

The four pillars and the digital literacy curriculum become the basis for the development of a national digital literacy program that includes a digital society, digital economy, and digital government.

Table 1. Indonesia Digital Literacy Index 2021-2022 (The Ministry of Communication and Informatics, 2022)

Digital Literacy Pillar	Digital Literacy Index 2021	Digital Literacy Index 2022
Digital Skill	3.44	3.52
Digital Culture	3.90	3.,84
Digital Ethics	3,.53	3.68
Digital Safety	3.10	3.12
Total	3.49	3.54

The National Digital Literacy Index in 2022 will increase by 0.05 points to 3.54 from index achievements in 2021. There are three pillars that have increased, namely Pillar 1 (Digital Skills) increased by 0.08 points, Pillar 3 (Digital Ethics) increased by 0.15 points, and Pillar 4 (Digital Safety) rose by 0.02 points. However, Pillar 2 (Digital Culture) decreased by 0.06 points.

Indonesia's digital literacy is included in the calculation of the Digital Literacy Index for 2021 categorized as moderate with an index score of 3.49. The pillar with the highest score is the pillar of digital culture, while the pillar with the lowest is Digital Safety. The score for the Digital Skill Pillar is 3.44, Digital Ethics 3.53, and Digital Safety Pillar 3.10.

The Digital Literacy Program held by MCI had a good impact. The Digital Literacy Index between those who followed and showed no difference being the biggest. Most of those who follow the program have a larger index, especially on the pillars of Digital Skill and Digital Security. However, activities are focused on improving the pillars of Digital Ethics and Digital Culture. Digital literacy outcomes are dependent on the inputs and outputs of ICT learning and teaching activities, as well as their interaction with activities designed to improve fundamental literacy skills.

DIGITAL SKILLS AND COMPETENCIES

The development of digital skills can hasten the fulfillment of national development objectives. Indonesia has enormous chances to reap maximum benefits by growing citizen participation in the digital economy ecosystem and experiencing a digital transformation of its labor force by promoting access to digital skill creation. The potential for a digitalized economy in Indonesia stems not only from digital sector activities, such as a digital services and platform economy, but also from non-digital sector adoption and utilization of ICT, such as the use of e-government, e-commerce, and Industry 4.0 solutions (Ministry of Communication and Informatics, 2020). To maximize the potential of a digital economy, it is essential to cultivate the digital competencies of the workforce. Having a talent group with digital competencies can extend the development and application of digital tools and processes to strengthen business models in various industries (Diningrat et al., 2022).

Digital skills are essential for society, government, and businesses to use, implement, and develop digital technology. To maximize the digital economic dividend and speed the fulfillment of development goals, a digital skills development framework is required. Digital skills must be included in the human development agenda, which necessitates the government's commitment to improving digital inclusion, literacy, and competency (Diningrat et al., 2022). Opportunities for digital upskilling must be expanded through formal education, vocational training, and on-the-job training. The smooth operation of these channels is controlled by the policy, financial, and infrastructural frameworks in place, which will allow Indonesia to build a favorable digital skills ecosystem. Digital competence is a broad concept involving diverse competency-based abilities and communication technology skills, with a focus on assessment abilities and software-integrated expertise and mindset. Digital competence includes four aspects such as basic skills, attitude and perspectives, background knowledge, and central competencies (Wahyuningtyas et al., 2021). The terms of competency concept include the knowledge, skills, and attitudes that enable people to convey work behavior to accomplish what an organization requires to deliver a specific performance for each workplace (Fernando and Wulansari, 2020). In organizations, the millennial generation is considered knowledgeable about technology (Wahyuningtyas et al., 2021).

Figure 2. Digital Skills Development Framework
Source: (Diningrat et al., 2022)

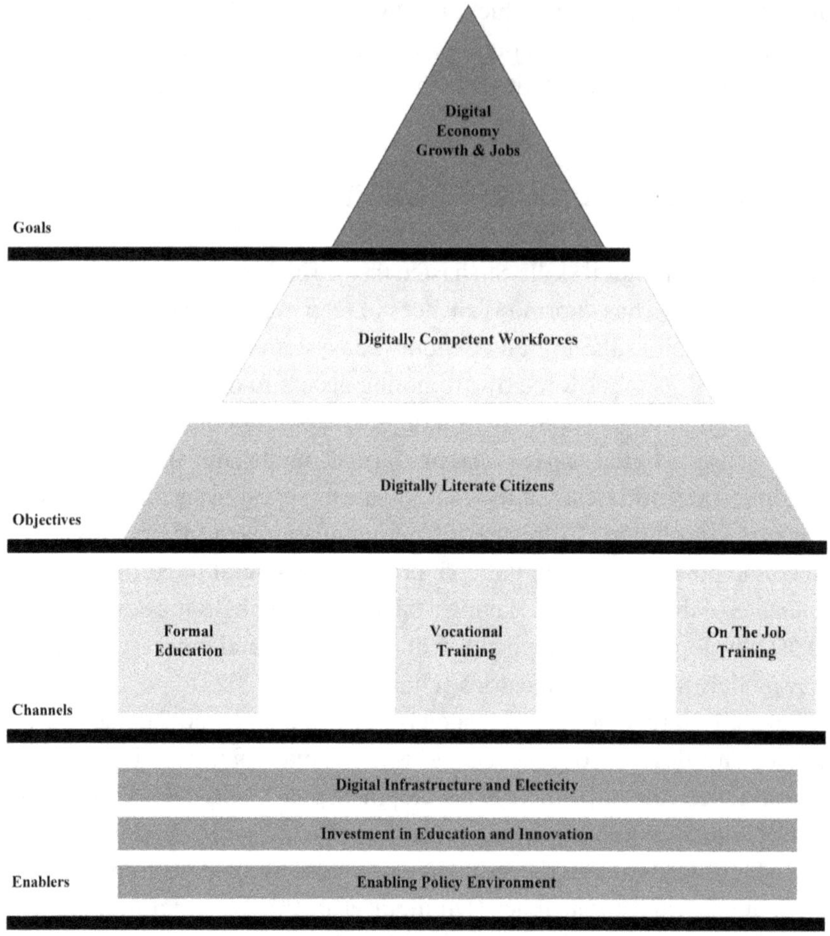

Based on Figure 2, the Indonesian government has made a digital skills development plan. The government mapping is based on the Goals, Objectives, Channels, and Enablers needed to support the program. A better understanding of Indonesia's digital skills landscape will assist the government in developing a strategy outline to handle the difficulties and will help to strengthen the Digital Nation Roadmap 2021-2024, which aims to produce digitally literate individuals and a digitally capable workforce. The development of digital capabilities in the workforce is critical for leveraging the potential of a digital economy (The Ministry of Communication and Informatics, 2020). Having a talent pool with digital skills can help to extend the creation and implementation of digital tools and processes to boost business models in a variety of industries, including platform economies, e-commerce, and Industry 4.0. An organization's talent consists of employees with the expertise and ability to drive the organization forward both in the present and in the future (Wahyuningtyas and Anggadwita, 2017).

Adapted from the World Bank's framework, Bashir and Miyamoto (2020) divided a digitally competent workforce into four categories unskilled, basic skilled, intermediate skilled, and advanced skilled. This approach identifies the ability of employees to use technology in helping their work. Based on the National Labor Survey Conducted by the Central Bureau of Statistics (BPS) in 2018 and 2020, the current workforce in Indonesia has a majority of unskilled workers (Table 2). Based on category, unskilled means a labor force that is unfamiliar with or has limited exposure to simple digital technologies. Basic skilled means simple digital technologies, such as Microsoft Office, Google Chrome for web perusing, and email, are mastered by the labor force. Intermediate skill means the labor force utilizes a variety of digital technologies applicable to their occupations. Journalists, for instance, should be able to compile, search, and analyze information using digital tools. Advanced skills mean the workforce applies more sophisticated analytic and theoretical abilities. This level of proficiency is typically required for jobs with a high ICT intensity, such as artificial intelligence specialists.

Table 2. Digital Skills of Labor Force (BPS, 2018 and 2020)

Category	2018	2020
Unskilled	55%	49%
Basic Skilled	32%	37%
Intermediate Skilled	12%	13%
Advanced Skilled	0.61%	0.63%

Approximately half of Indonesia's existing labor is digitally unskilled, with the number increasing to around two-thirds of the present workforce when individuals with basic IT skills are included. These figures show that the bulk of the workforce in Indonesia has limited ability to participate in, contribute to, and benefit from the rise of the digital economy. More workers with basic digital skills are employed in the informal sector, whereas those with intermediate digital skills can obtain formal employment due to their higher education levels. Only 16% of those with intermediate digital skills are employed in the informal sector, compared to 43% of those with fundamental digital skills (Diningrat et al., 2022) .

Industry 4.0 is associated with the reorganization of human-resource management in businesses, which will impact the need for skills or competencies. Obviously, it is associated predominantly with technological advancement and economic and social change (Kowal et al., 2022). Innovation in technology expands the requirement for skills, and investment in skills makes it possible to fulfill that demand. Employees function in an innovative setting because of business digitization. The workforce is highly responsive to change, necessitating the necessity to develop new competencies or skills (Bakhshi et al., 2017). Based on Deloitte (2017), the transformations in competencies will happen by 2030 because of the rapid pace of digitization and automation. As upcoming competencies, they identified those that deliver humans a benefit over machines in an environment with low automation capability. These competencies are in popular demand, leading to the development of fresh roles that require the acquisition of new skills or qualifications.

Future needs can be subdivided into numerous categories. The first of these (Table 3) is 'fundamental competencies,' which include reading comprehension, speaking, critical thinking, and active learning. The foundation for acquiring more advanced competencies, primarily creativity, and social intelligence, is fundamental skills (Deloitte, 2017). There are both 'mass competencies' and 'niche competencies,' which are required for an extensive range of positions and a lesser number of specialized positions, respectively (Deloitte, 2017).

Table 3. Future Requirement Knowledge and Skills (Deloitte, 2017)

Competencies	Mass competencies		Niche competencies
Basic Competencies	Reading comprehension Active Listening Speaking Active Learning	Writing Critical Thinking Languages Monitoring	Learning strategies Science Mathematics
Complementary Competencies	Creativity		
	Social Intelligence		Negotiation
	System Understanding		System analysis System evaluation
	Technology		Operational analysis Technology Design Programming
			Health Medicine Therapy Psychology
			Education and training
			Communications

Essentially there are three categories of niche competencies: health, training, and communications. Compared to mass competencies, will be particularly essential for a smaller group of jobs. Over the coming years, employment creation will be especially robust in the health sector (Kowal et al., 2022). In addition, the ICT competencies component of fundamental education must be modernized. In recent years, knowledge of computing and electronics has evolved from a niche competency to a widespread competency. Simultaneously, the significance of programming, systems analysis, and systems evaluation has increased significantly and will continue to rise in the future (Deloitte, 2017). Therefore, it is crucial that young people acquire ICT skills early in their education.

Based on research, Kowal et al. (2022) identified various types of competencies namely technical, digital, social, and personal. Technical competence needs expertise across processes and systems. In contrast, digital competence focuses primarily on data evaluation and analysis. Social competencies consist of basic abilities to communicate and teamwork. Personal competencies have been identified as a group that demands special consideration because the skills included in this group are exceedingly difficult to acquire through formal education or additional training. Currently, digital skills are focused on the requirements of the economy. Importantly, the primary objective is to provide people with sufficient digital literacy skills from online inquiry to digital content creation so that they can live and work in the digital economy (Stofkova et al., 2022).

According to research conducted by MCI in 2022, the main causes of the digital gap in Indonesia are 68% labor market competition, 65% unsupportive regulations, 41% lack of investment in R&D and job training, 85% lack of quality ICT graduates and opportunities for upskilling, and 56% limited access to higher education and resistance of workers to acquire new ICT skills. Overall, these digital competencies and skills are essential for anyone looking to succeed in Indonesia's labor market, and they are likely to become even more important in the future as technology continues to play a growing role in business and society.

THE ROLE OF HIGHER EDUCATION IN BUILDING DIGITAL SKILLS AND COMPETENCIES

In the modern, post-industrial economy of the twenty-first century, higher education is a vital source of specialized labor that ensures future prosperity. Students perceive employability as the capacity to find and keep employment in the future; however, the transition to further education and the labor market is an extended and complex process (Aljohani et al., 2022). Universities must place a larger emphasis on meeting students' expectations by analyzing the demands of the target employment markets and adapting their programs accordingly. Higher education plays a critical role in building digital skills and competencies in individuals, which are essential in today's digital world (Damuluri et al., 2020). The main priority of the Golden Indonesia Vision in 2045 is human development and mastery of science and technology. Based on the predictions of Bappenas (2020), the improvement in the quality of education for individuals who will continue their education in tertiary institutions will increase significantly from year to year. The following details are in Table 4.

Table 4. Education Quality Improvement (Bappenas, 2020)

Education	2015	2025	2035	2045
Higher Education	29.9%	35%	50%	60%

The role of higher education can support the Digital Skills Development Framework. Formal education, especially higher education, has core elements to focus on curriculum development and teaching quality. It is important to improve digital literacy and fundamental digital competency for work. The level and advancement of students' digital competency are directly influenced by the maturity and improvement of the lecturers' digital competency (Julio et al., 2021). ICT graduates must be able to master evolving digital tools, as well as have strong analytical, critical thinking, and problem-solving abilities. Graduates should be able to integrate data and information generated by technical capabilities into important insights for decision-making and business model enhancements, with the goal of increasing industrial productivity.

Another opportunity is provided by the Ministry of Education's implementation of the 'Kampus Merdeka' program. The "Kampus Merdeka is part of the Freedom to Learn policy by the Ministry of Education, Culture, Research, and Technology, which gives all students the opportunity to hone their abilities according to their talents and interests by going directly into the world of work as a career preparation step. The program has several alternative activities, such as certified internships in industry, student exchanges, and the Indonesian International Student Mobility Awards. This activity does not only focus on student development but also lecture development, such as teaching campus "Kampus Mengajar" activities and internships in industry. The program needs a higher commitment from educational institutions and industry, particularly to guarantee students' digital competencies for work are improved. ICT schools continue to experience hurdles in implementing the program, ranging from difficulties in establishing agreements to developing an internship curriculum that meets industry standards.

Primarily due to the transforming nature of work environments and the different needs of the market because of the introduction of new technologies, it has become more and more challenging to integrate academic curriculum with the job market. Graduates are typically uneasy at the beginning of their job search, wasting precious resources on positions that are not suitable for them (Aljohani et al., 2022). Table 5 presents a tool for obtaining the digital skills of individuals in the workforce based on four different factors: employment and unemployment status; the use of the Internet; the level of education; and ICT background.

Table 5. Classification of the digitally competent workforces (Diningrat et al., 2022)

Education		Employed		Unemployed
		Not using internet	Using Internet	
Undergraduate,	Non-ICT	Intermediate	Intermediate	Intermediate
	ICT	Advance	Advance	Advance
Postgraduate	Non-ICT	Intermediate	Intermediate	Intermediate
	ICT	Advance	Advance	Advance

Based on Table 5, graduates of undergraduate and postgraduate programs with non-ICT backgrounds are expected to have a minimum intermediate standard in ICT skills. Meanwhile, for ICT backgrounds, the minimum standard that must be achieved is at an advanced level. People with a background in ICT, a higher level of education, employment in the ICT industry, or work in occupations requiring moderate to high levels of skill experience earn greater premiums (Diningrat, 2022). Higher education must additionally develop an appropriate curriculum, but also enhance the ICT skills of lecturers and students. Improving ICT skills is crucial for lecturers, but motivation is also necessary so that they are more mindful of digital transformation (Tokovska, Zat'kova and Jamborova, 2022).

According to Tokovska, Zat'kova and Jamborova (2022), there are actual deficiencies and prospective difficulties on the part of both students and lecturers. Students are aware of the potential for combining digital technologies with the allure of educational content, and they are knowledgeable of the labor challenges created by social demands. Sakernas additionally acquires data on the highest level of education reached. It provides information on the field of study of university graduates (bachelor's, master's, and doctoral degrees), allowing us to determine whether the educational institutions attended by everyone provided an ICT-specific curriculum. People with extensive training in information technology will be more knowledgeable than those without such training. Although it would have been ideal to measure the extent of IT content in each program's curriculum, this was not possible. "ICT graduates" obtain more ICT-related training and knowledge than students in other subject areas and programs (Diningrat, 2022).

The ICT-specific education program empowers employees with the theoretical and practical knowledge necessary for advancement in ICT-intensive occupations. Increased exposure to the internet through the daily use of ICT tools at work and for other purposes would raise the requirement for complementary skills, which include information processing, self-direction, problem-solving, and communication. Combining these dimensions yields an estimate of the availability of digital skills through formal education and their implementation in the workplace (Aljohani et al., 2022).

As a response to industry needs, higher education has the following priority roles (Ministry of Communication and Informatics, 2020):

1. *Providing relevant coursework and programs*: Higher education institutions offer degree programs, certifications, and short courses that equip students with digital skills and competencies necessary for success in the digital age. These courses and programs are designed to meet the current demands of the labor market and ensure that students are equipped with the right knowledge, skills, and tools to succeed in the digital economy.
2. *Hands-on training and experience*: Higher education institutions provide students with hands-on training and experience in digital technologies, tools, and platforms. This hands-on experience enables students to apply the theoretical knowledge learned in the classroom to real-world scenarios, helping them develop practical skills that are valuable to employers.
3. *Encouraging innovation and creativity*: Higher education institutions foster an environment that encourages innovation and creativity. This environment provides students with opportunities to experiment, develop and implement new ideas, and learn from failures. This helps students to think creatively, which is critical in the digital age, where new technologies and digital business models are constantly emerging.
4. *Developing critical thinking and problem-solving skills*: Higher education institutions promote critical thinking and problem-solving skills that are essential in the digital age. Students are exposed to real-world challenges and are required to analyze problems, identify solutions, and make informed decisions using digital tools and technologies.

In conclusion, the role of higher education in building digital skills and competencies is crucial in preparing individuals for the digital economy. Higher education institutions must continue to develop programs and coursework that meet the current demands of the labor market and ensure that students are equipped with the right skills, knowledge, and tools to succeed in the digital age.

JOB OPPORTUNITIES FOR DIGITAL SKILLS AND COMPETENCIES

Advanced digital skills refer to the capacity to apply theoretical ICT knowledge and analytical skills to the resolution of digital problems (Diningrat, 2022). These qualitative findings are consistent with the findings of a survey conducted by the MCI in 2020 on 48 large industries from eight business sectors in Indonesia, which were compiled using data from a survey of 48 large industries. Full-stack programmers (27.18%), big-data analysts (16.77%), and network architects (10.04%) are the most popular ICT occupations in Indonesia, according to the 2020 survey. Digital technology helps people to manage their careers. Using online video, chats, and other online communication tools that support free calling and messaging, career management professionals were also able to provide career guidance and training (Wang et al., 2022).

In Indonesia's labor market, there are several important digital competencies and skills that are highly sought after by employers (Das et al., 2019). Based on recent research, there are several job opportunities for digital skills and competencies in Indonesia, including:

1. *Data analysis*: As businesses increasingly rely on data to inform their decision-making processes, data analysis has become a crucial skill in many industries. Proficiency in data analysis tools such as Excel, SQL, and Python can be an asset in the job market.
2. *Communication*: Good communication skills are essential in any job, but they are particularly important in a digital environment. Effective communication in a virtual setting requires proficiency in written and verbal communication, as well as the ability to use online collaboration tools.
3. *Cybersecurity*: With the rise of cyber threats and data breaches, employers are increasingly looking for candidates with knowledge of cybersecurity. This includes understanding the basics of online security, such as creating strong passwords and recognizing phishing scams.
4. *Digital marketing*: With more and more businesses going online, digital marketing skills have become highly valued in the job market. This includes knowledge of social media platforms, search engine optimization (SEO), and content creation.

5. *Software Development*: The software development industry is growing rapidly in Indonesia, with companies looking for skilled developers to create applications, websites, and other digital products. The most in-demand skills include programming languages such as Java, Python, and JavaScript, as well as web development frameworks like React and Angular.
6. *E-commerce*: As the e-commerce industry grows in Indonesia, there is a need for professionals with skills in e-commerce operations, such as logistics, payment systems, and customer service.
7. *Digital Design*: The demand for digital designers is also increasing in Indonesia, with companies seeking professionals with skills in graphic design, UI/UX design, and animation to create engaging digital content.

Overall, Indonesia's digital economy is rapidly growing, and there are many job opportunities for professionals with digital skills and competencies. According to Aljohani et al. (2022), the big data engineering sector garnered the most interest (56.59%), then developers and engineers (31.87%), and finally database systems (11.44%). Educational qualifications dominate bachelor's degrees with a percentage of 78.88%.

Table 5. Job Market (Aljohani et al., 2022)

Job Sector	Percentage
Big Data Engineer	56.69%
Developers and Engineers	31.87%
Database System	11.44%

Table 6. Academics Qualifications (Aljohani et al., 2022)

Qualifications	Percentage
Diploma	6.83%
Bachelor's degree	78.88%
Master's degree	4.97%
PhD	9.32%

In Indonesia, automation may have less of an effect on occupations involving people management, areas of expertise, and social interactions, where machines are still behind human performance. As many as 16% of present jobs in Indonesia could be automated by the year 2030, displacing a total of 23 million workers (Das et al., 2019). The latest study from the McKinsey Global Institute suggests that Indonesia could add $135 billion annually to its GDP by 2025 in the greatest-case scenario for the region. This is a 9% increase over the business as a normal scenario for 2025. Similar contributions are made by all three sources of improved economic value. The scale of future job creation due to the mentioned patterns and the impact of automation on the labor force differ significantly by country, based on four variables: wage levels, demand growth, demographics, and the mix of sectors and occupations. In broad terms, the present educational needs for jobs that may experience growth are greater than those for occupations that will be automated away. In Indonesia and other emerging economies, there is an increase in labor demand for all levels of education, with the greatest number of new positions requiring a secondary education. However, occupations currently requiring a bachelor's degree or higher will experience the greatest job growth.

THE CHALLENGE OF DIGITAL SKILLS AND COMPETENCIES

Worldwide competition now exists in an unpredictable and extremely complex condition; competition is not only based on the product but also on the strategy of creating knowledge (Wahyuningtyas and Anggadwita, 2017). Therefore, a company's future competitiveness and growth are heavily dependent on its employees as the organization's driving force. Despite new solutions and technologies that look to replace labor with automation, the human factor will continue to play a crucial role in industrial processes in the years to come. It is possible to create sustainable production systems if employees' competencies are appropriately developed (Kowal et al., 2022). Universities and corporations are too closely following present technological and social trends and collaborating with relevant entities. Companies should identify the skills that are deficient in the labor force, and universities ought to adjust the curriculum adequately. Employment expansion is likely to alter the sectoral composition of the Indonesian economy. Construction and manufacturing, as well as accommodation and food service, education, healthcare, and retail and wholesale trade, may experience an increase in labor demand. To reap the full benefits of the productivity increase brought on by automation, an integrated effort to resolve skill transitions will be required (Das et al., 2019).

Indonesia is the fourth most populous country in the world with a growing economy and a young population. However, the country is facing a challenge in terms of digital skills and competencies in the labor market. One of the main challenges is the lack of digital skills and competencies among the workforces. According to a survey by the World Bank (2021), only 22% of the Indonesian workforce has basic digital skills, and less than 10% have advanced digital skills. This is a major concern, especially since digital technology is rapidly transforming the global economy.

Another challenge is the mismatch between the skills that are in demand by employers and the skills that are possessed by job seekers. This is because the education system in Indonesia is not adapting quickly enough to the changing demands of the labor market (Ministry of Education, 2021). As a result, many young Indonesians are finding it difficult to secure well-paying jobs. Furthermore, there is a lack of investment in training and upskilling programs to help workers acquire the digital skills and competencies that are needed in the labor market. This is compounded by the fact that many small and medium-sized enterprises (SMEs) in Indonesia do not have the resources to invest in digital technology and digital skills training for their employees.

To address these challenges, there needs to be a concerted effort from the government, the private sector, and educational institutions. The government should prioritize investment in digital skills training programs and work with the private sector to identify the skills that are in demand in the labor market. The private sector should also invest in training and upskilling programs for their employees, and work with educational institutions to ensure that graduates have the necessary digital skills and competencies. Educational institutions should adapt their curriculums to incorporate digital skills and competencies that are relevant to the labor market.

Overall, the development of digital skills and competencies is essential for Indonesia to remain competitive in the global economy and to ensure that its young population has access to well-paying jobs.

CONCLUSION AND RECOMMENDATION

The economic development that might arise from AI and automation's productivity contributions is inspiring. AI and automation will not only contribute to dynamic economies that create employment but will also help create the economic advantages that will enable nations to prepare for the workforce transformations that will occur regardless. Consequently, an integrated effort will be required to address skills transitions. All stakeholders, including policymakers, academic institutions, non-governmental organizations, and corporate leaders, are required to get ready for significant changes in the future of work. Companies must start preparing for and changing into the future of work immediately, with long-term learning programs for those whose jobs are changing nowadays and those whose occupations will change in future generations. Developing digital competencies and skills is becoming increasingly important in Indonesia, as technology continues to transform the way people live, work, and learn. Here are some recommendations to help face this challenge:

1. *Invest in digital literacy programs*: The government, educational institutions, and private organizations should invest in digital literacy programs that equip people with the knowledge and skills needed to use digital tools and technologies. These programs can be delivered through online platforms, mobile applications, or face-to-face training.
2. *Encourage the use of online learning platforms*: Online learning platforms like Coursera, edX, and Udemy offer a wide range of courses in various digital competencies and skills. Encourage students, professionals, and individuals to use these platforms to upskill themselves and stay updated with the latest trends and technologies.
3. *Provide access to digital infrastructure*: Access to digital infrastructure is critical for developing digital competencies and skills. The government should focus on improving digital infrastructure in rural areas and remote locations to ensure that everyone has access to the internet and other digital tools.
4. *Foster a culture of innovation and experimentation*: Encourage people to experiment with digital tools and technologies to develop new solutions and improve existing processes. This can be done through hackathons, incubators, and other innovation programs.
5. *Develop partnerships with the private sector*: The private sector can play a crucial role in developing digital competencies and skills. Partnerships between the government, educational institutions, and the private sector can help to provide funding, expertise, and resources to support the development of digital skills.

Overall, developing digital competencies and skills is critical for Indonesia's future success, and it requires a multi-stakeholder approach involving the government, educational institutions, the private sector, and individuals.

REFERENCES

Aljohani, N. R., Aslam, M. A., Khadidos, A. O., & Hassan, S.-U. (2022). A Methodological Framework to Predict Future Market Needs for Sustainable Skills Management Using AI and Big Data Technologies. *Applied Sciences (Basel, Switzerland)*, *12*(14), 6898. doi:10.3390/app12146898

Ameliah, R., Negara, R.A. and Rahmawati, I. (2022). *Status Literasi Digital di Indonesia 2021*.Katadata Insight Center. Kementrian Komunikasi dan Informasi Indonesia.

APJII. (2020). Penetrasi dan Profil Perilaku Pengguna Internet Indonesia. Survei. [Online], Retrieved July 10, 2023, from.

Ayuningtias, H. G., & Anggadwita, G. (2015). Human capital development of information and communication technology industry in Indonesia. *Journal ofResearch and Applications: Accounting and Management*, *1*(2), 157–164. doi:10.18382/jraam.v1i2.15

Ayuningtias, H. G., Anggadwita, G., & Prasetia, A. Y. (2018). Global talent program as determinants of employees' performance on telecommunication company in Indonesia. *International Journal of Learning and Intellectual Capital*, *15*(3), 242–257. doi:10.1504/IJLIC.2018.094721

Bakhshi, H., Downing, J., Osborne, M., & Schneider, P. (2017). *The Future of Skills: Employment in 2030*. Pearson and Nesta.

Bappenas. (2020). *Berdaulat, Maju, Adil dan Makmur*. [Online] Available at: https://www.bappenas.go.id/files/Visi Indonesia 2045/Ringkasan Eksekutif Visi Indonesia 2045_Final.pdf [Accessed: July 10, 2023]].

Bashir, S., & Miyamoto, K. (2020). Digital Skills: Frameworks and Programs. World Bank. [Online], Retrieved July 10, 2023, from. doi:10.1596/35080

BPS. (2018). *Survey Angkatan Kerja Nasional 2018*.

BPS. (2020). Survey Angkatan Kerja Nasional 2020.

Damuluri, S., Islam, K., Ahmadi, P. and Qureshi, N. S. (2020). Analyzing Navigational Data and Predicting Student Grades Using Support Vector Machine. *Emerging Science Journal*, 4(4), pp,243–252. doi:10.28991/esj-2020-01227

Das, K., Wibowo, P., Chui, M., Agarwal, V., & Lath, V. (2019). *Automation and the future of work in Indonesia*. McKinsey & Company.

Deloitte. (2017). *What Key Competencies Are Needed in the Digital Age? The Impact of Automation on Employees, Companies, and Education*. [Online] Available at: https://www2.deloitte.com/content/dam/Deloitte/ch/Documents/innovation/ch-en-innovation-automation-competencies.pdf [Accessed: July 10, 2023]].

Diningrat, R. A., Rizku, M., Oley, J. D. B., Natih, P., & Astini, F. N. (2022). *Digital Skills Landscape in Indonesia*. The SMERU Research Institute.

Fernando, Y., & Wulansari, P. (2021). Perceived understanding of supply-chain integration, communication, and teamwork competency in the global manufacturing companies. European. *Journal of Management and Business Economics*, *30*(2), 191–210.

Government of Indonesia. (2020). Rencana Pembangunan Jangka Menengah Nasional (RPJMN) Tahun 2020–2024 [2020–2024 National Medium-Term Development Plan (RPJMN)]. Government of Indonesia. [Online], Retrieved July 10, 2023, from.

IMD World Competitiveness Center. (2022). *World Digital Competitiveness Rankings 2020*. [Online]. Available at: https://www.imd.org/centers/wcc/world-competitiveness-center/rankings/world-digital-competitiveness-ranking// [Accessed: July 10, 2023].

Indonesian Information Portal. (2023). Indonesia Profile. [Online]. Available at: https://indonesia.go.id/profil//] Accessed: July 10, 2023]

Julio, C. A., Guillén-Gámez, F. D., Ruiz-Palmero, J., & Palacios-Rodríguez, A. (2021). Digital competence of higher education professor according to DigCompEdu. Statistical research methods with ANOVA between fields of knowledge in different age ranges. *Education and Information Technologies*, *26*(4), 4691–4708. doi:10.100710639-021-10476-5 PMID:33758572

Kowal, B., Włodarz, D., Brzychczy, E., & Klepka, A. (2022). Analysis of Employees' Competencies in the Context of Industry 4.0. *Energies*, *15*(19), 7142. doi:10.3390/en15197142

Ministry of Communication and Informatics. (2020). *Roadmap Literasi Digital 2020-2024*.

Ministry of Education. Culture, Research, and Technology (Kemendikbudristek). (2021). *The Merdeka Campus Program Invites Indonesian Students to Become Creative and Adaptive Human Resources*. [Online] Available at: https://dikti.kemdikbud.go.id/kabar-dikti/kabar/program-kampus-merdeka-ajak-mahasiswa-indonesia-menjadi-sdm-kreatif-dan-adaptif/[Accessed: July 10, 2023].

Ramadhan, H. A., & Putri, D. A. (2018). Big Data, Kecerdasan Buatan, Blockchain, dan Teknologi Finansial di Indonesia. Direktorat Jenderal Apl. Inform. Kementeri. Komun. dan. *Inform (Silver Spring, Md.)*, 1–66.

Stofkova, J., Poliakova, A., Stofkova, K. R., Malega, P., Krejnus, M., Binasova, V., & Daneshjo, N. (2022). Digital Skills as a Significant Factor of Human Resources Development. *Sustainability (Basel)*, *14*(20), 13117. doi:10.3390u142013117

Tokovska, M., Zat'kova, T. S., & Jamborova, L. (2022). Digital Competencies Development in Higher Education Institutions: A Mixed Methods Research Study. *Emerging Science Journal*, *6*(Special Issue), 150–165. Advance online publication. doi:10.28991/ESJ-2022-SIED-011

UN. (2022). *UN Special Rapporteur on the Right to Education launches new report into the impact of digitalisation.*

UNESCO. (2018). *A Landscape Review: Digital Inclusion for Low-skilled and Low-literate People.*

UNESCO. (2018). *A Global Framework of Reference on Digital Literacy Skills for Indicator 4.4.2.* [Online]Available at: http://www.uis.unesco.org [Accessed: July 10, 2023].

Wahyuningtyas, R., Alamsyah, A., & Diliana, N. A. (2021). Mapping Digital Talent Based on Competency using Social Network Analysis.*2nd International Conference on ICT for Rural Development (IC-ICTRuDev)*, Jogjakarta, Indonesia, 2021, pp. 1-5, 10.1109/IC-ICTRuDev50538.2021.9656521

Wahyuningtyas, R., & Anggadwita, G. (2017). The perspective of Managing Talent in Indonesia: Reality and Strategy. In: Ordoñez de P, Rablos, P. and Tennyson, R.D. (eds.), Handbook of Research on Human Resources Strategies for the New Millennial Workforce. IGI Global.

Wahyuningtyas, R., Fajriani, V.M, Chaudhuri, M.R. and Yusoff, Y.M. (2021). Gap analysis of multigeneration digital competence. *Proceedings of Synergizing Management, Technology, and Innovation in Generating Sustainable and Competitive Business Growth.* Taylor Francis Group.

Wang, Y., Jiang, S., Wu, C., Cai, X., & Wang, F. (2022). Impact of the Global Megatrends, COVID-19, and Digital Economy on Professional Career Management Transformation in Asian Countries. *Sustainability (Basel)*, *14*(17), 10981. doi:10.3390u141710981

World Bank. (2021). Beyond Unicorns: Harnessing Digital Technologies for Inclusion in Indonesia. World Bank. [Online], Retrieved July 10, 2023, from.

Chapter 7
Developing Skills and Competencies for Educators' Transitions to Using Digital Technologies:
An E-School in Johannesburg, Gauteng, South Africa

Peter Kisembo
University of South Africa, South Africa

Leila Goosen
https://orcid.org/0000-0003-4948-2699
University of South Africa, South Africa

ABSTRACT

The purpose of the study reported on in this chapter was to investigate the extent to which educators were developing skills and competencies for their transitions to using digital or Information and Communication Technologies (ICTs) at a specific school in the greater Johannesburg area, Gauteng province, South Africa. Against the background of Developing Skills and Competencies for Digital and Green Transitions, the objectives of this study also included establishing to what extent the school was ready to implement such technologies in teaching and learning and hence be characterized as an e-school.

DOI: 10.4018/978-1-6684-9089-1.ch007

INTRODUCTION

This section will describe the general perspective of the chapter and end by specifically stating the **objectives**.

Developing Skills and Competencies for Digital and Green Transitions

As part of the book and like that of Van Heerden and Goosen (2023), this chapter aims at *fostering* international dialogue in terms of cross- and approaches to action research and action learning for e-schools, community engagement, and Information and Communication Technologies for Development (ICT4D), as well as methodologies and tools for diversity and inclusion through the *transformation* of education *in the digital era*, with a special focus on **green** education and **sustainability**. It will broaden the visibility of new digital technologies for education, create an international "scientific platform for **collaboration** on digital" and **green** education and accelerate innovation related to *educational technologies* (Ordóñez de Pablos, Zhang, & Almunawar, 2023, p. xxii).

The current technological advancements in the world have had a huge **impact** on almost all sectors of production and service delivery (Lin & Huang, 2011). This technological **impact** has led to a complete shift in methods in a bid for organizations to integrate technology into their operations, and meet their clients' needs (Raja & Nagasubramani, 2018). Reliable internet connectivity has also further enhanced this (Wright, Cilliers, Van Niekerk, & Seekoe, 2017). Some of the aspects that technology has been associated with include providing a faster means of communication, enabling accurate record keeping and data analysis (Raja & Nagasubramani, 2018).

In this research, a look was taken at e-schools' community engagement, focusing on various aspects relating to the extent to which *educational technologies* are being used to enhance teaching and learning at the school chosen for the study, as well as the support that the school offers to the community where it is based.

2030 Agenda and Sustainable Development Goals

Focusing on the **Sustainable Development Goals** (SDGs) and advocating in favor of "public accountability as the agenda proceeds towards 2030", Palmer (2015b, p. 262) provided **introductions** to the "**2030 Agenda** for Sustainable Development", as well as briefly explaining the United Nations (UN) process through which the SDGs had been formulated (Palmer, 2015a, p. 3).

Objectives

Resilience

This chapter will explore how **green** education and advanced information technologies can be applied in the education sector to create **inclusive** and increase the *resilience* of e-schools (Mautso & Goosen, 2023) as part of more climate neutral and **green** *economies* and societies. The chapter will discuss how **green** education and digital technologies can boost the creation of new jobs, build *circular economies* and societies and increase the **inclusiveness** of citizens.

BACKGROUND

In this section of the chapter, the authors will present broad **definitions** and discussion on the topic of **Developing Skills and Competencies for Educators' Transitions to Using Digital Technologies** at **an e-school in Johannesburg, Gauteng, South Africa**, and incorporate the views of others (in the form of a *literature review* for the study) into the discussion to support, refute, or demonstrate the authors' position on the topic.

Definitions

Terminology as used and interpretated in this study included:

- An e-school refers to a school that teaches students primarily online (Mqina & Goosen, 2021).
- E-learning refers to a system based on formalized teaching but with the help of electronic resources, teaching may take place in or out of the classroom (Sangrà, Vlachopoulos, & Cabrera, 2012).
- A School Governing Body (SGB) is a statutory body of parents from the community, learners, educators, teaching and non-teaching staff, who seek to work together to promote the well-being and effectiveness of the school community so as to promote teaching and learning (Van Wyk, 2004).
- A community is a group of people living in a particular area having certain common characteristics (Muzambi & Goosen, 2022).
- In terms of a definition, characteristics, architecture, enabling technologies and application, the Internet of Things (IoT) refers to a global network of computing resources, which support communication and information sharing using various communication protocols (Patel & Patel, 2016).

To carry out the *review* of existing *literature*, the researchers drew on resources that were provided as part of the course materials, as well as t with a focus on credible academic articles. The key themes of the research included:

- Integration of digital technologies in education
- Awareness and effective use
- The Substitution, Augmentation, Modification and Redefinition (SAMR) model.

The themes above were selected in line with the research **objectives** presented earlier.

Integration of Digital Technologies in Education

According to the then national South African Department of Education (2004), there was an increase in the number of computers acquired by schools. However, this did not improve e-learning, as there was still a lack of technical support for the educators and school management on how to use these technologies, and this links to effective use, which will be discussed later. School-based management was introduced in schools (Steyn, 2002), and this led to the establishment of the school governing body. The research that has been done has shown that SGB members lack the required knowledge regarding school operations (Van Wyk, 2004). Communities play a vital role in ensuring the security of the Information and Communication Technology (ICT) equipment at the schools and they must be involved in the implementation strategy of ICT usage in schools from the beginning (Mqina & Goosen, 2021). Community development initiatives must be put in place by the government, which may involve equipping themselves with **skills**, such as the use of the technologies and ICT resources at these schools. This was used to facilitate these initiatives especially on the weekends and after school.

Awareness and Effective Use

Infrastructure

In digital integration, awareness occurs when opportunities that digital tools and resources offer are known to the user. In the setting of this study, awareness will occur if all the stakeholders are aware of the tools available, and these opportunities are exploited appropriately within the context.

Effective use of technology is said to have occurred if digital tools and resources are used to add value to the learning process, student engagement and the achievement of the intended **objectives**. For example, if a digital tool is used to teach a concept, effective use occurs only if the concept has been taught and learners have understood the concept because of the digital tool. Effective use is not always measured in terms of more digital tools but rather in terms of how the tools are adding value. In this study, if the school has all the *infrastructure* required and all the resources, if they are not used effectively, then the school cannot be called an e-school.

The Substitution, Augmentation, Modification and Redefinition Model

The SAMR model was developed by Puentedura (2010). The framework was created to inform technology integration, and it is a good tool to use when analyzing a school that is transitioning from a traditional setting to an e-school. The model has four aspects as defined below.

- **Substitution** refers to replacing a process with another process. Technology substitutes other processes that were used before digital capabilities. The tasks remain unchanged, for example, using vodcasts to teach programming in an Open Distance Learning (ODL) environment. In this study, one way of identifying substitution would be to look at how smartboards have been substituted for blackboards.
- **Augmentation** refers to using technology enhance the process. Technology augments the task. The task is largely unchanged, for example, the digital report card generator augments the process of creating report cards for learners. The report card system is one of the processes that this study may investigate at the school during the process of data collection.
- **Modification** refers to changing a process. Technology makes it possible to modify the tasks, for example, recorded lessons make it possible for learners to listen to the lesson after it has been completed.
- **Redefinition** refers to the process of changing how a task is performed. Technology makes it possible to redefine the task, for example, an editing software makes it possible to redefine a writing task as a video production including written scripts.

Section 1: Green Education and Digital Education: State of the Art and Priorities for Inclusive Education

Green Education and Digital Education

Against the background of environmental science and pollution research, Lu, Chen, Huang and Usman (2023) investigated the impact of **education** on renewable energy demand, carbon intensity, and **green** growth, and asked whether **digital** financial inclusion and environmental policy stringency mattered in China.

"ATHENA is a European University aimed at nurturing the development of all-**inclusive green**" **education**, as well as "to advance **green education** and inclusion" in the European higher **education** area (Escudeiro, Cruz, & Escudeiro, 2021, p. 1).

According to Racsko and Troll (2020, p. 69), **education** "is increasingly relying on **digital** tools to become a major emitter of CO_2 itself." **Digital education** habits, in the light of their environmental impact in **education** systems, need to reconciled with **digital** sustainability and the role of **green** computing in **education**.

Mapping the practices in the field "of higher **education** assessment and" professional accreditation to harness the potential of **digital** technologies for the "**green and digital** *transformation* of higher **education**" was the focus of a bibliographic coupling and co-citation analyses by Chen and Liao (2020, p. 281).

Inclusive Education

Towards a more **inclusive** and resilient e-school, Mautso and Goosen (2023) were using applicable e-learning principles to promote effectiveness and quality *in the digital era* in the *handbook of research on education institutions, skills, and jobs.*

With a "special focus on higher **education** institutions", the book on **inclusivity** edited by Ordóñez de Pablos, Lytras, Zhang and Chui (2019, p. xv) also focused "on information technology and application" in terms of *educational technologies*, and analyzed the implications of these for **education**.

A paper presented at the South Africa international conference on *educational technologies* by Tshabalala and Ndeya-Ndereya (2015) offered criteria and guidelines for the selection and implementation of such *educational technologies* in high schools.

Section 2: Digital Skills and Competencies for a Changing Labor Market

Digital Skills and Competencies

"**Digital** technologies are transforming *economies* and societies around the world. As such, markets demand new types of **skills and competences** that students must learn in order to" contribute in terms of IT and the development of **digital skills and competences** in education (Ordóñez de Pablos, Lytras, & Zhang, 2020, p. xv).

Technology Platforms in Education

Ordóñez de Pablos (2022, p. 391) pointed out that *technology "platforms* offer opportunities to sectors like *education"*, as well as energy and financial services. Some keys for competitiveness in the post-pandemic era are thus linked to unlocking the potential of intellectual capital, innovation and higher education.

COVID-19 and Learning

An editorial on the **digital economy**, intellectual capital and competitiveness by Ordóñez de Pablos (2023) provided some lessons for the post-**COVID-19** era.

Ethical Use of Artificial Intelligence and Data in Education

Information on the "emergence of **artificial intelligence** … and its derivative technologies, such as machine learning (ML) and deep learning (DL)", heralded a new era in terms of the optimization "of knowledge management (KM) presentation and discovery" (Anshari, Syafrudin, Tan, Fitriyani, & Alas, 2023, p. 1).

Gamification

In the *Handbook of Research on Acquiring 21st Century Literacy Skills Through* ***Game-Based*** *Learning*, the chapter by Muzambi and Goosen (2022) looked at factors affecting the successful implementation of an e-education policy and community engagement towards learners acquiring 21st century skills through e-learning.

MAIN FOCUS OF THE CHAPTER

Issues, Challenges, Problems

This section of the chapter will present the authors' perspective on the **issues**, **challenges**, **problems**, etc., as these relate to the main theme of **Developing Skills and Competencies for Digital and Green Transitions** and arguments supporting the authors' position. It will also compare and contrast with what has been, or is currently being, done as it relates to the specific topic of the chapter, **Developing Skills and Competencies for Educators' Transitions to Using Digital Technologies** at **an e-school in Johannesburg, Gauteng, South Africa**.

Integration of Digital Technologies in Education

Infrastructure

The use of Information and Communication Technologies (ICTs) has become a significant aspect in the day-to-day operations of different organizations and many **developing** countries are taking up the usage of ICT serious so as to achieve the set developmental goals. In South Africa, various policies have been put in place to guide on the usage of ICT so as to support teaching and learning in the classroom, as well as e-learning (Goosen, 2015) and these include the White Paper on e-Education (Department of Education, 2004). Using ICTs to supplement the traditional teaching methods widely used leads to improved quality of education (Farikayi & Goosen, 2022). ICT helps to develop learners' ability with regards to self-learning, **problem** solving, as well as critical thinking (Wright et al., 2017). In South Africa today, more than ten years after the White Paper was released by the Department of Education (2004) was meant to ensure that all schools were able to have well established ICT *infrastructure* to support teaching and learning as well as management functions. It can be clearly seen that the extent to which ICTs had been incorporated into teaching and learning and differs from one institution to another due to the various factors, some of which may include lack of enough resources to purchase equipment, unsupportive leadership at the institution towards ICT development (Farikayi & Goosen, 2021). The various tools used for eLearning include computers, software systems that are able to support teaching and learning also education management systems used by the administrators of the learning institutions (Bhasin, 2012). For ICT integration to be a success educators are expected to play an important role and thus they need to have the appropriate training, **skills** and attitude (Jimoyiannis & Komis, 2007). According to Wang (2008), technology availability is not a major **problem** with regard to the effective use of ICT but the main **problem** is the teaching and social design strategies employed by the educators. The lack of knowledge and **skills** is a huge factor that limits integration of ICTs into teaching and learning (Khan, Hasan, & Clement, 2012). Various ICT-enabled ways of teaching and learning have ensured that learners are constantly engaged in with their work (Muzambi & Goosen, 2021). Assistive technology tools have been able to assist some learners with disorders, such as autism. Due to the current corona virus pandemic, various institutions resorted to e-learning but found a lot of **challenges** especially with regards to learner engagement. The South African government is responsible for ensuring that e-education is implemented through the Department of Education (DoE, 2004). In order to ensure quality teaching and learning various factors come into play, these also include the communities where the learners come from and the various **problems** faced by these communities must be solved as well for improved quality teaching and learning outcomes (Cloete, 2014).

The book edited by Ordóñez de Pablos, et al. (2020, p. xviii) described a chapter, which aimed to access different levels of education, and pointed out the **gap** related to "education in Indonesia by utilizing technological advancements including the skills needed to face the current digital era."

Metaverse and Education

The importance of the application of the **metaverse in education** is underlined by "the use of the word metaverse, which" one had surely heard of against the background of "Facebook and all its surroundings" (Contreras, et al., 2022, p. 34). However, "there are still many people who do not know exactly what a metaverse is".

A definition, as well as potential applications and research **issues** related to the **metaverse in education** from an artificial intelligence perspective were presented in the article on artificial intelligence, computers and education by Hwang and Chien (2022, p. 1). "Moreover, the roles of AI in the metaverse as well as metaverse-based education" were discussed.

An overview of "the implementation of **metaverse** in the **education** industry" was presented in the chapter by Singh, Malhotra and Sharma (2022, p. 135).

The use of **metaverse** technologies "can also be valuable in language education." The study reported on by Inceoglu and Ciloglugil (2022, p. 171) examined "the application examples of Metaverse technology in Korean language education".

Inclusive Education

According to Sandrone (2022), educational "leaders should work to **reduce** the digital divide and exploit the metaverse to provide *inclusive*" medical *education*, as the metaverse could "add a new dimension to medical education."

When analyzing education based on metaverse technologies, Mustafa (2022, p. 280) also advocated in favor of "the metaverse and its role in education and learning. With its qualitative and" analytic research methods, the study by the latter author investigated how the "metaverse exists in various forms of education."

Research on technology-supported teaching and learning methods for educators of learners with autism could also contribute towards more *inclusive education* (Goosen, 2019).

Section 2: Digital Skills and Competencies for a Changing Labor Market

Ordóñez de Pablos (2022, p. 391) further pointed out that education "policies and systems need to be prepared for the **challenges** and *opportunities* of the *transformation* of" **labor markets**.

Section 3: Digital Education, Gender Issues and Science, Technology, Engineering, the Arts and Mathematics

Gender Dimensions and Digital Skills

Cramer and Crocco (2010, p. 185) saw "**digital** technology both as content and tool in" urban social studies education. In the context of women and technology, upon reflection, the latter authors aimed at linking global women's **issues** to the **digital gender** divide, in order "to infuse technology, itself a gendered subject, into a course focused on **gender issues**".

Women and Science, Technology, Engineering, the Arts and Mathematics

Although Kahn and Ginther (2017, p. 6) further back in time saw that "women's representation in each separate STEM field increased rapidly in the 1970s", more than a decade ago, Beede, et al. (2011, p. 8) found that women were "underrepresented both in STEM jobs and STEM undergraduate degrees". "The relatively few women who" "do receive STEM degrees are less likely to work in STEM jobs than their male counterparts", contributing to a gender **gap** to innovation.

One would think that a lot has changed since…? However, reflections by Adams, Oduor, Wahome and Tondapu (2022, p. 23) on two years of teaching earth science at the Women in Science (WiSci) STEAM camp *STILL* report "fewer women participating in STEAM careers. An increase in women in STEAM fields would" require time and interventions at the WiSci "STEAM camp, inclusive of the arts".

A review of the literature 2007–2017 on the status of women in STEM in higher education by Blackburn (2017, p. 235) showed that STEM education in the United States have drastically increased the number of studies offering insight into the experiences of women in STEM programs in higher education."

The paper by Bhatia (2022) envisaged "insights into women's entrepreneurship in STEAM at the global level", "for promoting knowledge in STEM/STEAM for women and girls" and creating awareness of such entrepreneurship.

Although the study by Bloodhart, Balgopal, Casper, McMeeking and Fischer (2020, p. 1) showed "that women are outperforming men in both physical and life science undergraduate courses at the same institution," women's abilities are undervalued.

Section 4: Teaching and Learning in Biodiversity, Climate Change, *Sustainability*, and Other Issues: Educators, Learners and Institutions

Sustainability

The **sustainability** and **inclusiveness** of quality **education** can be improved through research-informed practice on information and communication technologies.

The White Paper released by the Department of Education (2004) was meant to address the various **issues** that existed with regard to the usage of technologies and the other main emphasis was to ensure that **educators, learners and institutions** were to be made ICT capable by the year 2013.

Policy Challenges to and Impacts of Digital and Green Transitions

Challenges for the Circular Economy

The *Handbook of Research on* ***Green****, Circular, and Digital Economies as Tools for Recovery and* **Sustainability** edited by Ordóñez de Pablos, Zhang and Almunawar (2022) indicated that societies across the globe had "simultaneously faced several unprecedented health, social, and economic **challenges**. Countries need to recover economic growth quickly, boost productivity and job creation, invest in smart healthcare systems and services, and work toward a climate-neutral and *circular economy*."

Section 5: Learning Analytics and Education

Learning Analytics

Learning analytics in an IT environment helps with understanding "the whole process of students' learning in groups, analyze" **problems** related to "**collaborative** learning and find ways to improve group learning effect." Zhang, Meng, Ordóñez de Pablos and Sun (2019, p. 625) "examined the role of Slack", a Software-as-a-Service (SaaS) application.

Section 6: University-Industry Collaboration for Green Education

Primary, Secondary and University Education

Learners at *primary and secondary* (basic) *education* (schools), as well as students at **university** (higher) **education** levels, had indicated that they want access to e-learning in an ICT4D Massive Open Online Course (MOOC). Computer lecturers are therefore using their institutional learning management systems for ICT education in the cyber world (Goosen & Naidoo, 2014).

Also in the *Handbook of Research on Education Institutions, Skills, and Jobs in the Digital Era* edited by Ordóñez de Pablos, et al. (2023, p. xxii) already mentioned, the editors indicated the importance "of **education** *in the digital era*" and broadening "the visibility of new technologies for **education"**, as well as "**for collaboration** on digital **education**, and" accelerating innovation in terms of *education technologies*.

Vorster and Goosen (2017) suggested a framework for **university collaboration** promoting continued support of e-schools in the proceedings of the 46th Annual conference of the Southern African Computer Lecturers' Association (SACLA).

Problem Statement

According to the White Paper on e-Education (Department of Education, 2004), e-schools can be characterized as learning institutions where students make use of technology to enhance their learning, and spaces where qualified and competent leaders use educational technology for planning, management and administration (Goosen, 2015). The White Paper on e-Education by the Department of Education (2004) had set certain targets that were to be achieved by the year 2013. A significant number of years have gone by since this date and some of the goals are yet to be achieved. The prospects of what has been achieved so far gives a positive indication that necessary steps need to be put in place to accelerate the implementation of technology usage in schools to enhance teaching and learning and hence reduce the limitations of implementing technology for teaching and learning (Vorster & Goosen, 2017). In this research, the focus was on an e-school's community engagement with a focus on a school in the north of Johannesburg, Gauteng province. The study was guided by the following research purpose and questions:

Assumptions and Limitations

The first assumption for the study was that the participants chosen from the school and the community where the school is based would answer the survey and interview questions honestly and factually.

The second assumption for the study was that the school selected would have ICT implemented, functional and integrated in the process of teaching and learning.

Research Design

This was a qualitative study. Brantlinger, Jimenez, Klingner, Pugach and Richardson (2005, p. 196) defined a qualitative **research design** as a "systematic approach to understanding qualities, or the essential nature, of a phenomenon within particular research". Qualitative research allows the researcher to view reality as multi-layered and interactive (Foreman-Peck & Winch, 2010) and it also enables the researcher to understand the social phenomenon from the perspective of the participants in the study. The study was conducted in a real-life setting, at a school.

Population and Research Context

The research was carried out at a public secondary school with Grade 8 to 12 learners. Public secondary schools are funded by the government in terms of resource allocation and *infrastructure* development, such as the installation of ICT equipment, which they require for e-learning. The school has a learner population of around 998 learners across all grades, 36 educators, one principal, two deputies and a school governing body comprising of parents, educators, and learners. The school management committee provides updates to the district office of the provincial Gauteng Department of Education (GDE) regularly regarding any new developments. Officials from the department of education come to the school to monitor the implementation of the teaching year plans set up by the educators.

The researcher focused the study on identifying the level at which all stakeholders mentioned were able to integrate ICT into their daily operations to achieve their roles at the school and to what extent that happens. The researcher also determined, after analysis of the data collected, whether the school qualifies to be characterized as an e-school.

Sampling and Data Collection

The researcher used a three-phase approach to collect data. The first phase of data collection was pre-observation interviews with various stakeholders in terms of the integration of ICTs in their various roles at the school. These stakeholders included members of the school management, educators, SGB, parents and community members. The second phase was observations of the stakeholders, and the third phase was post-observation stakeholder interviews.

For this research, purposive sampling was used (Cohen, Manion, & Morrison, 2018).

Phase 1: Pre-Observation Interviews

For this phase, three educators and one member of the School Management Team (SMT) (these included the principal, school administrator and bursar), the SGB chairperson, three classes and two active community members, who come to the school regularly, were selected for semi-structured interviews around the pre-observation schedule. The interviews were audio recorded and helped to provide information, which participants might not to display during observations (Foreman-Peck and Winch, 2010).

A pre-observation schedule was used for the pre-observation interviews as shown in Table 1.

Phase 2: Observation Interviews

The purpose of the study was to investigate the extent to which a particular school could be characterized as an e-school and to what extent it is connected to the community. The main method of data collection was observation, because the researcher sought to study the participants' behavior in situ. In a brief history of the previous decade (Goosen, 2008), Hammer (2003) argued that observation provides the researcher with a chance to look at what is taking place in situ. The researcher therefore visited the school for one week before the start of the research to familiarize himself with the school environment and to meet the various stakeholders at the school. The researcher also used this week visit to obtain the weekly timetables of the participants to plan the observations and interview times prior to the commencement of the research.

Table 1. School management and community pre-observation interview

Questions/Statements	Yes/ No	Comments
Does the school have a digital strategy to guide the usage of ICT to enhance teaching and learning?		
Have funds been allocated to the school to be used to procure the necessary ICT equipment required?		
Is there ICT training for the educators and the other relevant support staff at the school?		
Are there some members in the community where the school is based who use the school premises to share their ICT **skills** with others or help to support the school in maintaining and repair of the ICT infrastructure?		
Does the school allow members from the community to access ICT resources on days when learners are not there?		
Educator has adequate technology **skills**.		
Acknowledges technology and implements it in teaching		
Educator incorporates technology in their lesson using the available technology		

After the first week, the researcher carried out the research observations. The researcher used an observation schedule to take field notes during the observations to be able to capture as much information as possible with regards to how all the stakeholders, who had been identified above integrate ICT in their operations regarding completion of tasks.

The researcher used the observation schedule shown in Table 2 to obtain the necessary information. Observations were only done of the educators chosen with regards to teaching and handling their various administrative tasks.

Phase 3: Post-Observation Interviews

Post-observation interviews were conducted with selected participants at the school, and these included the questions arising from observations in *Phase 2*. The post-observation interviews were semi-structured and were created after data had been transcribed. As the researcher wanted to make sure that all the questions were in place before asking the participants to avail themselves, the questions were formulated in such a way that they help me confirm some **issues**, which were not clear during the observation. All the post-observation interviews will last about 15 minutes. The interviews were recorded. The post-observation interview questions were based on the data from the observations.

Table 2. Focus of observation

Focus	Comments
Classroom set up	
Teaching and learning resources available	
Incorporating digital technology in teaching and learning	
Learner`s reaction to educator's digital incorporation in the classroom?	

Data Analysis in Education

All audio recordings collected were transcribed in preparation for data analysis. The observation schedules notes were summarized and analyzed. The study drew on the typological analysis method (LeCompte, 2000) to analyze data. In typological analysis, data are divided into different categories, which were informed by literature, themes from the data and any theoretical frameworks used in the study. The steps followed during data analysis are outlined in Table 3.

Table 3. Typological analysis steps

Step	Activity
1	Classify the typologies, e.g., awareness; effective use; substitution; modification; redefinition; augmentation
2	Go through the data to identify the selected typologies
3	Get themes from the data
4	Code themes
5	Group themes within the codes
6	Identify relationships in the data
7	Describe patterns
8	Identify excerpts from the data to support the description

Ethical Use of Artificial Intelligence and Data in Education

Due to the sensitivity and nature of data collection procedures, the researcher ensured that ethical clearance had been obtained from the applicable ethics committee. The researcher also sought permission from the provincial Department of Education.

Consent and/or assent were obtained from all participants in the study.

The principal of the school was also informed with regard to permission obtained and provided with an information sheet indicating what the study was about. After the principal has agreed to take part in the study, other school stakeholders were contacted. The researcher then moved on to obtain permission from the educators using written consent forms (including consent for audio recording), again accompanied by an information sheet incorporating e.g., the purpose of the study. After the educators had consented to take part in the study and permission was received from the department of education, the researcher started with the pre-observation interviews. The parents of learners, who were below the age of 18, were requested to complete consent forms, and such learners requested to complete assent forms. It was stated clearly that participants would in no way be forced to take part in the study.

All consent and/or assent forms were collected prior to the start of data collection. The **issues** of anonymity and confidentiality were emphasized to all the participants. In the analysis and write-up of findings, pseudonyms were used.

Relevant measures were put in place to ensure that social stigmatization and victimization of respondents did not occur.

SOLUTIONS AND RECOMMENDATIONS

This section of the chapter will discuss **solutions and recommendations** in dealing with the **issues**, **challenges** and/or **problems** presented in the preceding section.

Solutions

Ethical Use of Artificial Intelligence and Data in Education

Resonating with some aspects of the theme of this book, the *Handbook of Research on Artificial Intelligence and Knowledge Management in Asia's Digital Economy* edited by Ordóñez de Pablos, Zhang and Almunawar (2023) indicated that *artificial intelligence* "and knowledge management can create innovative **digital solutions** and business opportunities in Asia from *circular* and **green** *economies* to technological disruption, innovation, and smart cities."

According to Semivolos and Anshari (2023, p. 1), *artificial intelligence* had also "become more widely used in many organizations, including logistics, such as driver assistance, parcel sorters, and inventory planners." The chapter in the book mentioned previously by the latter authors therefore discussed the application of *artificial intelligence* to the process of logistics during **COVID**-19.

"A digital marketplace is a digital area or *platform* where demand and supply sides communicate and transact through the digital space or *platform* controlled by a digital marketplace provider." Also in the book mentioned previously, the chapter by Anshari, Almunawar and Ordóñez de Pablos (2023, p. 366) offered **solutions** with regard to disruptive innovation and a multi-sided digital marketplace.

Ensuring research integrity and the *ethical* management of data in the context of e-schools and community engagement (Goosen, 2018a) and towards ICT4D **solutions** for MOOCs also shows the way (Goosen, 2018b).

Results

During the first week, the researcher conducted pre-observation visits at the institution. It was noted, at the school administration office and in some of the classrooms, that a school ICT user policy was available to guide the implementation of ICTs. The institution had computing equipment set up in different rooms around the school, and these included desktop computers in the administration block offices, internet connectivity at the main office block, a printer at the front office, smartboards in 80% of the classrooms, white boards in 20% of the classrooms, a functional intercom system and a computer center. All educators were allocated a new laptop by the department of education, except for the two new teachers, who had joined the institution at the beginning of the year. A non-staff member attached to the school to provide ICT training was also available at the school every Thursday to help train and support the school staff regarding the usage of ICT equipment.

Observations followed the week after, directed by the observation schedule designed to guide the researcher in obtaining the relevant information pertaining the research, and an integrated approach was used to present the findings from the research (LeCompte, 2000).

All participants in the research were observed in situ, and these included three teachers, one member from school management (the principal), the school governing body chair and two active community members that come to the school often.

Post-observation interviews were also carried out with the educators to obtain more information regarding some of the aspects that the researcher needed further explanations from during the observation.

Accessible Digital Education

Mobile Learning

One of the academic publishing companies has provided at no cost to the school a fully *developed mobile learning* application, which allows learners to practice and study various subjects from anywhere. One of the advantages of using this application is that it is zero rated and does not require any data for learners to access it and practice the various activities their educators can set up there. The application also supports self-study by the learners. It was, however, noted that very few learners continued to practice on the application after the first week when it was introduced earlier in 2021. In recent months, it was also noted that load shedding had become a major **problem**, especially for educators, as the institution did not have an alternative source of electricity to keep the technology devices running. According to the school policy on cell (*mobile*) phone usage by learners, it was noted that learners were not allowed to use cell phones during instruction time by their educators. It was also noted that some educators enforced this policy, while other educators let learners use their cell phones during instruction time without enforcing the institutional policy on cell phone usage. Some of the educators at the institution were part of various subject specific Professional Learning Communities (PLC) that were made up of educators from the various neighboring schools. These groups were set up using the WhatsApp *technology platform* and the educators used these to discuss amongst each other subject related content and also shared learner tests and activities among each other on the PLC groups. It was also further noted that these PLC groups were created by the department of education subject advisors to support all educators seamlessly from anywhere regarding the various subject-related content and communication regarding the various subject specific updates from the department of education.

How Information and Communication Technology Resources were Utilized by the School Management Team and Department of Education

The software that was installed on the devices of the educators and those used by the school administration officials at the institution (SASAMS) enabled most of the computing tasks required by the department of education from educators and school management team being completed in a shorter period of time, as they were able to compute the different records for the different attributes that were input by the educators and the school management. These included marks for the different tasks and subjects, as well as daily learner and educator attendance statistics. It was noted that this software had done a great deal in ensuring timely delivery of learner reports with marks that have correct conversions, as it was a huge **challenge** prior to using the software: errors, especially during the conversions of marks, were encountered and this led to delays, especially with the printing of reports. The head office was also able to log in and obtain the information without the school management travelling to submit hard copies at the head office. This helped the school administration with regard to saving on costs incurred in travelling to the head office to submit the different documents, as these could be accessed on the *technology platform* once the school administration clerk ensured that all the required information was captured.

It was also noted that the school was able to motivate for more resource allocation using the statistics from the software to the department of education. This included requesting more educators, laptops, classrooms and other learning support materials, as the department was able to monitor the capacity at the institution using this software and then make the necessary adjustment to ensure that the institution functioned as required.

Other Important Aspects to Note from the Findings

In recent months, load shedding has become a huge **problem** and it was noted that most computing operations that required the use of ICT equipment at the institution were affected by this **problem**. Educators have been forced to switch to traditional means of teaching and also postponing some of their lessons to a time when the electricity is available - this negatively affects curriculum coverage and task completion by the various educators at the institution.

From the above discussion of the results from the research at the institution, after looking at how the various tasks are carried out at the institution, it was noted that the school has the potential to operate as an e-school due to the various capabilities it has. These capabilities comprised ICT equipment, including the various software applications, an ICT educator trainer and educators able to use ICT equipment. It was, however, noted that some **issues** needed to be addressed for the school to fully benefit from the advantages of operating as an e-school. The researcher will discuss some of the **issues** that need to be addressed under **recommendations** in the next subsection of the chapter.

Section 7: Plans, Strategies and Recommendations

From the research conducted and the various findings that the researcher was able to make from the data collected during the research, the institution can be classified as an e-school. However, the following **recommendations** may be implemented so as to ensure that the institution is able fully benefit from the advantages that come as a result of using technology in teaching and learning at the institution (Mqina & Goosen, 2021).

More funds are required to purchase ICT equipment and the different academic software that will be used implement teaching and learning at the institution as some of the educators did not receive laptops and also some classrooms did not have smartboards installed, the funds will be used to ensure that all the required equipment is purchased at the institution to ensure the smooth functionality of the institution as an e- school as all the educators will have their own ICT enabled classrooms with all the necessary equipment required.

Quality ICT training should be conducted for all the educators and the various members on the school management committee (Khan, et al., 2012). This can be achieved by **developing** a fully accredited ICT course targeting the main topics that are relevant (Sangrà, et al., 2012) for the educators and school management committee on how to use technology to achieve the learning outcomes desired. There was ICT training at the institution but most of the concepts that were taught during this training were generic and did not directly address the topics or fields, which are very specific to helping educators to have the ability to be able to work well using ICT equipment (Muzambi & Goosen, 2021). It was noted that various educators used technology according to their own level of expertise. Once an accredited ICT in education course is developed, it should be made compulsory for all the educators to complete this course as it will set a standard across the board whereby the majority of the educators will have the capacity to use ICT (Tshabalala & Ndeya-Ndereya, 2015) to ensure that they achieved their learning outcomes.

The department of education should also ensure that the quality of eLearning materials that are supplied to schools will enhance the quality of teaching and learning as it was noted that the smartboards had PDF copies of the hard cover text books that were supplied to the learners. With technology advancement, there are different publishers in the country that are **developing** interactive e-books and other technologically advanced learning materials, which can improve learners' understanding of certain concepts in the classroom as well as short videos explaining certain steps in different subjects (Wang, 2008).

The department of education should standardize the process regarding teaching and learning with regard to the use of ICTs (Khan, et al., 2012), as well as handling of different records and learner statistics, as the researcher noted that some educators were still using hand written documents and also some smartboards in some classes were never used throughout the entire lesson as educators stated that they preferred using a chalkboard instead of a smartboard. Once the standardization with regards to the use of ICT is put in place with all the necessary preparations educators will be able to use technology to their advantage.

The institution needs to put in place the necessary steps towards improving the relationships with the members of the community (Muzambi & Goosen, 2021) where the institution is located as these relations will be very vital towards building and enhancing its self as well for example some members in the community may have certain **skills**, such as computer **skills** or some may be able to assist with installation or repairs of different ICT components while some may have the **skills** to fix or do certain repairs and renovations at the institution, once this relationships are established community members will ensure that they also support the institution.

The institution needs to put in place some funds to secure alternative energy sources to keep the computing devices running in case of load shedding; these may include generators and solar powered inverters. On a longer term, the government needs to find a **solution** to the energy crisis that is affecting almost all the sectors of the *economy*.

FUTURE RESEARCH DIRECTIONS

This section of the chapter will discuss **future** and **emerging trends** and provide insight about the **future** of the theme of **Developing Skills and Competencies for Digital and Green Transitions**, from the perspective of the chapter focus on **Developing Skills and Competencies for Educators' Transitions to Using Digital Technologies** at **an e-school in Johannesburg, Gauteng, South Africa**. The viability of a paradigm, model, implementation **issues** of proposed programs, etc., may also be included in this section. **Future research directions** within the domain of the topic will finally be suggested.

Technology advancements have also impacted the education sector globally, with South Africa embracing the **emerging** digital **trends** related to teaching and learning at all levels of education (Goosen, 2015). The technologies may include smartboards, learning management systems, e-books and many others.

For the "planning of digital green education", Soliman (2021, p. 3) suggested "digital green education skills for" students, and presented "a proposed **future** vision for" the planning required "to enhance digital green education skills" for students at applied technology schools.

Metaverse and Education

Looking to the **future**, the review article by Collins (2008) discussed "the implications of the emergence of" the **metaverse**. The latter author also discussed "the role of higher **education** in" the **metaverse** "in preparing students for the explosion of virtual worlds in the near future."

Singh, et al. (2022, p. 135) also indicted that the **metaverse** "is an enormous term that contains every digital aspect of the **future**."

CONCLUSION

This section of the chapter will provide a discussion of the overall coverage and concluding remarks.

From the research carried out at the institution, the main findings included that there was evidence of resource allocation by the Department of Education towards the purchase of ICT equipment and software at the institution as computing devices were installed in most of the classrooms and school administration block at the institution.

ICT training was conducted at the institution, an ICT educator trainer was allocated to the school by the Department of Education to train educators and school management in computer literacy.

The e-learning materials used at the school included mostly PDF copies of the hard cover text books, which were being used at the institution prior to the introduction of smartboards the PDF copies of the text boards did not offer advanced features that could be used by the educators to enhance teaching and learning such as interactive options (Khan, et al., 2012).

An ICT user policy (Department of Education, 2004) was evident at the institution however I noted that most of the educators used ICT according to their level of expertise, I also noted that some educators continued to teach without using the ICT devices that were installed in their classrooms.

The level at which members from the community engaged with the school was very low this was a big **problem** as various community members have different **skills**, which they would be able to share with the institution once this engagement happened some of these include ICT **skills**, computer repairs and other **skills** that may be vital to the school such as handyman services.

The **problem** of load shedding hugely impacted the usage of ICT resources, as the ICT devices are powered by electricity thus the educators had to use the traditional methods of teaching or even post pone their lessons to a later date when the power would be available.

Contribution of the Research

The research carried out should be able to contribute the following towards teaching and learning:

- The research was able to **highlight** the different attributes that are required at the institution for it to operate as an e-school.
- The research also **highlighted** the various **challenges** that need to be addressed for the institution to run more smoothly as an e-school.

- The research further provided an indication of how the institution is working towards becoming an e-school and also possible ways on how to improve different aspects at the institution.
- Although the research was carried out at only one institution, it can also be used as reference for **future challenges** by researchers to establish certain specific aspects regarding e-learning and e-schools (Patel & Patel, 2016).

Therefore, from the research carried out, it was established that the institution is currently running as an e-school with some **challenges** that have been highlighted. Once these had been addressed, more positive outcomes will be evident at the institution.

REFERENCES

Adams, E. C., Oduor, P., Wahome, A., & Tondapu, G. (2022). Reflections on two years teaching earth science at the Women in Science (WiSci) STEAM camp. *Journal of Women and Minorities in Science and Engineering*, *28*(1), 23–39. doi:10.1615/JWomenMinorScienEng.2021033536

Anshari, M., Almunawar, M. N., & Ordóñez de Pablos, P. (2023). Disruptive Innovation and a Multi-Sided Digital Marketplace. In *Handbook of Research on Artificial Intelligence and Knowledge Management in Asia's Digital Economy* (pp. 366–377). IGI Global.

Anshari, M., Syafrudin, M., Tan, A., Fitriyani, N. L., & Alas, Y. (2023). Optimisation of Knowledge Management (KM) with Machine Learning (ML) Enabled. *Information (Basel)*, *14*(1), 35. Advance online publication. doi:10.3390/info14010035

Beede, D. N., Julian, T. A., Langdon, D., McKittrick, G., Khan, B., & Doms, M. E. (2011). *Women in STEM: A gender gap to innovation.* Retrieved from https://files.eric.ed.gov/fulltext/ED523766.pdf

Bhasin, B. (2012). Integration of information and communication technology in enhancing teaching and learning. *Contemporary Educational Technology*, *3*(2), 130–140. doi:10.30935/cedtech/6073

Bhatia, M. (2022). *Women Empowerment with the Power of STEAM.* Retrieved from http://library.ediindia.ac.in:8181/jspui/handle/123456789/13212

Blackburn, H. (2017). The status of women in STEM in higher education: A review of the literature 2007–2017. *Science & Technology Libraries*, *36*(3), 235–273. doi:10.1080/0194262X.2017.1371658

Bloodhart, B., Balgopal, M. M., Casper, A. M., Sample McMeeking, L. B., & Fischer, E. V. (2020). Outperforming yet undervalued: Undergraduate women in STEM. *PLoS One, 15*(6), e0234685. Advance online publication. doi:10.1371/journal.pone.0234685 PMID:32584838

Brantlinger, E., Jimenez, R., Klingner, J., Pugach, M., & Richardson, V. (2005). Qualitative studies in special education. *Exceptional Children, 71*(2), 195–207. doi:10.1177/001440290507100205 PMID:15757375

Chen, K. S., & Liao, H. T. (2020, December). Mapping the Field of Educational Assessment and Professional Accreditation in Higher Education: Bibliographic Coupling and Co-Citation Analyses for Green and Digital Transformation. *Management Science Informatization and Economic Innovation Development Conference (MSIEID)* (pp. 281-284). IEEE. 10.1109/MSIEID52046.2020.00058

Cloete, N. (2014). The South African higher education system: Performance and policy. *Studies in Higher Education, 39*(8), 1355–1368. doi:10.1080/03075079.2014.949533

Cohen, L., Manion, L., & Morrison, K. (2018). *Research methods in education* (8th ed.). Routledge.

Collins, C. (2008). Looking to the future: Higher education in the Metaverse. *EDUCAUSE Review, 43*(5), 50–52. https://www.learntechlib.org/p/102249/

Contreras, G. S., González, A. H., Fernández, M. I., Martínez, C. B., Cepa, J., & Escobar, Z. (2022). The importance of the application of the metaverse in education. *Modern Applied Science, 16*(3), 1–34. doi:10.5539/mas.v16n3p34

Cramer, J., & Crocco, M. S. (2010). Women and technology, upon reflection: Linking global women's issues to the digital gender divide in urban social studies education. In Technology Implementation and Teacher Education: Reflective Models (pp. 184-202). IGI Global.

Department of Education. (2004, September 2). White Paper on e-Education: Transforming Learning and Teaching through Information and Communication Technologies (ICTs). *Government Gazette* (26734), pp. 3 - 46.

Escudeiro, N., Cruz, M., & Escudeiro, P. (2021, April). ATHENA: A Novel Higher Education Approach to Advance a Green Digital Europe. *Open Conference of Electrical, Electronic and Information Sciences (eStream)* (pp. 1-6). IEEE.

Farikayi, B., & Goosen, L. (2022). Designing Effective E-Learning Environments in Gweru, Midlands Province, Zimbabwe: An E-School's Community Engagement Case Study. In T. Driscoll (Ed.), *Designing Effective Distance and Blended Learning Environments in K-12* (pp. 258–281). IGI Global., doi:10.4018/978-1-7998-6829-3.ch015

Foreman-Peck, L., & Winch, C. (2010). *Using educational research to inform practice: A practical guide to practitioner research in universities and colleges.* Routledge. doi:10.4324/9780203847619

Goosen, L. (2008). A Brief History of Choosing First Programming Languages. In J. Impagliazzo (Ed.), *History of Computing and Education 3* (Vol. 269, pp. 167–170). Springer. doi:10.1007/978-0-387-09657-5_11

Goosen, L. (2015). Educational Technologies for an ICT4D MOOC in the 21st Century. In D. Nwaozuzu, & S. Mnisi (Ed.), *Proceedings of the South Africa International Conference on Educational Technologies* (pp. 37 - 48). Pretoria: African Academic Research Forum.

Goosen, L. (2018a). Ethical Data Management and Research Integrity in the Context of e-Schools and Community Engagement. In C. Sibinga (Ed.), *Ensuring Research Integrity and the Ethical Management of Data* (pp. 14–45). IGI Global., doi:10.4018/978-1-5225-2730-5.ch002

Goosen, L. (2018b). Ethical Information and Communication Technologies for Development Solutions: Research Integrity for Massive Open Online Courses. In C. Sibinga (Ed.), *Ensuring Research Integrity and the Ethical Management of Data* (pp. 155–173). IGI Global., doi:10.4018/978-1-5225-2730-5.ch009

Goosen, L. (2019). Research on Technology-Supported Teaching and Learning for Autism. In L. Makewa, B. Ngussa, & J. Kuboja (Eds.), *Technology-Supported Teaching and Research Methods for Educators* (pp. 88–110). IGI Global., doi:10.4018/978-1-5225-5915-3.ch005

Goosen, L., & Naidoo, L. (2014). Computer Lecturers Using Their Institutional LMS for ICT Education in the Cyber World. In C. Burger, & K. Naudé (Ed.), *Proceedings of the 43rd Conference of the Southern African Computer Lecturers' Association (SACLA)* (pp. 99-108). Port Elizabeth: Nelson Mandela Metropolitan University.

Hammer, P. E. (2003). The last decade. *History Today*, *53*(5), 53.

Hwang, G. J., & Chien, S. Y. (2022). Definition, roles, and potential research issues of the metaverse in education: An artificial intelligence perspective. *Computers and Education: Artificial Intelligence*, *3*, 100082. Advance online publication. doi:10.1016/j.caeai.2022.100082

Inceoglu, M. M., & Ciloglugil, B. (2022, July). Use of Metaverse in education. *Computational Science and Its Applications (ICCSA) Workshops: Malaga, Spain, Proceedings, Part I* (pp. 171-184). Cham: Springer International Publishing. Retrieved from https://link.springer.com/chapter/10.1007/978-3-031-10536-4_12

Jimoyiannis, A., & Komis, V. (2007). Examining teachers' beliefs about ICT in education: Implications of a teacher preparation programme. *Teacher Development*, *11*(2), 149–173. doi:10.1080/13664530701414779

Kahn, S., & Ginther, D. (2017). *Women and STEM*. National Bureau of Economic Research. Retrieved from https://www.nber.org/system/files/working_papers/w23525/w23525.pdf

Khan, M. S., Hasan, M., & Clement, C. K. (2012). Barriers to the introduction of ICT into education in developing countries: The example of Bangladesh. *International Journal of Instruction*, *5*(2), 61–80.

LeCompte, M. D. (2000). Analyzing qualitative data. *Theory into Practice*, *39*(3), 146–154. doi:10.120715430421tip3903_5

S. Lin, & X. Huang (Eds.). (2011, June). Preface. *Advanced Research on Computer Education, Simulation and Modeling (CESM): International Conference, Proceedings, Part I.* 175. Wuhan, China: Springer. 10.1007/978-3-642-21783-8

Lu, L., Chen, Q., Huang, R., & Usman, A. (2023). Education and its impact on renewable energy demand, carbon intensity, and green growth: Do digital financial inclusion and environmental policy stringency matter in China. *Environmental Science and Pollution Research International*, *30*(5), 12020–12028. doi:10.100711356-022-22759-6 PMID:36103074

Mautso, K., & Goosen, L. (2023). Using Applicable E-Learning Principles to Promote Effectiveness and Quality in the Digital Era: Toward a More Inclusive and Resilient E-School. In P. Ordóñez de Pablos, X. Zhang, & M. Almunawar (Eds.), *Handbook of Research on Education Institutions, Skills, and Jobs in the Digital Era* (pp. 359–381). IGI Global., doi:10.4018/978-1-6684-5914-0.ch021

Mqina, B. P., & Goosen, L. (2021). Investigating the Roles of School Management Teams in Curriculum Delivery Through ICTs in Education: e-Schools' Community Engagement. In A. Mawela, M. van Wyk, L. Lebeloane, & A. Mudau (Eds.), *Investigating the Roles of School Management Teams in Curriculum Delivery* (pp. 158–178). IGI Global., doi:10.4018/978-1-7998-7168-2.ch011

Mustafa, B. (2022). Analyzing education based on metaverse technology. *Technium Social Sciences Journal, 32*, 278–295. doi:10.47577/tssj.v32i1.6742

Muzambi, A., & Goosen, L. (2022). Factors Affecting the Successful Implementation of an E-Education Policy and Community Engagement: Acquiring 21st Century Skills Through E-Learning. In C. Lane (Ed.), *Handbook of Research on Acquiring 21st Century Literacy Skills Through Game-Based Learning* (pp. 580–605). IGI Global., doi:10.4018/978-1-7998-7271-9.ch030

Ordóñez de Pablos, P. (2022). Unlocking the potential of intellectual capital, innovation and higher education: Some keys for competitiveness in the post-pandemic era. *International Journal of Learning and Intellectual Capital, 19*(5), 391–394.

Ordóñez de Pablos, P. (2023). Digital economy, intellectual capital and competitiveness. Some lessons for the post-COVID-19 era. *International Journal of Learning and Intellectual Capital, 20*(1), 1–5.

Ordóñez de Pablos, P., Lytras, M. D., & Zhang, X. (Eds.). (2020). *IT and the Development of Digital Skills and Competences in Education*. IGI Global.

Ordóñez de Pablos, P., Lytras, M. D., Zhang, X., & Chui, K. T. (Eds.). (2019). *Opening Up Education for Inclusivity across Digital Economies and Societies*. IGI Global. doi:10.4018/978-1-5225-7473-6

Ordóñez de Pablos, P., Zhang, X., & Almunawar, M. (Eds.). (2023). *Handbook of Research on Education Institutions, Skills, and Jobs in the Digital Era*. IGI Global.

Ordóñez de Pablos, P., Zhang, X., & Almunawar, M. N. (Eds.). (2022). *Handbook of Research on Green, Circular, and Digital Economies as Tools for Recovery and Sustainability*. IGI Global., Retrieved from https://www.igi-global.com/book/handbook-research-green-circular-digital/280314 doi:10.4018/978-1-7998-9664-7

Ordóñez de Pablos, P., Zhang, X., & Almunawar, M. N. (Eds.). (2023). *Handbook of Research on Artificial Intelligence and Knowledge Management in Asia's Digital Economy*. IGI Global., Retrieved from https://www.igi-global.com/book/handbook-research-artificial-intelligence-knowledge/297912

Palmer, E. (2015a). Introduction: The sustainable development goals forum. *Journal of Global Ethics, 11*(1), 3–9. doi:10.1080/17449626.2015.1021091

Palmer, E. (2015b). Introduction: The 2030 agenda. *Journal of Global Ethics, 11*(3), 262–269. doi:10.1080/17449626.2015.1119928

Patel, K. K., & Patel, S. M. (2016). Internet of things-IOT: definition, characteristics, architecture, enabling technologies, application & future challenges. *International journal of engineering science and computing, 6*(5), 6122-6131.

Puentedura, R. R. (2010). *SAMR and TPCK: Intro to Advanced Practice.* Retrieved from http://hippasus.com/resources/sweden2010/SAMR_TPCK_IntroToAdvancedPractice.pdf

Racsko, R., & Troll, E. M. (2020). Our digital education habits in the light of their environmental impact: The role of green computing in education. *Teaching Mathematics and Computer Science, 18*(1), 69–86. doi:10.5485/TMCS.2020.0508

Raja, R., & Nagasubramani, P. C. (2018). Impact of modern technology in education. *Journal of Applied and Advanced Research, 3*(1), 33–35. doi:10.21839/jaar.2018.v3iS1.165

Sandrone, S. (2022). Medical education in the metaverse. *Nature Medicine, 28*(12), 2456–2457. doi:10.103841591-022-02038-0 PMID:36385158

Sangrà, A., Vlachopoulos, D., & Cabrera, N. (2012). Building an inclusive definition of e-learning: An approach to the conceptual framework. *International Review of Research in Open and Distance Learning, 13*(2), 145–159. doi:10.19173/irrodl.v13i2.1161

Semivolos, A., & Anshari, M. (2023). The Application of Artificial Intelligence to the Process of Logistic During COVID-19. In *Handbook of Research on Artificial Intelligence and Knowledge Management in Asia's Digital Economy* (pp. 1–11). IGI Global.

Singh, J., Malhotra, M., & Sharma, N. (2022). Metaverse in Education: An Overview. In D. Bathla & A. Singh (Eds.), *Applying Metalytics to Measure Customer Experience in the Metaverse* (pp. 135–142). IGI Global. doi:10.4018/978-1-6684-6133-4.ch012

Soliman, E. (2021). Planning requirements to enhance digital green education skills for students of applied technology schools (future vision). *Journal of Education, 91*, 2959–3017.

Steyn, G. M. (2002). The changing principalship in South African schools. *Educare (San José), 31*(1-2), 251–274.

Tshabalala, M., & Ndeya-Ndereya, C. (2015). How far are we with the implementation of blended learning? Implementation guidelines on offer. In D. Nwaozuzu, & S. Mnisi (Ed.), *South Africa International Conference on Educational Technologies* (pp. 27-36). Pretoria: African Academic Research Forum.

Van Heerden, D., & Goosen, L. (2023). The Implications of COVID-19 for Fostering Diversity and Inclusion Through Curriculum Transformation: Teaching Open and Distance E-Learning Students. In C. Tabane, B. Diale, A. Mawela, & T. Zengele (Eds.), *Fostering Diversity and Inclusion Through Curriculum Transformation* (pp. 73–100). IGI Global., doi:10.4018/978-1-6684-6995-8.ch005

Van Wyk, N. (2004). School governing bodies: The experiences of South African educators. *South African Journal of Education, 24*(1), 49–54.

Vorster, J., & Goosen, L. (2017). A Framework for University Partnerships Promoting Continued Support of e-Schools. In J. Liebenberg (Ed.), *Proceedings of the 46th Annual Conference of the Southern African Computer Lecturers' Association (SACLA)* (pp. 118 - 126). Magaliesburg: North-West University.

Wang, Q. (2008). A generic model for guiding the integration of ICT into teaching and learning. *Innovations in Education and Teaching International, 45*(4), 411–419. doi:10.1080/14703290802377307

Wright, G., Cilliers, L., Van Niekerk, E., & Seekoe, E. (2017). The next stage of development of elearning at UFH in South Africa. *International Conference on E-Learning* (pp. 32-38). Lisbon: International Association for Development of the Information Society.

Zhang, X., Meng, Y., Ordóñez de Pablos, P., & Sun, Y. (2019). Learning analytics in collaborative learning supported by Slack: From the perspective of engagement. *Computers in Human Behavior, 92*, 625–633. doi:10.1016/j.chb.2017.08.012

KEY TERMS AND DEFINITIONS

Community: A group of people living in a particular area having certain common characteristics.

E-learning: refers to a learning system based on formalized teaching but with the help of electronic resources, teaching may take place in or out of the classroom.

E-school: refers to a school that teaches students primarily online.

Internet: refers to a global network of computing resources, which support communication and information sharing using various communication protocols.

School Governing Body: A statutory body of parents from the community, learners, educators, teaching and non-teaching staff who seek to work together to promote the well-being and effectiveness of the school community so as to promote teaching and learning.

Section 2

Digital Technologies and Competitiveness in the Green Economy

Chapter 8
Developing Consumers' Competencies for Digital Marketing Transitions

Xuan Tran
https://orcid.org/0000-0001-9521-5722
University of West Florida, USA

ABSTRACT

The issues of mismatching between the feedback and the strategies have deteriorated most businesses in their transitions. The purpose of this chapter is thus to provide an instant digital marketing transition from consumers' feedback to providers' strategies. The three theories of action regulation, actor-network, and motivation are used to direct the solutions and recommendations whereby the feedback has been clustered by the digital Linguistic Inquiry and Word Count into 12 words and transited to 12 marketing strategies. Findings indicated that the feedback words related to "body", "health", "risk", "work", "friend", "social", "adjectives", "family", "biological" "drives", "affect", and "positive emotion" have been transited to physical handling, fact finding, packaging, product planning, distribution, display, advertising, personal selling, branding, servicing, pricing, and promotion, respectively. Solutions and recommendations have been discussed.

INTRODUCTION

The issues of mismatching between the feedback and the strategies have deteriorated most businesses in their transitions of life cycle. The separation between consumers' feedback and providers' marketing strategies has created critical issues for providers

DOI: 10.4018/978-1-6684-9089-1.ch008

to meet consumers' demand instantly. For example, an increase in guests' comments in social media has made hoteliers change hotel brands to confuse hotel guests in the COVID-19 pandemic era as follows. 31 brands of Marriott have been changed into Bonvoy, 19 brands of Hilton have been changed into Honors, 15 Intercontinental have been changed into IHG Rewards Club, 18 Hyatt brands have been changed into the World of Hyatt, and 17 brands of Choice hotels have been changed into Choice of Privileges.

After COVID-19, the essence of the change is from being to doing. Therefore, hoteliers should not use branding. Instead, they should use promotion to respond to consumers' positive emotion to attract more guests as socializers with high power and achievement needs. Otherwise, the hoteliers only change hotel brands to respond to consumers' biological process, when guests' comments are related to basic human physiological need such as eating or dwelling. This case related to consumers' feedback and hoteliers' strategies are the motivation for this study.

The purpose of this chapter is thus to provide an instant digital marketing transition from consumers' feedback to providers' strategies. The three theories of action regulation, actor-network, and motivation are used to direct the solutions and recommendations whereby the feedback has been clustered by the digital Linguistic Inquiry and Word Count (LIWC) into 12 words and transited to 12 marketing strategies through canonical correlation regression.

The background information of the study of the relationships among feedback, transition, and strategies is thus in the three theories: The action regulation theory (Frese and Keith, 2015), the Actor-Network theory (Latour, 2005), and the 3-needs theory (McClelland, 1985). The action regulation theory (ART) states that feedback would create strategies in a transition. The actor-network theory (ANT) states that digital transition is a nonhuman actant to combine human feedback in a network of strategies. The 3-needs theory (3NT) states that a human being has 3 unconscious needs for doing excellent performance in strategies (the need for achievement), for controlling other people in feedback (the need for power), and for making friends with other people in transition (the need for affiliation). The hypothesis of the study is that consumers' feedback would be transited to providers' marketing strategies instantly based on the digital marketing transition relationship.

In the ANT, Salter (2019) states that nonhuman actants play in processes to consolidate the network with human. The digital Linguistic Inquiry and Word Count (LIWC, 2022) is a nonhuman actant to count human writing words and classify them into specific clusters used for different marketing purposes of human in an actor-network of trading between providers and consumers. Ozuem, Ranfagni, Willis, Rovai, and Howell (2021) reported the impact of ANT was on customers to change their feedback during the Covid-19 pandemic transition resulting in different providers' strategies.

In the ART, Frese and Keith (2015) has found that feedback provides people with information to regulate their behavior to achieve the goal. The feedback can be timely, positive (e.g., praise) or negative (e.g., error). According to action regulation theory, positive feedback is for maintaining behavior, whereas negative feedback is for learning and personal development. Zacher (2017) reported that the impact of ART was on workers to change their goals with relevant strategies, transiting behavior, and feedback.

In the 3NT, McClelland (1985) has found the relationship between the need for achievement and goal which is now controlled by consumers' feedback in this study. Mertens, Theisen, and Funke (2022) identified the providers' strategies with less power and high affiliation needs would attract senior consumers coming in the transition.

In sum, the above three theories ART, ANT, and 3NT are the literature background information for this study to hypothesize that consumers' feedback would be transited to providers' marketing strategies instantly based on the digital marketing transition relationship.

The remaining paper would include the literature with Borden's (1964) marketing mix, Linguistic Inquiry and Word Count (LWIC), Action regulation theory (ANT), Actor-Network theory (ANT) and 3-need theory (3NT). The study has used the LIWC to cluster the feedback into 12 words whereby were transited to 12 marketing strategies through canonical correlation regression for the sample of 11,959 consumers' feedback comments to support the hypotheses. Finally, the discussion as well as concluding remarks would be at the end of the paper.

LITERATURE

Borden's (1964) Marketing Mix

Marketing mix is an art to mix ingredients to create different recipes. According to Borden (1964), the father of the concept "marketing mix", the marketing mix of manufacturers including all marketing procedures and policies could be an art and a science. The marketing mix is a science because it involves in observation and classification of facts in verifiable laws to predict results accurately through the 12 marketing mix elements of a manufacturer in the priority order as follows. Firstly with "fact finding" is for the manufacturers to find facts and ideas to meet the consumers' demand. Secondly with "physical handling" is for the manufacturers to prepare for transportation supply. Thirdly with "servicing" is for the manufacturers to serve any consumers who are energetic. Fourthly with "advertising" is for the manufacturers to provide all modifications for consumers' selection. Fifthly with "personal selling" for the manufacturers to sell to most familiar customers. Sixthly with "distribution" is for the manufacturers to bring their products or services to their known customers. Seventhly with "product planning" is for the manufacturers to offer a product or service to the consumers at work. Eighthly with "branding" is for the manufacturers to let consumers research the bio of the business it. Ninthly with "pricing" is for the manufacturers to fit the consumers' income affect. Tenthly with "packaging" is for the manufacturers to offer a combination to avoid any consumers' risk. Eleventhly with "promotion" is for the manufacturers to congratulate happy consumers. And finally, twelfthly with "displaying" is for the manufacturers to assure their product or service to social consumers.

The marketing mix is an art because the mix can be designed as a guide (process) of the 4 market forces as follows. "1. *Consumers' Buying Behavior,* as determined by their: a) motivation in purchasing, b) buying habits, c) living habits, d) environment (present and future, as revealed by trends, for environment influences consumers' attitudes toward products and their use of them), e) buying power, and f) number (i.e., how many). 2. *The Trade's Behavior*—wholesalers' and retailers' behavior, as influenced by a) their motivations, b) their structure, practices, and attitudes, c) trends in structure and procedures that portend change. 3. *Competitors' Position and Behavior,* as influenced by: a) industry structure and the firm's relation thereto., 1. size and strength of competitors, 2. number of competitors and degree of industry concentration. 3. Indirect competition—i.e., from other products, b) relation of supply to demand—oversupply or undersupply, c) product choices offered consumers by the industry—i.e., quality, price, service, d) degree to which competitors compete on price vs. nonprice bases, e) competitors' motivations, and attitudes—their likely response to the actions of other firms, f) trends technological and social, and portending change in supply and demand. 4. *Governmental Behavior—Controls over Marketing: a)* Regulations over products, b) regulations over pricing, c) regulations over competitive practices, d) regulations over advertising and promotion." (Borden, 1964, p. 10).

The above 12 marketing mix elements have become the solutions and recommendations when they are combined with the consumers' feedback that has been clustered by the digital Linguistic Inquiry and Word Count (LIWC) into 12 words and transited to 12 marketing strategies.

LINGUISTIC INQUIRY AND WORD COUNT (LIWC)

The digital transition in this study is the use of the linguistic inquiry and word count (LIWC, 2022). Studies measuring personality in participants through writing samples (Pennebaker, Boyd, Jordan, & Blackburn, 2015) and spoken dialogue (Mehl, Gosling, & Pennebaker, 2006) have shown that some LIWC categories connect words with motives and correspond with big-five personality traits. For example, Mehl and colleagues found that for both males and females higher word count and fewer large words predicted extraversion. Pennebaker and King (1999) showed that other LIWC categories showing complexity of language (such as articles, exclusive words, causal words, and negations) were less frequent in the writing of people who scored high on extraversion. Social and emotional language also differed with respect to extraversion; people who scored high on extraversion used more social words, more positive emotion, and less negative emotion. The findings from these two studies partially support traditional personality models. Models of extraversion would predict that extraverts engage in more social interaction and have a more positive response to that engagement. Also, these models would predict that people high in extroversion would be less inhibited in their language production, possibly leading to less complex language. Depressed and suicidal individuals are more self-focused, express more negative emotion and sometime use more death-related words. Studies on depression and suicide show that language features can be markers of mental health. Depressed patients are more likely to use more first-person singular and more negative emotion words than participants who have never been depressed in emotional writings (Rude, Gortner, Pennebaker, 2004). Suicidal poets in their published works compared with matched nonsuicidal poets use more first-person singular and more death-related words (Stirman & Pennebaker, 2001). This individual difference may show an attentional difference, that is, more self-focus in response to emotional pain, or it may indicate a thinking pattern that is a predilection for experiencing depression (see also work by Wolf, Sedway, Bulik, & Kordy, 2007, dealing with the language of anorexia). The study has used LIWC to examine the feedback of 11,959 hotel guests from social media to retrieve consumers' feedback and their motivation.

In sum, the 12 LIWC words have been used to support the relationships with the 12 marketing elements among feedback, transition, and strategies in the following 12 hypotheses.

Hypothesis 1: There would be a significant relationship between the LIWC word and the marketing mix element: Biologistic and Branding.

Hypothesis 2: There would be a significant relationship between the LIWC word and the marketing mix element: Family and Personal selling.

Hypothesis 3: There would be a significant relationship between the LIWC word and the marketing mix element: Adjectives and Advertising

Hypothesis 4: There would be a significant relationship between the LIWC word and the marketing mix element: Working and Planning

Hypothesis 5: There would be a significant relationship between the LIWC word and the marketing mix element: Friend and Distribution

Hypothesis 6: There would be a significant relationship between the LIWC word and the marketing mix element Social and Display.

Hypothesis 7: There would be a significant relationship between the LIWC word and the marketing mix element: Health and Finding

Hypothesis 8: There would be a significant relationship between the LIWC word and the marketing mix element: Body and Handling

Hypothesis 9: There would be a significant relationship between the LIWC word and the marketing mix element: Risk and Packaging

Hypothesis 10: There would be a significant relationship between the LIWC word and the marketing mix element: Drives & Servicing

Hypothesis 11: There would be a significant relationship between the LIWC word and the marketing mix element: Positive & Promotion

Hypothesis 12: There would be a significant relationship between the LIWC word and the marketing mix element: Affect & Pricing

ACTION REGULATION THEORY (ART)

The Action regulation theory (Frese and Keith, 2015) states that human action is goal-oriented behavior that unfolds in a sequence: the action sequence starts with the generation of a goal and people then have to process and collect information about their environment, plan how to reach the goal, monitor the execution of action plans, and process feedback that may help them to attain their goal (Frese et al. 2007; Frese and Zapf 1994; Zacher and Frese 2018).

The feedback reflects the emotional and analytical terms for the quality of plans in a long-range perspective with regard to future opportunities and potential hindrances. The more obstacles by challenged fact findings, the more anger human reflects by their degree of detailedness of sub-steps leading to goal achievement, and by the generation of back-up plans (Frese et al. 2007; Frese and Zapf, 1994).

The 12 hypotheses above have been supported in the action regulation theory, which reflect in the following articles.

Hypothesis 1: There would be a significant relationship between the LIWC word and the marketing mix element: Biologistic and Branding.

Cohen (2012) identifies a shift from a synchronic like brand to a diachronic perspective like biologistic demand. Keller (2019) reports that consumers would research the bio of the business for the pleasure and/or pain of brands, brand attachment and loyalty, consumer relevance and distinctiveness and branding, consumer communications about brands, and managerial branding considerations. While manufacturers are branding for consumers, would it be possible related to consumers' bio feedback? In the network of manufacturer and consumer including branding and bio, the branding of the manufacturer is thus positively related to the bio of consumer. Therefore, the hypothesis of the association between the marketing of branding policy and the consumers' bio feedback has been supported.

Hypothesis 2: There would be a significant relationship between the LIWC word and the marketing mix element: Family and Personal selling.

Cohen (2012) identifies a shift from a post-modern tendency like family to stress the de-differentiation personal selling between social domains. Azeez and Abeyratna (2022) using the sample of 322 undergraduate students report that personal selling of the providers is associated with buyer behavior when the contact of salesperson is comfortable to buyers. The people are most comfortable to the buyers are their family so personal selling would be associated with family. They also report there is no impact of presentation of commodities by the salespersons on buying behavior. In the network of manufacturer and consumer including personal selling and familiarity, the personal selling of the manufacturer is thus positively related to the family of consumer. Therefore, the hypothesis of the association between the marketing factor of personal selling policy and the consumers' family feedback has been supported.

Hypothesis 3: There would be a significant relationship between the LIWC word and the marketing mix element: Adjectives and Advertising

Cohen (2012) identifies a shift from structure like adjective to agency like advertising. Janiszewska (2013) report that consumer modifications though adjectives which has become more popular recently is one of the key product identities. The main reason of consumer adjective importance is dynamic changes in consumers' needs and behaviors, which determines their decisions. Segmentation process based on demographic data in context of postmodernism is not sufficient any more to differentiate the brand. Looking for advertising, manufacturer managers more often focus on consumer behavior, attitude, values, needs and expectations. In the network of manufacturer and consumer including adjective modifications and advertising, the advertising of the manufacturer is thus positively related to the adjectives of consumer. Therefore, the hypothesis of the association between the marketing factor of advertising and the consumers' adjective feedback has been supported.

Hypothesis 4: There would be a significant relationship between the LIWC word and the marketing mix element: Working and Planning

Cohen (2012) identifies a shift from the blurring of the border between reality like work to virtuality like product planning. We shall here discuss three important inter-related theoretical developments sharing this perspective, (11) the mobilities "paradigm" (desires-servicing), (12) the performativity approach (health-fact finding) and (13) actor-network theory (achievement-consumer, affiliation-trade, power-competitor, and reward-government) (ANT)" (p. 2177). Thomke, S. & Reinertsen, D. (2012) report that there would be an association between consumers' work and manufacturers' product planning. In product development many tasks are unique, project requirements constantly change, and the output resides in multiple places at the same time. Therefore, there are 6 myths on product planning as follows: High utilization of resources will improve performance (In fact, projects take longer when people are not working 100% of the time); Processing work in large batches improves the economics of the development process. (In fact, small batches allow manufacturers accelerate feedback); Our development plan is great; we just need to stick to it. (In fact, there is never a single product development project whose requirement remained stable); The sooner the project is started, the sooner it will be finished. (In fact, work is perishable, technologies become obsolete; The importance of reducing the amount of work in process is evident when we look at one of the classic formulas of queuing theory: Little's Law. It simply states that, on average, cycle time is proportional to the size of the queue divided by the processing rate.); The more features we put into a product, the more customers will like it. (In fact, adding features creates value for customers and subtracting them destroys value.); and We will be more successful if we get it right the first time. (In fact, requiring success on the first pass biases teams toward the least-risky solutions and teams have little incentive to pursue innovative solutions to customers' problems. A tolerance for "getting it wrong the first time" can be the better strategy as long as people iterate rapidly and frequently and learn quickly from their failures.). While manufacturers are planning a product for consumers to work diligently due to the 6 above myths, would it be possible related to consumers' work feedback? In the network of manufacturer and consumer including product planning and work, the planning of the manufacturer is thus positively related to the work of consumer. Therefore, the hypothesis of the association between the marketing of product planning policy and the consumers' work feedback has been supported.

Hypothesis 5: There would be a significant relationship between the LIWC word and the marketing mix element: Friend and Distribution

Cohen (2012) identifies a shift from the break-down of conventional binary concepts like friend and distribution. Tian and Jiang (2017) report that in recent years consumer-to-consumer product sharing or "friend" affects the distribution channel because of mobile communications technologies and online sharing. The manufacturer has built its production capacity which product sharing will increase the manufacturer's optimal capacity to increase the distribution of retailer's share of the gross profit margin in the channel. In the network of manufacturer and consumer including friends and distribution, the distribution of the manufacturer is thus positively related to the friend of consumer. Therefore, the hypothesis of the association between the marketing factor of distribution policy and the consumers' friend feedback has been supported.

Hypothesis 6: There would be a significant relationship between the LIWC word and the marketing mix element Social and Display.

Cohen (2012) identifies a shift of emphasis from permanence like display to flux like social. Szocs, Williamson, and Mills (2022) report that there is a significant association between social and display. They confirm that it will be better to display some products socially. While manufacturers are doing display for consumers, would it be possible related to consumers' social feedback? In the network of manufacturer and consumer including display and social, the display of the manufacturer is thus positively related to the social feedback of consumer. Therefore, the hypothesis of the association between the marketing of display policy and the consumers' social feedback has been supported.

Hypothesis 7: There would be a significant relationship between the LIWC word and the marketing mix element: Health and Finding

Cohen (2012) identifies the shift in the performativity approach like health being switched to fact finding. While manufacturers are anxious in looking for their consumer demand in their marketing goal of fact finding, would it be possible related to consumers' health feedback? Russel and Mehrabian (1974) report that healthy is negatively associated with anxiety by the level of dominance in the network: pleasure, arousal, and dominance from the sample of 348 undergraduate students. In the network of manufacturer and consumer including pleasure, arousal and dominance, the anxiety of the manufacturer high dominance is thus negatively related to the health of consumer low dominance. Therefore, the hypothesis of the association between the marketing factor finding policy and the consumers' health feedback has been supported.

Hypothesis 8: There would be a significant relationship between the LIWC word and the marketing mix element: Body and Handling

Cohen (2012) identifies the shift from sedimented social patterns like body to the process of their emergence like physical handling. During their physical handling, manufacturers must do warehousing, transportation, and inventories (Borden, 1964). The body manufacturers' physical handling occurs in the network of manufacturer and consumer. Would the process be possible related to consumers' body feedback? Dagan, Facompre, and Bernard (2018) reports that the unresolved attachment is associated with body through the sample of 4,386 studies. The separation is significantly correlated in the unresolved organized body attachment. In the network of manufacturer and consumer attachment, the unresolved separation of manufacturers' physical handling is thus associated with the consumers' body. The body separation coming from physical handling when the providers separate the product elementaries for inventories. When an infant is separated from a primary attachment body, as in the Harlow studies of rhesus monkeys (Harlow, Dodsworth, & Harlow, 1965), the result is not commonly despair and passivity, immune system is also gotten harm, which lead toward depressive illness (Hennessy, Schimlwebb, Deak, 2009). History of separations, rejections, and insecure body attachments can be associated with depressed people. Hammen (2009), Nolan, Flynn, & Garber (2003), Weissman, Markowitz, & Klerman (2000), and Wakefield, Schmitz, First, &Horwitz (2007) report that people with a history of happy and secure body attachments may also fall into prolonged depression because of the loss of a beloved lifelong partner. Joormann, Siemer & Gotlib (2007) report that when depressed and non-depressed people asked to recall happier times, nondepressed people cheer up. But depressed people feel even worse, as if the happy memory makes them feel that they will never be happy again. Therefore, the hypothesis of the association between the marketing physical handling policy and the consumers' body feedback has been supported.

Hypothesis 9: There would be a significant relationship between the LIWC word and the marketing mix element: Risk and Packaging

Cohen (2012) identifies a shift from the interpenetration between formerly opposite categories like risk versus packaging. Organizations often packaging for price bundle to increase the demand for products and services. This practice would be risk because it increases short-term demand—but it may also reduce consumption. In addition, it is far easier to identify and account for the cost of an individual product in an unbundled transaction than within a bundled transaction. The one-to-one relationship between price and benefits in an unbundled transaction makes the cost of that item obvious, creating a strong sunk-cost effect and a high likelihood of consumption. Therefore, stagger payments will smooth consumption. Buchmüller, Bearth, and Siegrist (2022) report that there is a significant association between packaging and risk. They found that consumers' risk perception can be influenced by the packaging design. While manufacturers are packaging for consumers, would it be possible related to consumers' risk feedback? In the network of manufacturer and consumer including packaging and risk, the packaging of the manufacturer is thus positively related to the risk of consumer. Therefore, the hypothesis of the association between the marketing of packaging policy and the consumers' risk feedback has been supported.

Hypothesis 10: There would be a significant relationship between the LIWC word and the marketing mix element: Drives & Servicing

Cohen (2012) identifies a shift from the mobilities "paradigm" like from desires to servicing. During the energetic period of consumers who are thriving to find something most interested, manufacturers will focus on servicing (Borden, 1964). Dixon, Freeman, and Toman (2010) report that when consumers are full of drives, manufacturers should increase servicing by anticipating, dealing with related issues, arming reps, using feedback from disgruntled or struggling customers, and focusing on problem solving, not speed. As a result of servicing, Customer Effort Score is a better predictor of loyalty than customer satisfaction. In the network of manufacturer and consumer including drives and servicing, the servicing of the manufacturers is thus positively related to the drives of the consumers. Therefore, the hypothesis of the association between the marketing factor of servicing and the consumers' drives feedback has been supported.

Hypothesis 11: There would be a significant relationship between the LIWC word and the marketing mix element: Positive & Promotion

Cohen (2012) identifies a shift from being like positive emotion to doing like promotion. Silton, Kahrilas, Skymba, Smith, Bryant, & Heller (2020) report that there is a significant relationship between positive emotions and promotion. Positive emotions of consumers are thus associated with a range of beneficial outcomes resulting in the manufacturers' promotion. While manufacturers are doing promotion for consumers, would it be possible related to consumers' positive emotions feedback? In the network of manufacturer and consumer including promotion and positive emotions, the promotion of the manufacturer is thus positively related to the positive emotions of consumer. Therefore, the hypothesis of the association between the marketing of promotion policy and the consumers' positive emotions feedback has been supported.

Hypothesis 12: There would be a significant relationship between the LIWC word and the marketing mix element: Affect & Pricing

Cohen (2012) identifies a shift from a focus on the more stable fixtures of social life like affect to the mobilities linking them like pricing. Gourville and Soman (2002) report that there is an association between manufacturers' pricing and consumers' affect. They gave an example of two friends to join the local health club with two different number of payments. One on an annual payment plan and the other on a monthly payment plan and ask who is more likely to work out on a regular basis. Their finding indicates that the one who selects more numbers of payment is much more likely to exercise at the club than her friend and they conclude that people are more likely to consume a product if they are aware of its cost. The same situation with the immediacy of payment can be critical for the consumption of a paid-for product. While manufacturers are pricing for consumers to affect consumptions, would it be possible related to consumers' affect feedback? In the network of manufacturer and consumer including pricing and affect, the pricing of the manufacturer is thus positively related to the affect of consumers. Therefore, the hypothesis of the association between the marketing of pricing policy and the consumers' affect feedback has been supported.

In sum, the ART has confirmed the association between consumers' feedback and providers' marketing goal.

The study has been then used actor-net work theory to integrate the variables in the above 12 hypotheses into 12 network solutions as follows.

ACTOR-NETWORK THEORY (ANT)

The study has applied the Actor Network Theory including human, non-human, and process to Marketing including consumers (human), trader (non-human), and government (process) including product. The horizontal axis is the vector of non-human (government – product, place, price, and promotion), The vertical axis is the vector of human (consumers – demand (motivation), convenience (habits), cost (buying power), communication (trends)). The diagnose axis is the vector of process (trade – wholesaler/retailer (industry), practices, price, trends).

Although it is best known for its controversial insistence on the capacity of nonhumans to act or participate in systems or networks or both, ANT is also associated with forceful critiques of conventional and critical sociology. Developed by science and technology studies (STS) scholars Michel Callon (1998), Madeleine Akrich (2006) and Bruno Latour (2005), the sociologist John Law (Adams, 2012), and others, it can more technically be described as a "material-semiotic" method.

This means that it maps relations that are simultaneously material (between things) and semiotic (between concepts). It assumes that many relations are both material and semiotic.

For the market motivations, Actor–network theory (ANT) (Latour, 2005) posits that social and natural are shifting in the same world and nothing exists outside the constantly shifting networks of social and natural worlds. That is, if there are some human motivations to integrate into inhuman forces, these human with inhuman ideas are seen equally important in creating new social forces. In addition, the ANT holds that the new social forces that associate inhuman forces with human motivations do not exist in themselves and considered only as an abstract theoretical concept, not something which genuinely exists in the world.

Applying the actor-network theory, the study has suggested 12 solutions for the 12 above hypotheses based on the recent research as follows.

For Biologistic and Branding in hypothesis 1, the first network solution is understanding and incorporating consumer biologistic into business branding. Martin, Clark, Ferrell, Stewart, and Pitt (2020) report that by integrating consumer biologistic concerns into business branding, companies can effectively communicate their commitment to health, sustainability, and ethical practices, ultimately attracting and retaining customers who prioritize these values.

For Family and Personal selling in hypothesis 2, the second network solution is strengthening the relationship between consumer family and provider through personal selling. Foster and Cadogan (2000) report that by employing personal selling techniques tailored to consumer families, providers can establish a stronger connection, build trust, and better meet the unique needs of families during the sales process

For Adjectives and Advertising in hypothesis 3, the third network solution is strengthening the relationship between consumer adjectives and provider advertising. Ebaid (2018) report that by aligning consumer adjectives with provider advertising, businesses can create a stronger connection with their target audience, influence perceptions, and effectively communicate the desired attributes and values associated with their brand.

For Working and Planning in hypothesis 4, the fourth network solution is enhancing the relationship between consumer working and provider product planning. Iheanachor, Umukoro, and West (2021) report that by incorporating work-friendly features into product planning, providers can better serve the needs of working consumers, enhance their working experience, and establish a positive and mutually beneficial relationship.

For Friend and Distribution in hypothesis 5, the fifth network solution is strengthening the relationship between consumer friends and provider distribution. Lu, Oh, Leahy, and Chopik (2020) report that by tailoring distribution policies to cater to consumer friends, providers can create a seamless and enjoyable experience for groups traveling together, ultimately strengthening the relationship and fostering positive word-of-mouth recommendations.

For Social and Display in hypothesis 6, the sixth network solution is enhancing the relationship between consumer social and provider display. Jamil, Dunnan, Gul, Shehzad, Gilani, and Awan (2022) report that by designing display policies that cater to consumer social needs and encourage social interactions, providers can create an engaging and community-oriented environment that enhances the overall experience and fosters long-lasting connections among consumers.

For Health and Fact finding in hypothesis 7, the seventh network solution is strengthening the relationship between consumer health and provider fact finding. Sun, Zhang, Gwizdka and Trace (2019) report that by actively seeking consumer feedback and conducting fact-finding exercises, providers can better understand consumer health-related needs and make informed decisions to enhance health-related services, ultimately strengthening the relationship and improving the overall health experience for consumers.

For Body and Handling, the eighth network solution is strengthening the relationship between consumer body and provider physical handling. Castelo, Boegershausen, Hildebrand, and Henkel (2023) report that by continuously evaluating and improving physical handling policies and practices, providers can establish trust and confidence with consumers, demonstrating their commitment to ensuring the well-being and safety of consumers' bodies during the handling of products or provision of services.

For Risk and Packaging, the ninth network solution is addressing the relationship between consumer risk and provider packaging. Buchmuller, Bearth, Zurich, Siegrist, and Zuich (2022) report that by addressing consumer risk through thoughtful packaging solutions, providers can instill confidence in consumers, mitigate concerns, and build a strong relationship based on trust and safety.

For Drives and Servicing, the tenth network solution is enhancing the relationship between consumer drives and provider servicing. Cheung, Pires, Rosenberger, Leung, and Sharipudin (2021) report that by aligning servicing policies with consumer drives, providers can create meaningful and engaging experiences that resonate with their target customers. This enhances customer satisfaction, fosters loyalty, and strengthens the relationship between the consumer and the provider.

For Positive and Promotion, the eleventh network solution is capitalizing on the relationship between positive consumers and providers promotion. Jamil, Dunnan, Gul, Shehzad, Gilani, and Awan (2022) report that by capitalizing on the positive emotions and experiences of consumers through promotional strategies, providers can strengthen the relationship with positive consumers, reinforce their loyalty, and attract new customers who are inspired by their positive testimonials.

For Affect and Pricing, the twelfth network solution is addressing the relationship between consumers' affect and providers' pricing. Zhao, Yao, Liu, and Yang (2021) report that by tailoring pricing strategies to evoke positive affect and effectively communicating value for money, providers can enhance the emotional connection with consumers and create a positive relationship based on perceived fairness, satisfaction, and trust.

MOTIVATION THEORY OR 3 NEEDS THEORY (3NT)

Mertens, Theisen, and Funke (2022) reported the significant impact of the 3 needs theory including the need for achievement, the need for power, and the need for affiliation on mobility of consumers that covers consumers' feedback, transition, and providers' strategies. Borden's (1964) manufacturer's market forces of consumer, trade, competitor, and government as inhuman forces must relate to consumers' motivations of achievement, affiliation, power, and reward as human natural ones as follows.

Achievement motivation is the human desire to exceed their normal performance (McClelland, 1985). A customer with a high achievement motivation always want to win over his/her provider in the market. According to the ANT, the consumer force will associate with consumers' achievement motivation, so a new social force of actor-network is "achievement-consumer".

Affiliation motivation is the human desire to make friends with other people (McClelland, 1985, Tran, Tran, & Tran, 2018). A trade of wholesalers or retailers often make customers happy in the market. A trade force will be thus associated with consumers' affiliation motivation. According to the ANT, the trade force will correlate consumers' affiliation motivation, so another new social force of actor-network is "affiliation-trade". If the affiliation-trade network focuses on people orientation, the achievement-consumer network will focus on task performance.

Contrary to previous networks focusing on the open relationship, the power motivation and the later reward motivation are focusing on reserved relationship. Power motivation is the human desire to control over other people (Winter & Barenbaum, 1985). Any competitor who wants to compete over other people will be thus associated with the power motivation. According to ANT, the competitor force will correlate competitor's power motivation, so another new social force of actor-network is "power-competitor". (4) Finally, the last motivation is reward motivation which is the reserved achievement motivation (McClelland, 1985). A government force including reserved government policies for product, place, price, and promotion (Borden, 1964) will correlate reward motivation. Therefore, another new social force of actor-network is "reward-government".

In addition to the 12 hypotheses, there are additional four hypotheses based on the empirical facts as follows. (1) the achievement-consumer guests need servicing, pricing, and promoting; (2) the affiliation-trade guests need work, friend, and social; (3) the power-competitor guests need advertising, personal selling, and branding; and (4) the reward-government guests need physical handling, fact finding, and packaging.

Hypothesis 13: There would be a significant relationship between servicing, pricing, and promoting and the achievement-consumer guests.
Hypothesis 14: There would be a significant relationship between product planning, distribution, and display and the affiliation-trade guests.
Hypothesis 15: There would be a significant relationship between advertising, personal selling, and branding and the power-competitor guests.
Hypothesis 16: There would be a significant relationship between physical handling, fact finding, and packaging and the reward-consumer guests.

METHODOLOGY

Sample

The respondents in the study were 11959 hotel guests (6090 females and 5869 males) posted their comments on 10 social media including TripAdvisor (42.1%), Booking (26.2%), Priceline (0.6%), Agoda (1.7%), Travelocity (0.3%), Hotels (9.8%), Orbitz (0.5%), Yelp (1.1%), Facebook (8.1%), and Google (10.1%) in 2 years (2015-2016).

These guests stayed in four hotel classes of different brands located in different location in the United States including Upscale (48.8%), Upper Midscale (29.2%), Midscale (10.1%), and Economy (11.8%). The 18 hotels include Best Western Premier, The Tides Hotel Orange Beach, Bikini Beach Resort Motel Panama, FL, Hampton Inn Orange Beach/Gulf Front, AL, Hampton Inn Pensacola Beach Gulf Front, FL, Hawthorn Suites Panama City Beach, FL, Hilton Garden Inn Orange Beach Beachfront, AL, Hilton Garden Inn Pascagoula, MS, Hilton Pensacola Beach Gulf Front, FL, Holiday Inn Express, Orange Beach On The Beach, AL, Holiday Inn Resort Fort Walton Beach, FL, Holiday Inn Resort Pensacola Beach Gulf Front, FL, Hotel Indigo Sarasota, Hyatt Place Pensacola Airport, FL, Inn of the Six Mountains Killington, VT, McCamly Plaza Hotel Battle Creak, MI, The Beachside Resort Panama City Beach, FL, Travelodge Pensacola Beach, FL, and West Bay Beach A Holiday Inn Resort Traverse City, MI).

CANONICAL CORRELATION ANALYSES

The two sets of variables (feedback comments and market forces) are continuous in the design. In order to examine the relationship between the two sets of continuous variables, canonical variate analysis has been conducted to relate the twelve feedback variables to the four market force variables simultaneously. The "canonical" term indicates the extracting technique from a square matrix.

Canonical variate is a form of latent variables. If the canonical correlation function is significant with p<.05, it represents an orthogonally separate pattern of relationships between the two latent variables (feedback comments and market forces) and indicates how well the groups of feedback factors could predict each of the four sets of market forces.

Wilks' Lamda (Λ) which an index from 0 to 1 to test the significance of the canonical variates has been used in this study. A Lamda close to 1 shows that the means of all the groups in the canonical correlation function have an approximate same value. A Lamda close to zero indicates the more different means of the group in the canonical correlation function.

Canonical correlation (canonical loadings) (Rc) has been used to assess how much one set of variables can be predicted by the other set. Thanks to the canonical correlation, this study would show the relationship among the four pairs of consumers' feedback and market forces by comparing the volume of correlation between any linear combination of the feedback variables and any linear combination of the market force variables. The most or least correlated index of feedback would predict the most or least correlated index of market force, respectively.

Eigenvalue has been used to measure the strength of relationship statistics for each canonical pair. Canonical coefficients (standard coefficients) measure the weight that is attached to 12 feedback comments that takes into account their relationships with the market force variables in the analysis.

RESULTS

The Cronbach's Alpha to test the reliability of the variables in this study that was 0.8 has been confirmed.

Stevens (1986) recommends that there should be at least 20 times as many cases as variables in the analysis to arrive at reliable interpretations of the canonical factor loadings. In this study the sample size was N= 11,959. As such, the standard recommended by Stevens was achieved (Figure 1)

Figure 2 indicates the canonical correlations using SPSS. Figures 3 shows the results of the analysis.

Figure 1. Descriptive Statistic of the study sample

Descriptive Statistics

	N	Minimum	Maximum	Mean	Std. Deviation
Adjective for Advertising	11959	.00	100.00	9.2515	9.40993
Affect for Pricing	11959	.00	100.00	13.1779	12.25620
Positive emotion for Promoting	11959	.00	100.00	10.4429	11.27311
Social for Display	11959	.00	100.00	7.3972	6.61304
family for Personal selling	11959	.00	33.33	.3349	1.19668
Friend for Distributing	11959	.00	50.00	.9206	2.48887
Bio for Branding	11959	.00	100.00	2.7893	4.36735
Body for physical handling	11959	.00	100.00	.2179	1.34661
Health for Fact finding	11959	.00	33.33	.1453	.78854
Drives for Servicing	11959	.00	100.00	13.9268	11.20989
Affiliation for Trade	11959	.00	75.00	3.9687	4.82721
Achievement for Consumer	11959	.00	100.00	1.3305	3.30595
Power for Competitor	11959	.00	50.00	2.7809	3.67370
Reward for Government	11959	.00	100.00	5.0168	8.07341
Risk for Packaging	11959	.00	66.67	1.4175	3.50429
Work for Product planning	11959	.00	100.00	3.0897	5.03333
Valid N (listwise)	11959				

Figure 2. Canonical Correlations of the sample study

Canonical Correlations

	Correlation	Eigenvalue	Wilks Statistic	F	Num D.F	Denom D.F.	Sig.
1	.985	32.127	.007	2551.575	48.000	46007.724	.000
2	.756	1.331	.223	707.782	33.000	35189.925	.000
3	.632	.663	.520	461.537	20.000	23890.000	.000
4	.367	.156	.865	206.408	9.000	11946.000	.000

H0 for Wilks test is that the correlations in the current and following rows are zero

The analysis of the 12 feedback comments compared with the market force results was completed utilizing canonical correlation. Only one of the squared canonical correlations for three pairs of canonical variates was found to be significant (Wilk's Lamda= 0.007, F=2551.5, $p < .01$).

The first pair of variates revealed a very strong relationship between three of the feedback variables (drives-servicing [.945], affect-pricing [.631], and positive emotion-promotion [.599]) and the reward-government force [.752]. Essentially, the first pair of variates suggested that hoteliers who select servicing, pricing, and promoting will seek reward-government related. The first hypothesis that there would be a significant between servicing, pricing, and promoting and reward-government related has been confirmed.

The second pair of variates revealed a strong relationship between three of the feedback variables (work-product planning [.353], friend-distribution [.333], and social-display [.291]) and the affiliation-trade force [.466]. Essentially, the second pair of variates suggested that hoteliers who select product planning, distribution, and display will seek affiliation-trade related. The second hypothesis that there would be a significant between product planning, distribution, and display and affiliation-trade related has been confirmed.

The third pair of variates revealed a normal relationship between three of the feedback variables (adjective-advertising [.271], family-personal selling [.107], and bio-branding [.078]) and the power-competitor force [.423]. Essentially, the third pair of variates suggested that hoteliers who select advertising, personal selling, and branding will seek power-competitor related. The third hypothesis that there would be a significant between advertising, personal selling, and branding and power-competitor related has been confirmed.

The fourth pair of variates revealed a weak relationship between three of the feedback variables (body-fact finding [-.057], health-physical handling [-.015], and risk-packaging [.052]) and the achievement-consumer force [.254]. Essentially, the fourth pair of variates suggested that hoteliers who select physical handling, fact finding, and packaging will seek achievement-consumers related. The fourth hypothesis that there would be a significant between physical handling, fact finding, and packaging and achievement-consumers related has been confirmed.

Figure 3. Set 1 and Set 2 of the Canonical loadings

Set 1 Canonical Loadings

Variable	1	2	3	4
Advertising	.271	-.262	.052	-.283
Pricing	.631	-.382	.141	.093
Promotion	.599	-.436	.207	.272
Display	.291	.862	.191	.214
Personalselling	.107	.325	.151	.212
Distribution	.333	.503	-.177	-.407
Branding	.078	.082	.078	.156
Physicalhandling	-.057	-.028	-.027	-.078
Factfinding	-.015	-.033	.011	-.012
Servicing	.945	-.092	-.049	-.045
Packaging	.052	-.256	.060	-.189
Productplanning	.353	.022	-.857	.283

Set 2 Canonical Loadings

Variable	1	2	3	4
Trade	.466	.870	.135	.092
Consumer	.254	-.131	-.618	.732
Competitor	.423	.141	-.701	-.557
Government	.752	-.564	.334	-.065

DISCUSSION

The LIWC feedback words related to "body", "health", "risk", "work", "friend", "social", "adjectives", "family", "biological" "drives", "affect", and "positive emotion" have been transited to physical handling, fact finding, packaging, product planning, distribution, display, advertising, personal selling, branding, servicing, pricing, and promotion, respectively. Identifying the words in the feedback, managers can apply related marketing strategies to specific consumers to match their expectations. The study has suggested 12 recommendations to the 12 above solutions as follows.

For strengthening the relationship between consumer family and provider through personal selling, managers should understand the family unique needs, adapt personal selling techniques to address concerns and preferences of both parents and children, design physical spaces, such as play areas or kid-friendly corners, within the sales premises to make families feel comfortable and cater to children's needs during the selling process, and anticipate common questions and concerns from families and equip sales personnel with the necessary knowledge to provide accurate information and guidance on family-related services, facilities, and policies. For example, a travel consultant engages with a family by showing genuine interest in their preferences and understanding the ages of their children. They recommend family-friendly destinations with age-appropriate activities and accommodations that cater to the needs of both parents and children, highlighting safety measures and kid-friendly amenities.

For strengthening the relationship between consumer adjectives and provider advertising, managers should conduct market research to understand the adjectives that consumers associate with your product or service. These adjectives can be related to quality, reliability, affordability, luxury, innovation, sustainability, or any other attribute that resonates with your target audience. A luxury hotel creates an advertising campaign that highlights the adjectives "opulent," "sumptuous," and "unforgettable." The visuals showcase elegant interiors, impeccable service, and stunning locations, evoking a sense of luxury and indulgence. The hotel features testimonials from satisfied guests who use adjectives like "pampered," "exquisite," and "magical" to describe their stay. These testimonials are integrated into the advertising materials, further reinforcing the desired perception of the hotel as a luxurious and unforgettable destination.

For enhancing the relationship between consumer working and provider product planning, managers should research and gather insights to understand the specific needs, preferences, and challenges of consumers who work. Use the gathered insights to inform product planning decisions and integrate work-friendly features into the offerings. Recognize the diverse work arrangements and requirements of consumers. Provide flexibility in terms of workspace configurations, booking options, and accessibility to accommodate different working styles and schedules. The hotel for business incorporates well-designed workspaces within guest rooms, including ergonomic desks, comfortable chairs, ample lighting, and easily accessible power outlets. They ensure reliable and high-speed internet connectivity throughout the property, including common areas, meeting rooms, and restaurants. The hotel offers business-oriented amenities such as business centers, conference rooms, and on-site tech support to cater to the working needs of business travelers. They provide concierge services that assist guests with printing, scanning, and other business-related requirements. Additionally, they offer flexible check-in/check-out options to accommodate varying work schedules.

For strengthening the relationship between consumer friends and provider distribution, managers should consider factors such as group size, social activities, shared accommodations, tailor distribution policies to cater to group travelers, offer group booking discounts, flexible room configurations, and dedicated services that enhance the social aspect of their trip. The agency designs adventure travel packages specifically for group travelers, offering activities such as hiking, white-water rafting, and zip-lining that cater to the social and adrenaline-seeking nature of the group. They collaborate with accommodations and local guides to provide comfortable and flexible options suitable for groups of varying sizes. They provide user-friendly booking options, allowing group travelers to easily communicate and coordinate their bookings, select desired activities, and make reservations together. Additionally, they offer group discounts and incentives to make the experience more appealing for friends traveling together.

For enhancing the relationship between consumer social and provider display, managers should research networking opportunities, collaborative environments, and a sense of community. They can organize events, workshops, or themed activities that bring consumers together and provide opportunities for networking, idea-sharing, and collaborative experiences. The hotel creates designated networking spaces within their business facilities, such as a dedicated lounge or a communal work area. They provide interactive displays with information about the professionals staying at the hotel, facilitating connections and encouraging collaboration. The displays may showcase attendees of conferences or events happening in the hotel, fostering networking opportunities. The hotel organizes themed evenings or social events in their common areas, such as cocktail receptions or live music performances, to create an inviting and social atmosphere. They incorporate display screens that showcase user-generated content, such as

For strengthening the relationship between consumer health and provider fact finding, managers should establish feedback channels, such as online surveys, suggestion boxes, or customer service hotlines, to gather insights and feedback from consumers regarding their health-related needs and concerns. Based on consumer feedback and fact-finding exercises, make necessary improvements to health-related services, facilities, and amenities. This may include upgrading fitness centers, providing healthier dining options, implementing wellness programs, or partnering with healthcare providers to offer additional services. The resort implements a feedback system where guests can provide input on their wellness experiences and suggest areas for improvement. They also conduct regular wellness-focused workshops and seminars, inviting guests to share their health-related experiences, challenges, and expectations. The restaurant identifies the need for a more comprehensive menu catering to specific dietary requirements, such as vegan, gluten-free, or low-sodium options. They revise their menu, ensuring a diverse range of healthy choices, and provide clear nutritional information to help customers make informed decisions.

For strengthening the relationship between consumer body and provider physical handling, managers should seek feedback from consumers regarding their physical well-being during their interactions with your products or services. This can be done through surveys, feedback forms, or customer reviews to understand their specific needs and preferences. Managers evaluate and assess your physical handling policies, such as warehousing, transportation, and inventory management, to ensure they align with consumer expectations for their well-being. The healthcare provider assesses their physical handling policies in terms of patient transfers, ensuring that they align with ergonomic guidelines and prioritize patient comfort and safety. They provide training to their staff, including nurses and caregivers, on proper lifting techniques and the use of assistive devices to minimize the risk of injuries or discomfort during patient transfers. The provider improves their inventory management processes to ensure a smooth and efficient flow of medical supplies. They prioritize proper storage conditions for sensitive items, such as medications or vaccines, and implement protocols for safe handling and distribution to minimize the risk of damage or contamination.

For addressing the relationship between consumer risk and provider packaging, managers should research and gather insights to understand how consumers perceive risk, especially related to packaging safety and security. Develop packaging solutions that address consumer concerns and provide reassurance. The food delivery service invests in packaging solutions that maintain food temperature and integrity during transportation. They use insulated containers or thermal bags to ensure hot foods remain hot and cold foods remain cold, minimizing the risk of foodborne illnesses due to improper temperature control. The service introduces tamper-evident seals on their food packaging to provide visual indicators of safety. This reassures consumers that their food has not been tampered with during delivery and builds trust in the service's commitment to quality and hygiene.

For enhancing the relationship between consumer drives and provider servicing, managers should identify the key factors that motivate and energize them, such as adventure, relaxation, exploration, personal growth, or social connections. Develop servicing policies that align with the specific drives of your target consumers. Customize the service offerings, amenities, and experiences to cater to their unique motivations and preferences. The adventure company offers personalized trip planning services, taking into account each traveler's specific adventure goals and preferences. They provide customized itineraries that align with different drives, such as adrenaline junkies seeking extreme activities or nature enthusiasts desiring eco-friendly experiences. The company's guides are trained to understand the diverse motivations and drives of adventure travelers. They provide expert advice, share knowledge about local cultures and natural wonders, and offer tailored guidance to cater to the unique interests and drives of each traveler.

For capitalizing on the relationship between positive consumers and providers promotion, managers should identify the key factors that contribute to their positive experiences, such as satisfaction, delight, happiness, or excitement. Managers should craft promotional strategies that capitalize on and further enhance the positive emotions of consumers. Use engaging storytelling, appealing visuals, and uplifting messages that evoke positive emotions and resonate with their positive experiences. Show appreciation to positive consumers by providing exclusive promotions, rewards, or loyalty programs. This not only reinforces their positive experience but also encourages them to continue engaging with your brand and sharing their positive experiences with others. The hospitality chain capitalizes on the relationship with positive consumers through promotion. The chain features testimonials and reviews from positive guests in their promotional materials, showcasing their memorable experiences and the positive emotions they felt during their stays. These testimonials are used in online advertisements, brochures, and website content to create an emotional connection with potential customers. The chain creates an online community where positive consumers can share their travel stories, tips, and recommendations. They engage with the community through social media platforms, organizing contests or challenges that encourage positive consumers to share their experiences. This fosters a sense of belonging and creates a platform for positive consumers to interact and inspire others.

For addressing the relationship between consumers' affect and providers' pricing, managers should understand how pricing can evoke positive or negative emotions, such as joy, satisfaction, disappointment, or frustration. Clearly communicate the value proposition and benefits associated with the pricing of your products or services. Highlight features, quality, uniqueness, or additional perks that justify the price and evoke positive emotions in consumers. Managers ensure transparency in pricing by clearly presenting information about costs, fees, and any additional charges. Avoid hidden costs or unexpected fees that may lead to negative affect in consumers. The luxury spa offers special packages that combine multiple services at a discounted price, such as a massage, facial, and body treatment bundle. This creates a sense of value and excitement for consumers, evoking positive affect and making them feel like they are getting a luxurious experience at a reasonable cost. The spa prominently displays the detailed benefits and unique features of each treatment along with their prices. This transparency helps customers understand the value they receive for the price paid, creating a positive affect by highlighting the quality and exclusivity of the spa services.

Methodological Limitations

LIWC (2005) is a content analysis that saves meticulous labor. The result of the method called "an X-Ray picture of the inner self" (Murray, 1943, p. 1) is more spontaneous and less self-critical than in self-report. Since this study utilized a random sample of hotel guests who may not represent the entire travelers in the United States.

In addition, the study has assumed that external factors were constant. When changing hotel guests' finances, education and gender could affect hotel experiences but were beyond the scope of this study. Time is the last constraint that might influence motives and travel preferences since "behavior in the same situation may vary during a period of time" (Smith, Feld, & Franz, 1992, p. 15).

Future Research

Although the results cannot be generalized to a specific hotel destination or the entire traveling population, these findings contribute to a greater understanding of the relationship between hotel guests' comments and hoteliers' policies. They also demonstrate that further research is needed in this area. It is evident that the findings from this research should be utilized when replicating the study using a random sample of hotel guests to a particular hotel. An analysis of the guests' comments and the market forces would be able to predict the hotel service choice of the general population.

Marketing Implications

To apply the results of this study to a current situation of hotels, it is appropriate to examine the position of the UNWTO promotes tourism as a driver of economic growth, inclusive development and environmental sustainability and offers leadership and support to the sector in advancing knowledge and tourism policies worldwide (UNWTO, 2023). Once a hotel has identified its positioning in the tourism network, it could research the target market through guests' comments to better understand their unconscious needs and subsequently policies appropriate to attract different guests to their hotel. The analysis of unconscious needs does provide a potentially effective means of the selections of hotel services.

Theory and research regarding affiliation, achievement and affiliation motives were found to be related to hotel rooms, value, location, cleanliness, service, and sleep quality. These findings will provide helpful information for marketers and managers in developing promotional materials for hospitality and tourism industry that will attempt to attract tourists from two important market segments: achievers, CEOs and presidents, and celebrity people. In addition, the application of the McClelland's implicit motives in text comments for the target markets would effectively arouse the needs for achievement, power, and affiliation when used in promotional materials. Based on the needs for achievement, power, and affiliation, marketers and managers could provide necessary information to satisfy the demands of hotel guests.

As a result, the study has discovered consumers' feedback that can provide providers' strategies instantly. The 4 market forces gathering from consumers' feedback would create the 4 types of customers including socializer, thinker, director, and relator (Abbey, 2005) as follows: (1) The director customers focusing on direct tasks and socially reserved might prefer value and cleanliness in a hotel. (2) The socializer customers focusing on indirect task and socially open might prefer hotel location. (3) The relater customers focusing on indirect tasks and socially open might prefer rooms and hotel service. Finally, the thinker customers focusing on indirect tasks and socially reserved might prefer hotel sleep quality.

CONCLUSION

To date, no model of marketing mix has been proved to be appropriate. This study has formulated the Trip Advisor categories to measure the model of the marketing process from manufacturing to consuming in hospitality industry. Using these findings, hoteliers can find the solutions for any issues in marketing from the beginning of product planning to the ending of the selling process of displaying due to the being to doing perspective. No matter what the limitation of the data prior COVID-19, the study has contributed to the relationship between marketing mix and consumers' roles in academia and provided hoteliers with essential marketing strategies to meet consumers' satisfaction.

REFERENCES

Abbey, J. (2005). *Hospitality Sales and Marketing*. American Hotel & Lodging Educational Institute Lansing.

Adams, J. (2012). Letters to John Law. Newton Page. pp. xiv, xxi, liii. ISBN 978-1934619087.

Akrich, M., Latour, B., & Callon, M. (2006), Sociologie de la traduction: textes fondateurs, Paris, Mines Paris, les Presses, "Sciences sociales,". ISBN 2-911762-75-4 doi:10.4000/books.pressesmines.1181

Azeez, M. S. A., & Abeyratna, W. A. M. I. (2020). The Impact of Personal Selling on Buying Behavior with Special Reference to Clothing Stores in Sri Lanka. *IUP Journal of Marketing Management*, *19*(4), 35–54.

Borden, N. (1964). The concept of the marketing mix. In Schwartz, G. (1964), Science in Marketing, New York: John Wiley. Available online at file:///C:/Users/xtran/Desktop/Borden-1984_The-concept-of-marketing-mix.pdf

Buchmuller, K., Bearth, A., & Siegrist, M. (2022). The influence of packaging on consumers' risk perception of chemical household products. *Applied Ergonomics*, *100*, 103676–103681. doi:10.1016/j.apergo.2021.103676 PMID:35021135

Buchmuller, K., Bearth, A., Zurich, E., Siegrist, M., & Zuich, E. (2022). The influence of packaging on consumers' risk perception of chemical household products. *Applied Ergonomics*, *100*(5), 103676. doi:10.1016/j.apergo.2021.103676 PMID:35021135

Callon, M. (Ed.). (1998). *The Laws of the Markets*. Blackwell Publishers.

Castelo, N., Boegershausen, J., Hildebrand, C., & Henkel, A. (2023). Understanding and Improving Consumer Reactions to Service Bots. *The Journal of Consumer Research*, ucad023. Advance online publication. doi:10.1093/jcr/ucad023

Cheung, M., Pires, G., Rosenberger, P., Leung, W., & Sharipudin, M. (2021). The role of consumer-consumer interaction and consumer-brand interaction in driving consumer-brand engagement and behavioral intentions. *Journal of Retailing and Consumer Services*, *61*, 102574. doi:10.1016/j.jretconser.2021.102574

Cohen, E., & Cohen, S. A. (2012). Current sociological theories and issues in tourism. *Annals of Tourism Research*, *39*(4), 2177–2202. doi:10.1016/j.annals.2012.07.009

Dagan, O., Facompre, C., & Bernard, K. (2018). Adult attachment representations and depressive symtoms: A meta-analysis. *Journal of Affective Disorders*, *236*, 274–290. doi:10.1016/j.jad.2018.04.091 PMID:29751243

Dixon, M., Freeman, K., & Toman, N. (2010). Stop trying to delight your customers. *Harvard Business Review*, (July-August). Advance online publication. doi:10.4236/ojml.2018.86022

Ebaid, N. (2018, December). Adjectives as Persuasive Tools. *Open Journal of Modern Linguistics*, *8*(6), 262–293. doi:10.4236/ojml.2018.86022

Foster, B., & Cadogan, J. (2000). Relationship selling and customer loyalty: An empirical investigation. *Marketing Intelligence & Planning*, *18*(4), 185–199. doi:10.1108/02634500010333316

Frese, M., & Keith, N. (2015). Action errors, error management and learning in organizations. Annual Review of Psychology, 66, 21.1-21.27.

Frese, M., Krauss, S. I., Keith, N., Escher, S., Grabarkiewicz, R., Luneng, S. T., & Friedrich, C. (2007). Business owners' action planning and its relationship to business success in three African countries. *The Journal of Applied Psychology*, *92*(6), 1481–1498. doi:10.1037/0021-9010.92.6.1481 PMID:18020791

Frese, M., & Zapf, D. (1994). Action as the core of work psychology: A German approach. In H. C. Triandis, M. D. Dunnette, & L. Hough (Eds.), *Handbook of industrial and organizational psychology* (Vol. 4, pp. 271–340). Consulting Psychologists Press.

Gourville, J., & Soman, D. (2002). Pricing and the psychology of consumption. *Harvard Business Review*, (September), 3–9. PMID:12227149

Hammen, C. (2009). Adolescent depression. *Current Directions in Psychological Science*, *18*(4), 200–204. doi:10.1111/j.1467-8721.2009.01636.x PMID:20161119

Harlow, H. F., Dodsworth, R. O., & Harlow, M. K. (1965). Total social isolation in monkeys. *Proceedings of the National Academy of Sciences of the United States of America*. Retrieved from https://www.ncbi.nlm.nih.gov/pmc/articles/PMC285801/pdf/pnas00159-0105.pdf

Hennessy, M., Schiml-Webb, P., & Deak, T. (2009). Separation, sickness, and depression: A new perspective on an old animal model. *Current Directions in Psychological Science*, *18*(7), 227–231. doi:10.1111/j.1467-8721.2009.01641.x PMID:20221300

Iheanachor, N., Umukoro, I., & West, O. (2021). The role of product development practices on new product performance: Evidence from Nigeria's financial services providers. *Technological Forecasting and Social Change*, *164*, 120470. doi:10.1016/j.techfore.2020.120470 PMID:33664533

Jamil, K., Dunnan, L., Gul, R., Shehzad, M., Gilani, S., & Awan, F. (2022). Role of Social Media Marketing Activities in Influencing Customer Intentions: A Perspective of a New Emerging Era. *Frontiers in Psychology*, *12*, 808525. Advance online publication. doi:10.3389/fpsyg.2021.808525 PMID:35111111

Janiszewska, K. (2013). The strategic importance of consumer insight in defining place brand identity and positioning. *Journal of International Students*, *6*, 2, 9–19.

Joormann, J., Siemer, M., & Gotlib, I. H. (2007). Mood regulation in depression: Differential effects of distraction and recall of happy memories on sad mood. *Journal of Abnormal Psychology*, *116*(3), 484–490. doi:10.1037/0021-843X.116.3.484 PMID:17696704

Keller, K. (2019). Consumer research insights on brands and branding: A JCR curation. *The Journal of Consumer Research*, *46*(5), 995–1001. doi:10.1093/jcr/ucz058

Latour, B. (2005). *Reassembling the Social: An Introduction to Actor-Network-Theory*. Oxford UP.

LIWC. (2022). Linguistic Inquiry and Word Count, Available online at https://www.liwc.app/

Lu, P., Oh, J., Leahy, K., & Chopik, W. (2020). Friendship Importance Around the World: Links to Cultural Factors, Health, and Well-Being. *Frontiers in Psychology*, *11*, 570839. Advance online publication. doi:10.3389/fpsyg.2020.570839 PMID:33536962

Martin, T., Clark, T., Ferrell, O., Stewart, D., & Pitt, L. (2020). Marketing's theoretical and conceptual value proposition: Opportunities to address marketing's influence. *AMS Review*, *10*(3-4), 151–167. doi:10.100713162-020-00176-7

McClelland, D. C. (1985). How motives, skills, and values determine what people do. *The American Psychologist*, *40*(7), 812–825. doi:10.1037/0003-066X.40.7.812

Mehl, M. R., Gosling, S. D., & Pennebaker, J. W. (2006). Personality in its natural habitat: Manifestations and implicit folk theories of personality in daily life. *Journal of Personality and Social Psychology*, *90*(5), 862–877. doi:10.1037/0022-3514.90.5.862 PMID:16737378

Mertens, A., Theisen, M., & Funke, J. (2022). Measuring Achievement, Affiliation, and Power Motives in Mobility Situations: Development of the Multi-Motive Grid Mobility. *Frontiers in Psychology*, *12*, 765627. doi:10.3389/fpsyg.2021.765627 PMID:35046868

Murray, H. A. (1943). *Thematic apperception test manual*. Harvard University Press.

Nolan, S. A., Flynn, C., & Garber, J. (2003). Prospective relations between rejection and depression in young adolescents. *Journal of Personality and Social Psychology*, *85*(4), 745–755. doi:10.1037/0022-3514.85.4.745 PMID:14561127

Ozuem, W., Ranfagni, S., Willis, M., Rovai, S., & Howell, K. (2021). Exploring customers' responses to online service failure and recovery strategies during Covid-19 pandemic: An actor–network theory perspective. *Psychology and Marketing*, *38*(9), 1440–1459. doi:10.1002/mar.21527 PMID:34539054

Pennebaker, J. W., Boyd, R. L., Jordan, K., & Blackburn, K. (2015). *The development and psychometric properties of LIWC2015*. University of Texas at Austin.

Pennebaker, J. W., & King, L. A. (1999). Linguistic styles: Language use as an individual difference. *Journal of Personality and Social Psychology*, *77*(6), 1296–1312. doi:10.1037/0022-3514.77.6.1296 PMID:10626371

Rude, S., Gortner, E., & Pennebaker, J. (2004). Language use of depressed and depression-vulnerable college students. *Cognition and Emotion*, *18*(8), 1121–1133. doi:10.1080/02699930441000030

Russell, J. A., & Mehrabian, A. (1974). Distinguishing anger and anxiety in terms of emotional response factors. *Journal of Consulting and Clinical Psychology*, *42*(1), 79–83. doi:10.1037/h0035915 PMID:4814102

Salter, M. (2019). Security Actor-Network Theory: Revitalizing Securitization Theory with Bruno Latour. *Polity*, 51, 2. The University of Chicago Press: Journals https://www.journals.uchicago.edu/doi/10.1086/701885

Silton, R. L., Kahrilas, I. J., Skymba, H. V., Smith, J., Bryant, F. B., & Heller, W. (2020). Regulating positive emotions: Implications for promoting well-being in individuals with depression. *Emotion (Washington, D.C.)*, *20*(1), 93–97. doi:10.1037/emo0000675 PMID:31961185

Smith, C. P., Feld, S. C., & Franz, C. E. (1992). Methodological considerations: Steps in research employing content analysis systems. In C. P. Smith (Ed.), *Motivation and personality: Handbook of thematic content analysis* (pp. 515–536). Cambridge University Press. doi:10.1017/CBO9780511527937.038

Stevens, J. (1986). *Applied Multivariate Statistics for the Social Sciences*. L. Erlbaum Associates.

Stirman, S. & Pennebaker, J. (2001). Word use in the poetry of suicidal and nonsuicidal poets. *Psychosomatic Medicine*, Jul-Aug;63(4):517-22. . doi:10.1097/00006842-200107000-00001

Sun, Y., Zhang, Y., Gwizdka, J., & Trace, C. (2019). Consumer Evaluation of the Quality of Online Health Information: Systematic Literature Review of Relevant Criteria and Indicators. *Journal of Medical Internet Research*, *21*(5), e12522. Advance online publication. doi:10.2196/12522 PMID:31045507

Szocs, C., Williamson, S., & Mills, A. (2022). Contained: Why it's better to display some products without a package. *Journal of the Academy of Marketing Science*, *50*(1), 131–146. doi:10.100711747-021-00800-3 PMID:34465928

Thompke, S., & Reinertsen, D. (2012). Six myths of product development. *Harvard Business Review*, (May), 4–10.

Tian, L., & Jiang, B. (2017). Effects of consumer-to-consumer product sharing on distribution channel. *Production and Operations Management*, *27*(3), 350–367. Advance online publication. doi:10.1111/poms.12794

Tran, X., Tran, H., & Tran, T. (2018). Information Communications Technology (ICT) and Tourism Experience: Can Serotonin become a measurement for tourism experience? *Ereview of Tourism Research*, *9*, 20–24.

UNWTO. (2023). Available online at https://www.unwto.org/who-we-are#:~:text=building%20resilience%20and%20promoting%20safe,communities%20both%20economically%20and%20socially

Wakefield, J. C., Schmitz, M. F., First, M. B., & Horwitz, A. V. (2007). Extending the bereavement exclusion for major depression to other losses: Evidence from the National Comorbidity Survey. *Archives of General Psychiatry*, *64*(7), 433–440. doi:10.1001/archpsyc.64.4.433 PMID:17404120

Weissman, M. M., Markowitz, J. C., & Klerman, L. (2000). *Comprehensive guide to interpersonal psychotherapy*. Basic Books.

Winter, D. G., & Barenbaum, N. B. (1985). Responsibility and the power motive in women and men. *Journal of Personality*, *53*(2), 335–355. doi:10.1111/j.1467-6494.1985.tb00369.x

Wolf, M., Sedway, J., Bulik, C. M., & Kordy, H. (2007). Linguistic analyses of natural written language: Unobtrusive assessment of cognitive style in eating disorders. *International Journal of Eating Disorders*, *40*(8), 711–717. doi:10.1002/eat.20445 PMID:17683092

Zacher, H. (2017). Action Regulation Theory. DOI:. In book: Oxford Research Encyclopedia of PsychologyPublisher: Oxford University PressEditors: Oliver Braddick doi:10.1093/acrefore/9780190236557.013.1

Zacher, H., & Frese, M. (2018). Action regulation theory: Foundations, current knowledge, and future directions. In D. S. Ones, N. Anderson, C. Viswesvaran, & H. K. Sinangil (Eds.), *The SAGE handbook of industrial, work, and organizational psychology* (Vol. 2, pp. 80–102). Sage. doi:10.4135/9781473914957.n7

Zhao, H., Yao, X., Liu, Z., & Yang, Q. (2021). Impact of Pricing and Product Information on Consumer Buying Behavior With Customer Satisfaction in a Mediating Role. *Organizational Psychology*, 12.

Chapter 9
New Trends in Organizational Learning and Its Impact on Business

Ayşe Meriç Yazıcı
Istanbul Gelisim University, Turkey

ABSTRACT

This study examines emerging trends in organizational learning and their profound impact on modern businesses. In an era marked by rapid technological developments, globalization, and changing customer expectations, businesses must constantly adapt to remain competitive. Organizational learning as a strategic tool plays a crucial role in fostering employee development, supporting innovation, and improving overall business performance. The paper begins by providing an overview of traditional approaches to learning in organizations and highlights the limitations that hinder their effectiveness in the dynamic business world. In conclusion, this study shows how adopting the latest trends in organizational learning can increase competitive advantage and foster a culture of innovation in businesses. By empowering employees with personalized, collaborative, and data-driven learning experiences, businesses can adapt more effectively to change and unlock their full potential in an ever-evolving business environment.

1. INTRODUCTION

In today's rapidly evolving business landscape, organizations face a perpetual need to adapt, innovate, and remain competitive. Amidst these dynamic changes, the importance of continuous learning and development has taken center stage.

DOI: 10.4018/978-1-6684-9089-1.ch009

Organizational learning, once considered a peripheral aspect of business operations, has emerged as a strategic imperative to navigate the complexities of the modern marketplace successfully.

As businesses strive to keep pace with technological advancements, changing consumer preferences, and global economic shifts, they must foster a learning culture that encourages employees to embrace knowledge acquisition as an integral part of their roles (Do, Budhwar, Shipton, Nguyen & Nguyen, 2022). In recent years, new trends in organizational learning have emerged, revolutionizing traditional approaches and revolutionizing the way companies impart knowledge, skills, and expertise to their workforce (Blaique, Ismail, & Aldabbas, 2023).

Organizational learning is a key concept in the field of management and organizational studies. Organizations need to constantly acquire new knowledge, skills, and abilities that enable them to adapt to changing environmental conditions so as to improve their performance (Rupcic, 2023). Typically, the process in organizational learning is dynamic based on knowledge sharing via knowledge acquisition, transfer and integration at different levels across individuals, groups, and organizations (Pham & Hoang, 2019). Such process involves the collective effort of all members in the organizational hierachy, from top managers to frontline employees, as it also takes place at different levels and forms in individual, group, or organizational learning (Basten & Haamann, 2018).

It is believed that operating in a dynamic, unpredictable, complex, and discontinuous environment leads to seeking the sources of competitive advantage not in specific products, Technologies, or resources, but in the ability to develop newer and more useful knowledge (Beer, Voelpel, Leibold & Tekie, 2005; Wu, 2010). As claimed by Santos-Vijande, López-Sánchez & Trespalacios (2012), the survival and growth of a business normally requires faster and more effective learning than it competitors.

To be competitive, firms need to learn to run their business effectively. Organizations should make maximum efforts to exercise their corporate knowledge internally and externally while empowering their employees to be more efficient and innovative.

In modern business, it is imperative that firms continuously adapt to changes and constantly learn to gain competitive advantage and sustainability. Learning should be an ongoing process during the workflow. Therefore, employees of the firm need to constantly learn new knowledge and use new technologies for work improvement or to overcome challenges (Hanson, Niqab & Bangert, 2021). In such mindset, employees must be equipped with skillset which empowers new learning and training, including artificial intelligence, online learning, blended learning, microlearning, continuous learning, cloud learning, gamification, adaptive learning, technological learning, m-learning, modular learning, and smart working.

The ability of employee learning in these mechanisms is deemed to siginificantly determine and influence the success of a firm which in turn supports and invests in employee's human capital for mutual growth (Do & Mai, 2022).

This paper examines the latest trends shaping organizational learning. Focusing on adaptability, agility, and employee empowerment, these trends are not just temporary innovations, but fundamental shifts in the way businesses harness the potential of their most valuable assets: their people.

2. ORGANIZATIONAL LEARNING

Organizational learning is an ability to create new knowledge and new perspective by transforming one's ability to create, acquire and transfer knowledge to behavior (Garvin, 1993). It is the acquisition of potentially usable knowledge by an organization (Huber, 1991). It differs from individual learning in that it has a common component involving group-interaction and social learning while it is a process leading to a commonly shared conceptual model, as it also involves and considers the needs, motives, and values of various members in the organization (Schein, 1993; Kim, 1993). Accordingly, organizational learning is the process of developing behavior through knowledge assimilation (Fiol & Lyles, 1985). It is the process of changes in individual and organizational thoughts and actions, and it endorses enhancement in organizational capacity for a fruitful business future (Senge, 1990; Vera & Crossan, 2004).

Conventionally, a learning organization endows with characteristics such as leaders who (i) strategize measuable risk-taking and experimentation in business, (ii) decentralize decision-making and promote employee empowerment, (iii) reserve skill inventories for talent sharing, using, and learning, (iv) reward and structure positive employee initiatives, (v) aim at long-term business outcome and ease the impact on the work of others, (vi) frequently utilize cross-functional work teams, (vii) create learning opportunities routinely, and (viii) cultivate a corporate culture for feedback and transparency (Watkins & Marsick, 1993).

Knowledge creation is necessary for learning (Bhatt & Zaver, 2002). Hence, organizational learning is essential in the ever-changing, dynamic, and complex business environment (Prahalad & Hamel, 1990). Learning organizations supporting effective learning should be made creative, flexible, contemporary, inclusive and open to communication and information-exchange while maintaining low or minimal control. In such system, each individual would be valued independently for his/her status and actions are to be taken to prevent and solve, rather than inviting, problems. Corporate missions trargeting in increased adaptation to the changing environment should integrate and align with employees' collective vision (İhtiyaroğlu, Karakuş

Yıldız & Yılmaz, 2022). The study of organizational learning is critical which promotes effective corporate management in the time of challenges and embolden businesses for long-term survival. Organizational learning is therefore a source of heterogeneity and a potential source of sustainable, competitive advantages (Lei, Slocum & Pitts, 1999; Easterby-Smith, Crossan & Nicolini, 2000).

Organizational learning can occur at various levels within an organization, ranging from individual learning to team learning and, ultimately, organizational-wide learning. It is not only about reacting to challenges or changes but also proactively seeking new opportunities for improvement and growth (Xu & Pero, 2023). A learning organization is one that actively fosters a culture of learning and encourages its members to engage in continuous learning and knowledge sharing. Such organizations tend to be more adaptive, innovative, and resilient in the face of challenges and uncertainties (Shaheen, Hayyat, Aslam & Shahbaz, 2023).

3. FUTURE LEARNING FUNCTION

In modern business, developing technologies or technological advancement may well change one's teaching and learning environment. In the process of technological development, access to information and its usage are often varied, as they affect the way we think, interact and work with others, and self-develop. Such technology-embedded approach reveals a new learning ecosystem and is deemed the key to future learning. It reflects the transformation from disconnected and episodic experiences to a continuum of lifelong learning tailored to individuals and delivered across different locations, media, and timeframes. Improved metrics and analytics help optimize such transformation while encouraging continuous adaptation across the system (Haleem, Javaid, Qadri & Suman, 2022). As compared with the traditional tech-absent approach, the technology-based learning in business is groundbreaking not only allowing ubiquitous access to wide spectrum of knowledge but also providing ways to stimulate individual and workforce development at an unprecedented pace.

4. FUTURE LEARNING TRENDS IN AN ORGANIZATION

With the recent shift to hybrid working or home officing due to the pandemic, coupled with an increasingly tight labor market and the relentless pace of digitalization, almost every firm faces the need to revisit its strategic business goals which affect employee learning and performance. The focus of online learning among employees becomes a new trend when work can be carried and performed remotely while many employees need to re-learn or newly adopt to new ways of tasking in different learning and working environments, including artificial intelligence, online learning, blended learning, microlearning, continuous learning, cloud learning, gamification, adaptive learning, technological learning, m-learning, modular learning, and smart working.

4.1. Artificial Intelligence

Artificial intelligence refers to the simulation of human intelligence in machines that are programmed to think, learn, and perform tasks that typically require human intelligence (Wamba-Taguimdje, Wamba, Kamdjoug & Wanko, 2020). These machines use algorithms, data, and computational power to analyze and solve complex problems, recognize patterns, and make decisions (Haenlein & Kaplan, 2019). While artificial intelligence and the relevant technologies have significantly affected the way businesses and firms operate, they have transformed the way businesses work and interact with their environment (Lee, Davari, Singh & Pandhare, 2018). Artificial intelligence brings a new approach to knowledge management. It is proposed to use artificial intelligence throughout the value chain of an organization, allowing the integration of all elements such as R&D, marketing, sales, operations, production planning and services (Kuzey, Uyar & Delen, 2014). Artificial intelligence can analyze vast amounts of data quickly, identifying patterns and trends that humans might miss. Organizations can leverage artificial intelligence to make data-driven decisions, enabling them to respond faster to market changes and improve their decision-making processes (Perifanis & Kitsios, 2023).

While there are certain concerns with the increase in the application of artificial intelligence, such as job loss and labor replacement, process transparency, and decreased customer loyalty, among others, we need to be more adaptable and keen to the upcoming change following such tech-induced moverment. With the rapid proliferation of artificial intelligence tools, organizations in emerging markets have started to adopt artificial intelligence functionalities to improve their performance and replicate their business to find and adjust to different directions (Bughin, Hazan, Ramaswamy, Chui, Allas, Dahlström, Henke & Trench, 2017).

Artificial intelligence can provide personalized learning experiences for employees based on their strengths, weaknesses, and learning styles. This approach can enhance the effectiveness of training programs and increase employee engagement and retention (Tapalova & Zhiyen, 2022). By automating mundane and repetitive tasks, artificial intelligence frees up human resources for more strategic and creative endeavors, allowing employees to focus on higher-value activities that contribute to organizational learning and growth (Ahmed, Krishna, Ganeshkumar, Anthiyur & Manivel, 2023).

4.2. Online Learning

Today, technology is the driving force of workplace learning (Bervell & Umar, 2018). Online learning consists of a wide range of application processes such as web-based learning, computer-based learning, and digital collaboration. Its primary methods include visual graphics, texts, animations, videos, and audio (Basar, Mansor, Jamaludin & Alias, 2021). With its beneficial features brought about cost-effectiveness, delivery efficiency, self-directed learning, on-demand training and unrestrictive time and space flexibility and availability, online learning is ncreasingly recognized as an important supporting structure for both formal and informal learning (Rosenberg, 2006). In business, it plays important roles in an organization as a tool to support just-in-time information access, knowledge management, peer collaboration, staff training, and online performance support (Brandenburg & Ellinger, 2003; Ong & Tan, 2022). Nonetheless, it is reported that in practice an adverse barrier to the success of online learning is the lack of adequate evaluation of the employee engagement in online learning at workplace (Admiraal & Lockhorst, 2009).

Online learning enables firms to provide consistent training to all employees, update training content when necessary, reduce the cost of traveling to external training facilities, and provide on-demand, anytime, and anywhere training to target employees (Burgess & Russell, 2003). Since it is rapidly changing due to tech-innovation, it is necessary for businesses to keep up with its latest developments and to upgrade and examine its applications regularly.

The importance of workplace learning as an effective way for employees to acquire knowledge and develop skills for individual professional competence to collectively improve organizational effectiveness has been widely recognized (Doornbos, Simons & Denessen, 2008). Workplace learning can be summarized as the tools, processes, and activities through which employees learn from basic skills to advanced technology and management practices that are immediately applicable to their jobs, tasks, and roles. Workplace learning includes both formal and informal learning. Formal learning refers to formally planned, structured, and institutionally supported activities, while informal learning is predominantly defined as unstructured, experiential and, non-institutionalized processes (Ellinger, 2005; Wannewitz, Petzold & Garschagen, 2023). The nature of workplace learning, such as its integration with work and daily office routines, has led to an understanding of the importance of contextual and situational factors in online learning practice.

Online learning is a form of modern organizational learning. Its potential as a tool for workplace learning can be utilized in an appropriate work and learning culture and environment (Tynjälä & Häkkinen, 2005). Work environment factors are also important in the design of online learning. For example, a performance-based learning approach should be adopted to align employees' learning with organizational and social contexts (Collis & Margaryan, 2004).

4.3. Blended Learning

Blended learning is an educational approach that combines traditional face-to-face teaching with online learning methods. It integrates both face-to-face teaching and technology-mediated learning experiences to create a more flexible and effective learning environment (Tong, Uyen & Ngan, 2022). Blended learning is a strategic integration of online and face-to-face education. Typically, it includes direct and indirect instruction, collaborative instruction, and individualized computer-assisted learning (Motteram & Sharma, 2009; Lalima & Dangwal, 2017).

The characterization of blended learning is as important as characterizing a domain, giving it importance, specifying it, and increasing its application in that branch of learning. In blended learning, participants create an atmosphere that can produce a strong sense of connectedness in both face-to-face and online learning (Comey, 2009). In an organization, it is believed to help employees to feel more comfortable interacting with each other, while the online interactions are practical as blended activities where employees can apply their experience. With the increasing evolution of most careers becoming practical and technology-orientated over the online environment, blended learning is deemed to enhance one's career readiness. The movement in blended learning is to create a new paradigm in education across all aspects of subjects (Duffey & Fox, 2012).

The future of blended learning and its implications relevant to the design and delivery of business training should be enhanced to allow business practitioners to respond to rapid changes in their fields so as to improve organizational performance. Proper use of blended learning increases job satisfaction, job performance, and productivity of employees (Kim, Bonk & Oh, 2008). It is expected to be increasingly used in the future across industries until it potentially becomes a standard as a common workplace setting; therefore, more business analysts and trainers have recommended blended learning as part of the human-resource training and development in their organizations (Mohanty, 1988).

4.4. Microlearning

Microlearning is a term used in the field of learning and education, and it usually makes sense in the context of educational technologies and learning methods. The prefix micro means small and scalable, and microlearning is an approach to learning in which learning is achieved through small and customized learning units (Fitria, 2022).

By definition, microlearning is an application system that enables the development of specific teaching-related behaviors and teaching practices under controlled conditions (Allen & Eve, 1968). It is a learning style emphasizing brevity, and used when the learning time is short and measurable, the content is small and simple, the curriculum is in modules and chapters or other small units, the process is simultaneous and iterative, and the medium is cyber-based or traditional. It is a form of learning where both the source and the units of knowledge are becoming increasingly important in areas of rapid development or with a high degree of change (such as technology). In the contemporary business world, it is clear that workplace learning seems to be undergoing a transformation toward microlearning where technological advancement allows learning to take place anywhere and anytime on demand (Bell, Tannenbaum, Ford, Noe & Kraiger, 2017).

Microlearning is a short-term learning method. It is a holistic approach based on e-learning, as its training is normally associated with small learning units. It involves a short-term strategy specifically designed for topic-understanding, learning, and skills-based training. It also enables real-time discovery of relevant data, photos and videos in mobile environments and can shorten the time required to find specific themes and contents (Park & Kim, 2018).

Microlearning emphasizes learning in a short timeframe such as that in online courses, training, seminars, and workshops. Under the format of microlearning, the content produced and presented is short and condensed, practical, while accessible anytime and anywhere. It is said that microlearning may most likely make one's learning result more effective in four reasons: (i) small learning content, (ii) topic- or theme-specific, (iii) fast process, (iv) learning content relatively and more easily understood and memorable for a long time (Fitria, Afdaleni, Sibolon & Suamba, 2022).

4.5. Continuous Learning

Continuous learning, also known as lifelong learning, is essentially the concept of continuously seeking and acquiring knowledge and skills throughout one's life for personal development or professional growth. It involves a proactive approach to learning where individuals take the initiative to extend their knowledge beyond formal education (Buranova & Latipova, 2023).

Continuous learning has some characteristics, listed as follows (Kluge & Schilling, 2003):

- Continuous learning is a learning and development process in an organizational context. It has no specific beginning or end.

- Continuous learning can be formal and informal. It can be all day, every day, and all the time that a person consciously thinks, contemplates, and learns.
- Continuous learning takes place from the individual to the group to the organizational level and vice versa.
- Continuous learning is generally difficult to measure because it is an ongoing process and can be about anything.
- In an organization, an example of continuous learning is when an employee learns about social networking sites, takes the initiative to discuss them further, and shares information.

Continuous learning is the process of continuously acquiring new knowledge, skills, and competencies throughout one's life or career. It goes beyond formal education and involves keeping abreast of the latest developments, adapting to changes, and seeking opportunities to grow personally and professionally. In the context of organizations, continuous learning plays a crucial role in fostering a culture of innovation, adaptability, and improvement (Lin, Tsai & Zhong, 2022).

Continuous learning helps the organization adapt its strategies and processes accordingly by keeping employees informed about new trends, technologies, and best practices. When employees participate in continuous learning, they can acquire new skills and improve their existing skills. Continuous learning opportunities demonstrate that an organization is investing in the development of its employees, leading to higher levels of engagement and job satisfaction (Akbari, Danesh, Rezvani, Javadi, Banihashem & Norozzi, 2023). Engaged employees are more likely to be productive and committed to the success of the organization. To foster a culture of continuous learning, organizations can implement a variety of strategies such as providing training programs, workshops, seminars, mentoring opportunities, and e-learning platforms, and encouraging employees to set personal development goals. It is crucial to create an environment where learning is celebrated and mistakes are seen as opportunities for growth rather than failures (Alarabiat, Hujran, Soares & Tarhini, 2023). Overall, continuous learning is a powerful tool that not only benefits individual employees but also contributes to the long-term success and sustainability of organizations.

4.6. Cloud Learning

All activities supported by modern technology are becoming easier, faster, and cheaper, while different information is abundant and widely and easily accessible. As technological advancement aims to make things easier, the world is entering the age of cloud computing.

Cloud computing is a computing system supported by computer network services, where the scale of computing can be dynamically changed and resources are provided over the internet services by service providers (Chan, Walker & Gleaves, 2015). The system runs in such a way that resources including computing power, storage, network, and software are operated as a service over a network environment while it can be accessed from anywhere as long as it is available and connected to the internet (Maimunah Yohanes & Puspitasari, 2012).

Thanks to cloud computing, the installation, updating, and security challenges related to computer hardware and software are overcome with the on-demand and flexible service options and the opportunities to contribute to the development of innovative learning environments have emerged (Özcan & Emiroğlu, 2020). In general, cloud-based online learning offers participants the opportunity to access text-based, audio, and video content without worrying about the storage capacity or computational capabilities of their mobile devices (Wang, Chen & Khan, 2014). Given the ever-changing technological innovation, cloud learning is an important management skill that reshapes the global business environment and firms need to understand and adapt to such learning and development for continuous growth. The adoption of cloud learning is therefore believed to be a needed employee skill to help an organization improve and streamline its training activities (Ratten, 2012).

4.7. Gamification

Gamification is the use of game design elements, mechanics, and principles in non-game contexts to encourage and motivate people to achieve specific goals or tasks (Oliveira, Hamari, Shi, Toda, Rodrigues, Palomino & Isotani, 2023). Gamification capitalizes on people's desire for achievement, recognition, and competition, making tasks more fun and increasing motivation. It has been successfully applied in a variety of contexts, such as educational platforms that promote learning through interactive games, fitness apps that encourage physical activity, and loyalty programs that reward customers for their interactions with a brand (Lampropoulos, Keramopoulos, Diamantaros & Evangelidis, 2022).

The literature on games and organizational management is mostly oriented towards improving processes and performances by using previous experience through gamification. This is because one of its major advantages is that it anticipates potential situations and conflicts and their resolution before they arise in the real world. Understood as a tool for designing behaviors, developing skills, and ultimately enabling innovation processes, gamification can improve employee performance, job learning, and customer engagement when combined with other technologies (Burke, 2020). The relevance and interest of gamification applied to the management context are increasing due to the use of organizational simulation methodologies such as project-based learning. These simulations are expected to foster the qualities of professionals with the capacity for self-inquiry while enabling them to develop systemic thinking (Versiani & Fachin, 2007). In addition, gamification could be a teaching strategy with positive effects on learning processes and outcomes when employees are required to solve problems, make decisions, and perform complex and challenging tasks (Gómez-Pablos & Muñoz- Repiso, 2019).

In sum, gamification is an innovative solution to increase one's motivation. It uses game design elements to solve organizational problems via user engagement so that one can effectively achieve business goals. These users are stakeholders internally or externally from within the business, consumers of the company's products or services, specific or nonspecific individuals and groups, or other community members (Gartner, 2012). Gamification is said to increase the effectiveness of teamwork in a company and is essential for human and social relations (Bang, Fuglesang, Ovesen & Eilertsen, 2010). By applying the game mechanics to training and other business-related processes, organizations can potentially increase employee engagement which improves the overall productivity and operational performance (Dulskaia, Bellini, Zabal, Garcia, Pañella & D'ascenzo, 2017).

4.8. Adaptive Learning

Adaptive learning is an approach to education that uses technology and data to personalize the learning experience. Adaptive learning aims to optimize and optimize outcomes and the overall educational experience by tailoring instruction, content, and assessments to the specific needs and abilities of each student (Kem, 2022). Adaptive learning platforms use data and algorithms to analyze learning preferences and knowledge gaps. Based on this information, the platform can deliver customized content and activities to meet the unique learning needs of the individual. The adaptive learning system ensures that the learner is appropriately challenged and engaged by continuous assessment, allowing the platform to adjust the difficulty and type of content presented to the learner (Wang, Christensen, Cui, Tong, Yarnall, Shear & Feng, 2023). Adaptive learning systems provide immediate feedback to learners, allowing them to understand their mistakes and areas for improvement. The system can also offer targeted remediation by suggesting specific resources or activities to help the learner overcome their difficulties. Students can learn at their own pace as the adaptive learning platform adapts content to match their level of understanding. This flexibility allows students to progress faster in areas where they excel and spend more time on challenging concepts (Li, Xu, Zhang & Chang, 2022).

Adaptive learning is a technique that breaks traditional designs and enables employees to understand their specific leaps. In the workforce, adaptive learning can be monitored individually while in real-time it can determine which learning method might best suit employees' needs. It has significant benefits for employees who form performance expectations on flexibility and interactivity, as it helps to increase productivity (Teece, Pisano & Shuen, 1997).

4.9. Technological Learning

Technological learning, also known as technological capability building or technological development, refers to the process through which individuals, organizations, or societies acquire and develop their knowledge, skills, and capacity to understand, adapt, and utilize technology effectively. This learning process can occur at various levels, including personal, organizational, and national levels, and it plays a crucial role in driving innovation, economic growth, and societal progress (Haas, Sayer, Ajanovic & Auer, 2022).

Technological learning entails the process of accumulating and developing technical knowledge and capabilities that can result in innovations, transforming into a firm's technological know-how and production capacity (Huang, 2013; Adelowo, Ilori, Siyanbola & Oluwale, 2015). It indicates the extent to which a firm's structure, system, and resources are aligned with technology, as it also shows the firm's willingness to use technology as a competitive advantage. Therefore, it is elemental whether a firm can transform technology into a competitive advantage, either from the external or internal sources (Kaya & Taşgit, 2013). In most developing countries, technological learning is often tied with technology transfer, as the latter is generally seen as the main source of the former (Ignatius, Leen, Ramayah, Kah-Hin & Jantan, 2012; Ghazinoory, Ali, Hassanzadeh & Majidpour, 2019). Given that technological development becomes the main basis for modern business, a firm's capacity to using technological learning in its production can often be a determinant in business success (Susan & Novianti, 2019).

4.10. Mobile Learning

Mobile learning, refers to the use of mobile devices, such as smartphones, tablets, and other handheld devices, for the purpose of delivering educational content and facilitating learning experiences. Mobile learning takes advantage of the ubiquity and portability of mobile devices to provide learners with access to educational materials and resources anytime and anywhere (Hartley & Andújar, 2022). Mobile learning has gained popularity due to the widespread adoption of mobile devices and the increasing need for flexible and accessible learning options. It has been utilized in various educational settings, including schools, universities, corporate training, and informal learning contexts (Lebedeva, Taranova & Beketov, 2022).

Contextual factors for mobile learning cover various dimensions such as time, spatial issues and environment, the usefulness of learning content and ease of use of the technology, individual learning styles and predispositions, extrinsic and intrinsic motivations, accessibility and cost, as well as integration with other learning approaches. In modrn business, mobile learning technologies can be used for the training and development of employees, with which training organizations can encourage course participants to engage in self-learning and development through formal, informal, or microlearning contexts or interactions. Corporate trainers and providers of continuing vocational training and development services can use mobile learning applications to enhance employees' skills and competencies, which can potentially lead to increased organizational productivity and competitiveness (Sarrab, Elgamel & Aldabbas, 2012). What makes mobile learning different from other types of learning in business is that it is a form of technology that designs learning experiences that capitalize on the opportunities that mobility can offer to workers. Mobile learning focuses on the mobility of the learner, interacting with portable technologies, and learning that focuses on how society and businesses would accommodate and support the increasingly mobile population given the features and functionality of mobile devices enabling one's learning.

4.11. Modular Learning

Modular learning, also known as modular education or modular instruction, is an educational approach that involves breaking down the learning content into smaller, self-contained units or modules. Each module typically focuses on a specific topic or skill, allowing learners to study and master one concept at a time before progressing to the next (Bustillo & Aguilos, 2022). The primary goal of modular learning is to offer flexibility and customization to learners, allowing them to learn at their own pace and focus on areas that interest them or that they find challenging. This approach is especially prevalent in online and distance learning settings, where students may have diverse needs and schedules (Tugano, Tria & Tonio, 2022).

Modular learning is defined as a learning element that shows the learning objectives to be achieved behaviourally, determines the activities necessary to achieve the objectives, allows controlling the extent to which the objectives are achieved, and has the feature of being used effectively in the teaching of individuals in different learning conditions while having integrity within itself (Özkan, 2005). With modular programming, an educational environment based on individual learning and competence development is created in the learning-teaching process, in which the content is organized as small wholes based on certain analyses. In general, a modular system has different features compared to other systems with respect to flexibility, progress, and function within the system, and the difference in learning speed. In business learning, motivating employees and establishing realistic network with professional practices are deemed two most important features and benefits under the modular system (Kaykı, 2008).

4.12. Smart Working

Smart working is a concept closely related to teleworking. It is defined as a new management philosophy based on giving employees the freedom and flexibility to choose the place, time, and means they use in their work, under conditions of greater responsibility and accountability for anticipated results. Smart working is seen as an evolved form of teleworking with additional mobility and versatility allowing flexible work scheduling and amiable or convenient workplace either from home or elsewhere. It not only means teleworking at home or in offices, part-time or full-time, but it also includes new approaches to the structuring of working time and workspace, as well as the emergence of flexible or smart workers in the organization (Stratige & Giaoutzi, 2000).

Smart working refers to a way of working characterized by the absence of time or space constraints and an organization by stages, cycles, and objectives. It illustrates a new way of working and functioning of enterprises, as it also fosters an organizational effort to increase or improve employees' work flexibility (Rapisarda, Ghersetti, Girardi, De Carlo & Corso, 2021). In sum, smart working emphasizes an office activity that follows an a-spatial concept, centering on the flexibility and autonomy of the employee rather than on one's predefined roles and tasks. Compared with teleworking which allows employees to perform their tasks at different places and time, smart working uses and relies on the most or advanced modern technologies. In other words, the former refers to the use of electronic information and telecommunication tools for remote work, whereas the latter depends on new or existing technologies to improve a worker's performance (Marino & Capone, 2021).

CONCLUSION

In modern business, organizational learning is important for building or improving employee skills and reshaping the corporate culture. In a growth mindset and knowledge-sharing culture of an organization, employee attitude and the ability to learn and adopt new skills and training are often paramount, which ties to the organization's potential success and its competitive advantage.

Organizational learning is key to business growth. It is the DNA in business management, promoting employees' self- and mutual professional development. Increasingly, firms realize that nowadays changing the way employees learn at work can be one of the biggest organizational challenges, especially when such change involves tech- or digital investment and transofrmation while not every firm is ready due to limited resources. During the transformation process, learning and development professionals are often expected to prove the value of learning, as they also need to justify the efficiency of the scheme when new learning paradigms, learning styles, and learning technologies are introduced or adopted.

In this study, it aims to provide perspectives regarding organizational learning in a broadly-defined online or hybrid environment, and how firms by facing a new learning trend use various novel learning approaches to enhance their business growth and empolyment investment and management. Along the line of literature, concepts of the organizational learning in this study are analyzed as follows: Artificial intelligence, online learning, blended learning, microlearning, continuous learning, cloud learning, gamification, adaptive learning, technological learning, mobile learning, modular learning, and smart working.

Given that many of the new learning methods in business are tech-embedded or tech-oriented, a firm's ability to the adoption or adaptation of new organizational learning is elemental and often a determinant for business success. As many of these new learning are complementary and mutually reinforcing, employees and executives in an organization are proposed to support each other by working closely, whether top-down or bottom-up mutual learning or being trained individually or as a group. The organization would then be able to integrate employees' new learning into its business operation to achieve production growth and efficiency.

RECOMMENDATIONS

Adopt learning management systems and other learning technologies that support microlearning, personalized learning paths, and data analytics to deliver effective and efficient training. A culture that values continuous learning and encourages employees to self-learn, share knowledge, and experiment should be fostered and learning initiatives should be aligned with the organization's strategic objectives, with a focus on skills development that directly contributes to business success. Managers play a crucial role in supporting and encouraging learning activities. It should be ensured that managers allocate time and resources to employee learning and recognize those who actively participate. Regularly assess the impact of learning programs on employee performance and business results to identify areas for improvement and adjust learning strategies accordingly.

FUTURE RESEARCH

Neurological mechanisms of learning should be investigated to optimize instructional design and improve knowledge retention. Investigate the potential of artificial intelligence driven adaptive learning systems that can dynamically adjust content and difficulty depending on individuals' progress and explore how emotional intelligence and psychological factors influence the effectiveness of learning programs. Examine how cultural diversity affects learning preferences and explore culturally responsive learning approaches. Longitudinal studies should be conducted to understand the long-term effects of adopting these new trends on employee performance, career development, and organizational success.

By adopting these new trends, implementing recommended solutions, and directing future research efforts, organizations can create a culture of continuous learning that empowers employees, improves business performance, and prepares them for the challenges of the future.

REFERENCES

Adelowo, C., Ilori, M. O., Siyanbola, W. O., & Oluwale, B. (2015). Technological learning mechanisms in Nigeria's technology incubation centre. *African Journal of Economic and Management Studies*, *6*(1), 72–89. doi:10.1108/AJEMS-10-2014-0071

Admiral, W., & Lockhorst, D. (2009). E-learning in small medium-sized enterprises across Europe. *International Small Business Journal*, *27*(6), 743–767. doi:10.1177/0266242609344244

Ahmed, S. F. A., Krishna, S. H., Ganeshkumar, K., Anthiyur, U., & Manivel, R. (2023). Explorıng the impact of artificial intelligence in business decision making. *Journal of Data Acquisition and Processing*, *38*(3), 686.

Akbari, M., Danesh, M., Rezvani, A., Javadi, N., Banihashem, S. K., & Noroozi, O. (2023). The role of students' relational identity and autotelic experience for their innovative and continuous use of e-learning. *Education and Information Technologies*, *28*(2), 1911–1934. doi:10.100710639-022-11272-5 PMID:35971410

Alarabiat, A., Hujran, O., Soares, D., & Tarhini, A. (2023). Examining students' continuous use of online learning in the post-COVID-19 era: An application of the process virtualization theory. *Information Technology & People*, *36*(1), 21–47. doi:10.1108/ITP-02-2021-0142

Allen, D., & Eve, A. (1968). Microteaching. *Theory into Practice*, *7*(5), 81–185. doi:10.1080/00405846809542153

Bang, H., Fuglesang, S. L., Ovesen, M. R., & Eilertsen, D. E. (2010). Effectiveness in Top Management Group Meetings: The Role of Goal Clarity, Focused Communication, and Learning Behavior. *Scandinavian Journal of Psychology*, *51*(3), 253–261. doi:10.1111/j.1467-9450.2009.00769.x PMID:20132458

Basar, Z. M., Mansor, A. N., Jamaludin, K. A., & Alias, B. S. (2021). The Effectiveness and Challenges of Online Learning for Secondary School Students- A Case Study. *Asian Journal of University Education*, *17*(3), 119–129. doi:10.24191/ajue.v17i3.14514

Basten, D., & Haamann, T. (2018). Approaches for Organizational Learning: A Literature Review. *SAGE Open*, *8*(3). doi:10.1177/2158244018794224

Beer, M., Voelpel, S., Leibold, M., & Tekie, E. (2005). Strategic management as organizational learning: Developing fit and alignment through a disciplined process. *Long Range Planning*, *38*(5), 445–465. doi:10.1016/j.lrp.2005.04.008

Bell, B., Tannenbaum, S., Ford, K., Noe, R., & Kraiger, K. (2017). 100 Years of training and development research: What we know and where we should go. *The Journal of Applied Psychology*, *102*(3), 305–323. doi:10.1037/apl0000142 PMID:28125262

Bervell, B., & Umar, I. N. (2018). Utilization decision towards LMS for blended learning in distance education: Modeling the effects of personality factors in exclusivity. *Knowledge Management & E-Learning*, *10*(3), 309–333.

Bhatt, G., & Zaveri, J. (2002). The enabling role of decision support systems in organizational learning. *Decision Support Systems*, *32*(3), 297–309. doi:10.1016/S0167-9236(01)00120-8

Blaique, L., Ismail, H. N., & Aldabbas, H. (2023). Organizational learning, resilience and psychological empowerment as antecedents of work engagement during COVID-19. *International Journal of Productivity and Performance Management*, *72*(6), 1584–1607. doi:10.1108/IJPPM-04-2021-0197

Brandenburg, D., & Ellinger, A. (2003). The future: Just-in-time learning expectations and potential implications for human resource development. *Advances in Developing Human Resources*, *5*(3), 308–320. doi:10.1177/1523422303254629

Bughin, J., Hazan, E., Ramaswamy, S., Chui, M., Allas, T., Dahlström, P., Henke, N., & Trench, M. (2017). Artificial intelligence the next digital frontier? ***McKinsey and Company Global Institute***, Discussion Paper.

Buranova, D., & Latipova, D. (2023). Involvement of a medical student in the process of learning foreign languages. ***British medical, 3(3)***.

Burgess, J., & Russell, J. (2003). The effectiveness of distance learning initiatives in organizations. *Journal of Vocational Behavior*, *63*(2), 289–303. doi:10.1016/S0001-8791(03)00045-9

Burke, B. (2020). Gamification 2020: What is the future of gamification. ***Gartner Research,*** https://www.gartner.com/en/documents/2226015 Accessed on: 11.01.2023.

Bustillo, E., & Aguilos, M. (2022). The Challenges of Modular Learning in the wake of COVID-19: A Digital Divide in the Philippine Countryside Revealed. *Education Sciences*, *12*(7), 449. doi:10.3390/educsci12070449

Chan, N. N., Walker, C., & Gleaves, A. (2015). An exploration of students' lived experiences of using smartphones in diverse learning contexts using a hermeneutic phenomenological approach. *Computers & Education*, *82*, 96–106. doi:10.1016/j.compedu.2014.11.001

Collis, B., & Margaryan, A. (2004). Applying activity theory to CSCL and work-based activities in corporate settings. *Educational Technology Research and Development*, *52*(4), 38–52. doi:10.1007/BF02504717

Comey, W. (2009). Blended learning and the classroom environment: A comparative analysis of students' perception of the classroom environment across community college courses taught in traditional face-to-face, online and blended methods. (Unpublished Doctoral Disertation), The George Washington University, The Faculty of The Graduate School of Education and Human Development.

Do, H., Budhwar, P., Shipton, H., Nguyen, H.-D., & Nguyen, B. (2022). Building organizational resilience, innovation through resource-based management initiatives, organizational learning and environmental dynamism. *Journal of Business Research*, *141*, 808–821. doi:10.1016/j.jbusres.2021.11.090

Do, T. T., & Mai, N. K. (2022). Organizational Learning and firms performance: A systematic review. *International Journal of Productivity and Performance Management*, *71*(4), 1230–1253. doi:10.1108/IJPPM-02-2020-0051

Doornbos, A., Simons, R.-J., & Denessen, E. (2008). Relations between characteristics of workplace practices and types of informal work-related learning: A survey study among Dutch police. *Human Resource Development Quarterly*, *19*(2), 129–151. doi:10.1002/hrdq.1231

Duffey, D., & Fox, C. (2012). National educational technology trends: 2012. State leadership empowers educators, transforms teaching and learning. ***Glen Burnie, MD: State Educational Technology Directors Association (SETDA).*** https://files.eric.ed.gov/fulltext/ED536746.pdf Access date: 25.02.2023.

Dulskaia, I., Bellini, F., Zabal, R., Garcia Pañella, O., & D'ascenzo, F. (2017). Gamification as an instrument for organizational behaviour change during the meeting: case study «ROBATIEMPOS». *Proceedings of the International Conference on Business Excellence,* 11(1). DOI: 10.1515/picbe-2017-0039

Easterby-Smith, M., Crossan, M., & Nicolini, D. (2000). Organizational Learning: Debates Past, Present and Future. *Journal of Management Studies, 37*(6), 783–796. doi:10.1111/1467-6486.00203

Ellinger, A. (2005). Contextual factors influencing informal learning in a workplace setting: The case of "Reinventing Itself Company. *Human Resource Development Quarterly, 16*(3), 389–415. doi:10.1002/hrdq.1145

Fiol, M., & Lyles, M. (1985). Organizational learning. *Academy of Management Review, 10*(4), 803–813. doi:10.2307/258048

Fitria, T. N. (2022). Microlearning in Teaching and Learning Process: A Review. ***CENDEKIA: Jurnal Ilmu Sosial, Bahasa dan Pendidikan, 2(4).***

Fitria, T. N., Afdaleni, Sibolon, N. E., & Suamba, I. B. P. (2022). Online Learning Implementation and Challenges during Covid-19 Pandemic: English Lecturers' Perspective in Indonesia. ***SOSHUM : Jurnal Sosial Dan Humaniora, 12(2)***, 171-183. https://doi.org/ doi:10.31940/soshum.v12i2.171-183

Gartner. (2012). Gamification: engagement strategies for business and IT. ***Gartner,*** https://www.gartner.com/en/documents/2246217 Accessed on: 22.01.2023.

Garvin, D. (1993). Building a learning organization. *Harvard Business Review, 71*(4), 78–92. PMID:10127041

Ghazinoory, S., Ali, A. A., Hassanzadeh, A. R., & Majidpour, M. (2019). Examining systematic technological learning of Syrian textile industry. *Journal of Science and Technology Policy Management, 10*(1), 116–142. doi:10.1108/JSTPM-10-2017-0049

Gómez-Pablos, V. B., & Muñoz-Repıso, A. G.-V. (2019). Opinión del profesorado hacia proyectos colaborativos con Tecnologías de la Información y Comunicación: Un estudio psicométrico. *Educação e Pesquisa, 45*, e213768. doi:10.15901678-4634201945213768

Haas, R., Sayer, M., Ajanovic, A., & Auer, H. (2023). Technological learning: Lessons learned on energy technologies. *Wiley Interdisciplinary Reviews. Energy and Environment*, *12*(2), e463. doi:10.1002/wene.463

Haenlein, M., & Kaplan, A. (2019). A brief history of artificial intelligence: On the past, present, and future of artificial intelligence. *California Management Review*, *61*(4), 5–14. doi:10.1177/0008125619864925

Haleem, A., Javaid, M., Qadri, M. A., & Suman, R. (2022). Understanding the role of digital technologies in education: A Review. *Sustainable Operations and Computers*, *3*, 275–285. doi:10.1016/j.susoc.2022.05.004

Hanson, J., Niqab, M., & Bangert, A. (2021). Determination and Validation of the 'Organizational Learning Capabilities' Scale in a New Context of Educational Instituties in Pakistan. *Journal of Education and Educaitonal Development*, *8*(2), 270–295. doi:10.22555/joeed.v8i2.408

Hartley, K., & Andújar, A. (2022). Smartphones and learning: An extension of M-learning or a distinct area of inquiry. *Education Sciences*, *12*(1), 50. doi:10.3390/educsci12010050

Huang, H-C. (2013). Technological learning and technological innovation creation: An Empirical analysis of biotechnology R&D teams. ***Journal of Engineering Science & Technology Review, 6(2)***, 120-124. https://doi.org/. 25 doi:10.25103/jestr.062

Huber, G. (1991). Organizational learning: The contributing processes and the literatures. *Organization Science*, *2*(1), 88–115. doi:10.1287/orsc.2.1.88

Ignatius, J., Leen, J. Y. A., Ramayah, T., Kah-Hin, C., & Jantan, M. (2012). The impact of technological learning on NPD outcomes: The moderating effect of project complexity. ***Technovation, 32(7-8)***, 452-463. https://doi.org/. technovation.2012.03.003 doi:10.1016/j

İhtiyaroğlu, N., Karakuş Yıldız, B., & Yılmaz, F. (2022). Okullarda Pozitif Psikolojik Sermaye ve Örgütsel Öğrenme İlişkisinin İncelenmesi. *Sinop Üniversitesi Sosyal Bilimler Dergisi*, *6*(2), 315–352. doi:10.30561inopusd.1139670

Kaya, N., & Taşgit, Y. E. (2013). Teknolojik Öğrenme, Örgütsel Öğrenme ve Ar-Ge Yeteneklerinin Entellektüel Sermaye Üzerine Etkisi. *Düzce Üniversitesi Sosyal Bilimler Dergisi*, *3*(2), 1–12.

Kaykı, İ. E. (2008). ***Modüler Öğretim Yöntemi ve Uygulamalı Dersler.*** (Unpublished Master Thesis), Beykent Üniversitesi, Sosyal Bilimler Enstitüsü, İstanbul.

Kem, D. (2022). Personalised and adaptive learning: Emerging learning platforms in the era of digital and smart learning. *International Journal of Social Science and Human Research*, *5*(2), 385–391. doi:10.47191/ijsshr/v5-i2-02

Kim, D. (1993). The link between individual and organizational learning. *Sloan Management Review*, *35*(1), 37–51.

Kim, K.-J., Bonk, C., & Oh, E. G. (2008). The present and future state of blended learning in workplace learning settings in the United States. *Performance Improvement*, *47*(8), 5–16. doi:10.1002/pfi.20018

Kluge, A., & Schilling, J. (2003). Organizational learning and learning organizations. *The Psychologist Manager Journal*, *6*(1), 31–50. doi:10.1037/h0095917

Kuzey, C., Uyar, A., & Delen, D. (2014). The impact of multinationality on firm value: A comparative analysis of machine learning techniques. *Decision Support Systems*, *59*, 127–142. doi:10.1016/j.dss.2013.11.001

Lalima, & Dangwal, K. L. (2017). Blended Learning: An Innovative Approach. ***Universal Journal of Educational Research, 5(1)***, 129-136. DOI: . doi:10.13189/ujer.2017.050116

Lampropoulos, G., Keramopoulos, E., Diamantaras, K., & Evangelidis, G. (2022). Augmented reality and gamification in education: A systematic literature review of research, ***Applications, and empirical studies. applied sciences, 12(13)***, 6809.

Lebedeva, M., Taranova, M., & Beketov, V. (2022). Assessment of academic achievements in m-learning. *Education and Information Technologies*, *1*, 21. PMID:36406788

Lee, J., Davari, H., Singh, J., & Pandhare, V. (2018). Industrial Artificial Intelligence for Industry 4.0-based manufacturing systems. *Manufacturing Letters*, *18*, 20–23. doi:10.1016/j.mfglet.2018.09.002

Lei, D., Slocum, J., & Pitts, R. (1999). Designing organizations for competitive advantage: The power of unlearning and learning. *Organizational Dynamics*, *27*(3), 24–38. doi:10.1016/S0090-2616(99)90019-0

Li, X., Xu, H., Zhang, J., & Chang, H.-H. (2022). Deep Reinforcement Learning for Adaptive Learning Systems. *Journal of Educational and Behavioral Statistics*, *48*(2), 220–243. Advance online publication. doi:10.3102/10769986221129847

Lin, W. C., Tsai, C. F., & Zhong, J. R. (2022). Deep learning for missing value imputation of continuous data and the effect of data discretization. *Knowledge-Based Systems*, *239*, 108079. doi:10.1016/j.knosys.2021.108079

Maimunah Yohanes, A. K. Y., & Puspitasari, N. (2012). Konsep dan penerapan cloud computing untuk meningkatkan mutu pembelajaran. *CSRID Journal*, *4*(3), 220–230.

Marino, L., & Capone, V. (2021). Smart Working and Well-Being before and during the COVID-19 Pandemic: A Scoping Review. *European Journal of Investigation in Health, Psychology and Education*, *11*(4), 1516–1536. doi:10.3390/ejihpe11040108 PMID:34940386

Mohanty, S. B. (1988). *Lifelong and adult education*. Ashish Publishing House.

Motteram, G., & Sharma, P. (2009). Blended learning in a Web 2.0 world. *International Journal of Emerging Technologies and Society*, *7*, 83–96.

Oliveira, W., Hamari, J., Shi, L., Toda, A. M., Rodrigues, L., Palomino, P. T., & Isotani, S. (2023). Tailored gamification in education: A literature review and future agenda. *Education and Information Technologies*, *28*(1), 373–406. doi:10.100710639-022-11122-4

Ong, E. C., & Tan, C. L. (2022). Soft TQM, agility, and knowledge management deliver organizational performance: A study of Malaysian manufacturing organizations in the electrical and electronics sector. *Global Business and Organizational Excellence*, *41*(4), 28–47. doi:10.1002/joe.22155

Özcan, H., & Emiroğlu, B. G. (2020). Bulut Tabanlı Öğrenme Yönetim Sistemi Seçiminde Bulanık Çok Kriterli Karar Analizi Yaklaşımı. *Bilişim Teknolojileri Dergisi*, *13*(1), 97–111. doi:10.17671/gazibtd.589757

Özkan, H. H. (2005). Öğrenme Öğretme Modelleri Açısından Modüler Öğretim. *Atatürk Üniversitesi Sosyal Bilimler Enstitüsü Dergisi*, *6*(2), 117–128.

Park, Y., & Kim, Y. (2018). A Design and Development of micro-Learning Content in eLearning System. *International Journal on Advanced Science, Engineering and Information Technology*, *8*(1), 56. Advance online publication. doi:10.18517/ijaseit.8.1.2698

Perifanis, N.-A., & Kitsios, F. (2023). Investigating the influence of artificial intelligence on business value in the digital era of strategy: A literature review. *Information (Basel)*, *14*(2), 85. doi:10.3390/info14020085

Pham, L. T., & Hoang, H. V. (2019). The relationship between organizational learning capability and business performance: The case of Vietnam firms. *Journal of Economic Development*, *21*(2), 259–269. doi:10.1108/JED-10-2019-0041

Prahalad, C. K., & Hamel, G. (1990). The core competence of corporation. *Harvard Business Review*, *68*, 79–93.

Rapisarda, S., Ghersetti, E., Girardi, D., De Carlo, N. A., & Corso, L. D. (2021). Smart working and online psychological support during the covid-19 pandemic: Work-family balance, well-being, and performance. *International Psychological Applications Conference and Trends*, pp.301-306. DOI: 10.36315/2021inpact062

Ratten, V. (2012). Implementing cloud learning in an organization: A training perspective. *Industrial and Commercial Training*, *44*(6), 334–336. doi:10.1108/00197851211254761

Rosenberg, M. (2006). *Beyond e-learning: Approaches and technologies to enhance organizational knowledge, learning, and performance*. Pfeiffer.

Rupcic, N. (2023). Means to improve organizational learning capability. *The Learning Organization*, *30*(1), 101–109. doi:10.1108/TLO-01-2023-289

Santos-Vijande, M. L., López-Sánchez, J. Á., & Trespalacios, J. A. (2012). How organizational learning affects a firm's flexibility, competitive strategy, and performance. *Journal of Business Research*, *65*(8), 1079–1089. doi:10.1016/j.jbusres.2011.09.002

Sarrab, M., Elgamel, L., & Aldabbas, H. (2012). Mobile Learning (M-Learning) and Educational Environments. *International Journal of Distributed and Parallel Systems*, *3*(4), 31–38. Advance online publication. doi:10.5121/ijdps.2012.3404

Schein, E. (1993). On Dialogue, Culture, and organizational learning. *Organizational Dynamics*, *22*(2), 40–52. doi:10.1016/0090-2616(93)90052-3

Senge, P. (1990). The Leader's New Work: Building learning organizations. *Sloan Management Review*, *32*(1), 7–23.

Shaheen, M. A., Hayyat, A., Aslam, S., & Shahbaz, M. A. (2023). Developing a culture of continuous learning for sustainable job performance: TEVT sector as learning organization, Pakistan. *Journal of Social Sciences Review*, *3*(2), 817–827. doi:10.54183/jssr.v3i2.321

Stratigea, A., & Giaoutzi, M. (2000). Teleworking and virtual organization in the urban and regional context. *NETCOM*, *14*(3-4), 331–357. doi:10.3406/netco.2000.1479

Susan, A., & Novianti, W. (2019). *Benefits of Technology for Business.* ***IOP Conf. Series: Materials Science and Engineering***, S.662, 032036. Doi:10.1088/1757-899X/662/3/032036

Tapalova, O., & Zhiyenbayeva, N. (2022). Artificial intelligence in education: AIEd for personalised learning pathways. *Electronic Journal of e-Learning*, *20*(5), 639–653. doi:10.34190/ejel.20.5.2597

Teece, D., Pisano, G., & Shuen, A. (1997). Dynamic Capabilities and Strategic Management. *Strategic Management Journal, 18*(7), 509–533. doi:10.1002/(SICI)1097-0266(199708)18:7<509::AID-SMJ882>3.0.CO;2-Z

Tong, D. H., Uyen, B. P., & Ngan, L. K. (2022). The effectiveness of blended learning on students' academic achievement, self-study skills and learning attitudes: A quasi-experiment study in teaching the conventions for coordinates in the plane. *Heliyon, 8*(12), e12657. doi:10.1016/j.heliyon.2022.e12657 PMID:36643330

Tugano, M. S., Tria, J. Z., & Tonio, J. Z. (2022). Modular learning amidst COVID-19 pandemic: Satisfaction among students in a higher education institution. ***International Journal of Professional Development, Learners and Learning, 4(2)***, ep2206.

Tynjälä, P., & Häkkinen, P. (2005). E-learning at work: Theoretical underpinnings and pedagogical challenges. *Journal of Workplace Learning, 17*(5/6), 318–336. doi:10.1108/13665620510606742

Vera, D., & Crossan, M. (2004). Strategic leadership and organizational learning. *Academy of Management Review, 29*(2), 222–241. doi:10.2307/20159030

Versiani, Â., & Fachin, R. (2007). Avaliando aprendizagem em simulações empresariais. Cadernos EBAPE.BR, Rio de Janeiro, 5, 1-13.

Wamba-Taguimdje, S.-L., Wamba, S. F., Kamdjoug, J. R. K., & Wanko, C. E. T. (2020). Influence of artificial intelligence (AI) on firm performance: The business value of AI-based transformation projects. *Business Process Management Journal, 26*(7), 1893–1924. doi:10.1108/BPMJ-10-2019-0411

Wang, M., Chen, Y., & Khan, M. J. (2014). Mobile cloud learning for higher education : A case study of Moodle in the cloud. *International Review of Research in Open and Distance Learning, 15*(2), 255–267. doi:10.19173/irrodl.v15i2.1676

Wang, S., Christensen, C., Cui, W., Tong, R., Yarnall, L., Shear, L., & Feng, M. (2023). When adaptive learning is effective learning: Comparison of an adaptive learning system to teacher-led instruction. *Interactive Learning Environments, 31*(2), 793–803. doi:10.1080/10494820.2020.1808794

Wannewitz, M., Petzold, J., & Garschagen, M. (2023). What makes people adapt together? An empirically grounded conceptual model on the enablers and barriers of collective climate change adaptation. *Frontiers in Climate, 5*, 1213852. doi:10.3389/fclim.2023.1213852

Watkins, K., & Marsick, V. (1993). *Sculpting the Learning Organization*. Jossey-Bass.

Wu, L.-Y. (2010). Applicability of the resource-based and dynamic-capability views under environmental volatility. *Journal of Business Research*, *63*(1), 27–31. doi:10.1016/j.jbusres.2009.01.007

Xu, J., & Pero, M. E. P. (2023). A resource orchestratipon perspectiv of organizational big data analytics adoption: Evidence from supply chain planning. *International Journal of Physical Distribution & Logistics Management*, *53*(11), 71–97. doi:10.1108/IJPDLM-04-2022-0118

Chapter 10
Volume Growth Advocacy and Sustainability Options in Sikkim Tourism in the Post-Pandemic Scenario

Supriya Dam
Netaji Subhash Mahavidyalaya, India

ABSTRACT

This study explores a dichotomy between the volume growth advocacy and survival options available to stakeholders in the post-pandemic scenario, applying sustainability principles in tourism in Sikkim. Pandemic inflicted respite from the volume growth advocacy model provided much-needed time to restore sanity, given the ecological fragility of the state. Scenario analysis of actual space on offer for tourism and host-guest ratio suggests the destination has embraced tourists beyond the acceptable limits. Per capita, municipal solid waste generation per day has left a lasting impact on land and masses. Mushrooming of concrete structures across urban centers exposes poorly planned arrangements for tourism facilities over sustainability considerations. Again, the projected demand for accommodation infrastructure and adjoining services is inadequate to meet spiraling demand. The study advocates for micro tourism besides practicing high-value and low-impact activities and spot-specific plans to strike a chord between volume growth advocacy and sustainability options.

DOI: 10.4018/978-1-6684-9089-1.ch010

INTRODUCTION

The upsurge of the pandemic from late 2019 brought a complete decline in people's movement away from home and badly hit the tourism sector more than ever. According to UNWTO estimates, till January-September, 2022, the tourism sector recovered around 62 percent to the pre-pandemic levels, recording the movement of 700 million international tourists (UNWTO, 2022) across the globe. However, people's perception of safety and cautious behavior deterred the sector's progress to match pre-pandemic performance (Orindaru et al, 2021). Consequently, a break from regular tourism activity provides time to reset and rethink along the sustainability lines (Seabra and Bhatt, 2022), given the ecological fragility of the destination.

During post-Covid 19, India registered a mixed bag of International Tourist Arrivals (ITAs), comprising Foreign Tourist Arrivals (FTAs) and Arrivals of Non-Resident Indians (ANRIs). FTAs recorded a negative growth rate of 44.5 percent, whereas ANRIs posted a positive growth of 52.62 percent in 2021 over 2020; and an expansion of Foreign Exchange Earnings (FEEs) of 8797 million USD in 2021, registering a 26.4 percent increment over 2020. Domestic tourism too documented an impressive rise of 11.5 percent in 2021 over 2020 to kick-start the resumption of tourism activity (ITS, 2022).

"Sustainability principles integrate environmental, economic, and socio-cultural aspects of tourism development, and striking a balance between these three dimensions to guarantee its long-term sustainability"(UNWTO, 2005,11). The outbreak ought to provide a fresh lease of life for volume-growth practitioners and the role played by national tourism organizations (NTOs) before, during, and after Covid 19. In this regard, Gossling et al (2020,16) cautioned against returning to comprehensive tourism activity during post-Covid 19, instead capitalizing on an opportunity for changed global tourism in line with Sustainable Development Goals (SDGs). The stoppage of tourism activity provides a much-needed occasion to relook at the structural changes in the tourism sector in agreement with the long-term vision (Romagosa, 2020). Ioannides and Gyimothy (2020, 630) stressed the urgency of exploiting the opportunity to repair an otherwise imperfect global system.

Sikkim has put up a brave front to fight the pandemic outbreak and progressively returning to the post-Covid level. Tourism entrepreneurs, primarily micro-entrepreneurs, find it hard to keep themselves afloat for a considerable period and are regaining their confidence with the resumption of tourism activity. Despite Covid inflicted losses, the destination got a much-needed respite from volume growth advocacy sustained during pre-pandemic and a scope for recovery on the sustainability front.

The thrust on the volume-growth model of tourism for substantial period results in anthropogenic pressure upon land and masses beyond the acceptable limit, especially during peak tourism season, coupled with consequent leakage of tourists' spending impaired the trickling down of derived benefits to small entrepreneurs at large. Again, the habitable area confined to a range of 2100 meters, constituting a mere 20 percent of the total landscape (GoS, 2011), offers limited scope for expansion of the accommodation sector. The pressure upon fragile landscapes for commercial purposes disregards environmental sustainability considerations (Rubita, 2012) for leisure. Anthropogenic pressure of this sort upon land and masses leads to the recurrence of landslides, and frequent degradation in hill slopes alongside tourism sites, calling for a revisit at the entire gamut of tourism offerings. Again, the stoppage of tourism brings a more aggressive growth strategy during post-Covid 19 (Gossling et al, 2020), in effect leading to disequilibrium in sustainability for this tiny hill state. A situation of this kind prompted us to go for the present study.

Therefore, the aim of the study is threefold: i) Conduct an in-depth analysis of tourists influx, pre and post-pandemic, in Sikkim concerning its contribution to the economy; ii) Examine the geographical space available to tourists from the perspective of sustainable tourism development; and iii) Explore the gaps and opportunities for the revival of sustainable tourism practices in post-Covid 19 scenarios.

To better understand the situations cited above, the main parts of this study focus on five aspects. The Section II draws heavily on the review of existing literature. The Section III enunciates the methodology adopted for the study. The subsequent section (Section IV) describes a detailed report on the status of tourism during pre and post-Covid 19 vis-a-vis sustainability issues concerning Sikkim. Section V sheds light on the discussion on matters that cropped up. The Section VI comments on the emerging issues, if any, about this study.

LITERATURE REVIEW

Niti Ayog (NA) (2018) observed that tourism development should not only aim at creating jobs or deriving economic output by destroying the natural resource base upon which it depends; but rather by strict adherence to sustainability principles and their implementation. Malik et al (2016) opined that, apart from tourism and adventure-related activities, the construction of dams, tunnels, and hydroelectric projects incur insurmountable damage to the mountain ecosystem coupled with unplanned developmental works like road constructions, mining causes several natural calamities (Bhattacharya, 2019). Insufficiency of rules and laws against illegal collection of forest items and travelers in high altitude areas causes perennial destruction of the flora and fauna of those locations (Chand et al, 2016). Addressing such biotic pressures upon fragile ecosystems is presumed to be an uphill task in protected zones (Maan and Choudhury, 2019) than ever before.

Space on offer for tourism activities got significance particularly from sustainability point of view. Tourism physical space is equated with destinations geographical area on offer for tourism covering attractions and adjunct facilities available to tourists (Butowski et al, 2018). In other words, the centre of activity do not confine to the physical space itself, rather the objects and subjects surrounding it (Lisowski, 2007). Tourisms' territorial implications with spatial dimensions shape up spacio-temporal pattern of growth (e Silva, 2018), needs a controlled intrusion and directed at sustainability, particularly for hill regions (Bhutia, 2015). This obviously calls for sustainable tourism infrastructure planning, in destination setting, in order to combine sustainability indicators in tourism planning (Boers and Cottrell, 2007). On the contrary, unplanned growth of tourism is often associated with low tourist satisfaction coupled with undesirable shocks to resources (Hultsman et al. 1999; Swarbrooke, 1999; Mowforth & Munt 2003). This again boils down to site-specific plans, depending on the landscape, by keeping sight on sustainability (Johnston and Braden, 2008).

Pandemic inflicted existential crisis for the tourism sector across the globe due to restrictions on the movement of tourists and its magnitude till the impact mitigates. Nientied and Shutina (2020) termed it a transitional phase that necessitated building resilience in coexistence with sustainable tourism practices for creating an integrated destination. Santos-Roldan et al (2020) emphasized the need for tourists' awareness in post-Covid 19 and the role of tourism establishments in its accomplishment.

Hall et al (2020) observed tourism destinations will now focus on the local and sustainable form of tourism with little governmental support and centered on promoting domestic tourism. Sigala (2020) outlined it as a transformation in tourism studies considering the end-user benefits and the need for stakeholders' resilience to accept the reconstruction induced (Ulak, 2020). Further, Haywood (2020) outlined the necessity of collaborative innovations with a proper assessment of destinations through intense scrutiny of past developments in shaping desirable future outcomes. In this regard, the preservation of traditional and physical environments depends on responsible tourism practices on a local scale to strike a balance between tourist inflows and sustainability concerns (Koscak and O'Rourke, 2021).

Again, Nepal (2020) sensed the opportunity for sustainable growth of adventure travel in place of haphazard one by re-conceptualizing the entire development process, given the scarcity of destination-specific indicators to measure (Kumari et al, 2006). Rubita (2012) opined that the adverse impact of mass tourism effectively threatened fragile ecosystems in mountainous destinations. Gossling et al (2020) called for a 'critical re-examination of volume growth model for tourism pursued over the years and its contribution to climate change' juxtaposing sustainability principles. Nyaupane and Timothy (2016) advocated for a low volume of high-value tourism practices with spatial restrictions, perhaps, a productive policy, especially for fragile mountainous regions. Considering these, tourists' arrival per square kilometer area emerged as a suitable tool to estimate the tourism density index (TDI) of destinations (Manera and Valle, 2018) and establish a negative relationship between TDI and environmental performance index (EPI) (Yassin et al, 2020); thereby helps in assessing adverse impact upon the environment aspect of tourism destinations.

Marques Santos et al (2020) stressed the need for utilizing existing policy initiatives with an innovative approach to safety and health-related measures in the face of change. Tourism policy initiatives in India also called for vertical collaboration between Union and State Ministries for tourism for its promotion in each state and union territories across the country (FICCI, 2020). The OECD (2020) professed vertical coordination between the national and sub-national governments for creating an effective response to face pandemic-related issues.

As a sizeable portion of tourism firms represent micro, small, and medium enterprises, measures to support these for survival should accord due attention along with digital literacy to enable these to make use of innovative technologies, big data as well as artificial intelligence to facilitate tourism experiences (Cepal, 2020). Social media can play a decisive role in creating a positive image of tourist sites' and demonstrate a case for virtual tourism (Vaisakh et al, 2020) across destinations.

Khazami et al (2020) stressed the need for the safety and security of hosts and called for flexible tourism services while rebuilding the destination. Brouder et al (2020) expressed optimism concerning peoples' awareness about visiting lesser-known yet safe destinations to assist local communities. Studies emphasized the need for nursing the quality of attraction of tourist sites during the post-pandemic situation (Hong et al, 2020).

The present study looks into possible fundamental changes in the overall approach towards tourism practices in Sikkim in the face of volume growth advocacy and sustainability options.

MATERIAL AND METHODS

This paper banks on secondary materials available in print and electronic media and encompasses the internet-published reports of governmental agencies and the studies carried out to date highlighting the issues concerned. Tourism reports of the Department of Tourism, Government of Sikkim, along with the reports of the Ministry of Tourism, Government of India forms the core of this analysis. The information so collected is processed and analyzed using statistical techniques as per the need of the study.

The study exploits select sustainability parameters, including tourism's contribution to Gross State Domestic Product (GSDP), physical space availability, tourism density, total density, host-guest ratio, projected accommodation status, and consultation with select stakeholders undertaken to know their perspective during post-Covid 19. The broad outline of the study will be limited to the geographical periphery of Sikkim alone, and the reference period is 2011-22. Considering this, the limitations of using secondary information would apply.

RESULTS

The 'tourist-centric' economy took center stage in Sikkim, engaging seven out of ten people in similar activities. The essence of the "Athithi Deva Bhava" principle, treating tourists as the god, has an enormous impact on local culture, followed by locals across Sikkim. Stakeholders' friendly attitude towards 'guests' led the way for a complete tourism experience. Despite vibrant organic farming in place for agriculture, tourism, and its multiplier effect has emerged as the prime money spinner for the Sikkimese people over the years.

During the last two decades, Sikkim has embraced all the visitors wishing to tour the tiny Eastern Himalayan State for a touristic reason leaving positive results on the state's employment generation, accommodation, and income-earning opportunities for the locals (Tamang and Solanki, 2020). The sector's contribution to Gross State Domestic Products (GSDP) peaked in percentage terms during 2016-17 (Niti Ayog, 2018). The annual Tourism Report of Sikkim for 2019-20 suggests 1,33,388 foreign tourists visited the state for the year, precisely 62216 tourists higher than the same for 2018-19 (DoT, 2019-20). In addition, the influx of over 13 lakh domestic tourists' exceeded the total population of Sikkim by two times. The peak season is prominent during the months of September-October and March-April. Further, the state's more than 12,000 rooms in the accommodation sector with over 1200 homestays usually register a 90 percent occupancy rate during peak season (Sarkar, 2020). The government of Sikkim unleashed a series of fiscal measures to keep the sector vibrant over the years before the onset of Covid 19. This can be evinced from the following figure 1 given below:

The tiny hill state, alongside Goa, accounts for the highest tourism contribution to Gross State Domestic Product (GSDP) during 2018-19. Master plans, Perspective Plans, and Plans for Trekking and tourism policies suggest the state's eagerness to view tourism as one of the vibrant sectors of the financial system. The Swadesh Darshan Project at Zero Point, Gangtok, started for Rs. 98.05 crores (US$ 14.02 million). The Eco Adventure Resort at Temi Tea Estate began in September 2019 as a part of the first public-private partnership (PPP) initiative between the State Government and private operators engaged in the tourism sector (IBEF, 2019).

Figure 1. Tourism Expenditure percentage of States' Budgetary Allocation during 2018-19
Source: The Economic Times online edition dated 13th Dec, 2019. https://economictimes.indiatimes.com/industry/services/travel/where-do-tourists-prefer-to-go-in-india/who-spends-the-most-on-tourism/slideshow/72516703.cms.

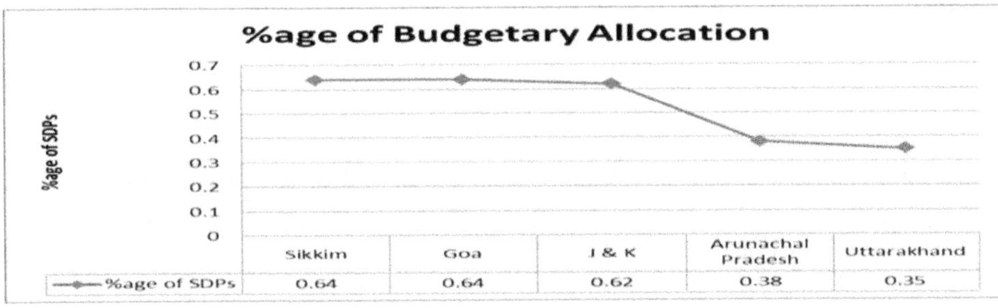

The state house's 6.19 lakh native population has over 12,000 rooms expected to generate a lodging business of 3 Crore INR a day, with over 90% accommodation occupancy rate, during peak season March-May and September-November each year (DoT, 2019-20). In the accommodation sector, homestays need a special mention primarily because of the involvement of locals and the source of providing a wholesome tourism experience. The Himalayan state has several registered tour operators, with 25000 registered local taxis plying across the destination and a sizeable number of tourist guides and porters serving visitors as per requirements (DoT, 2019-20).

Domestic tourism remains the mainstay of Sikkim as it accounts for little over 91 percent of the total visitors traveled to the state during the last decade.

Table 1. Inflow of Domestic Tourists during 2011-21

Year	2011	2012	2013	2014	2015	2016	2017	2018	2019	2020	2021
Mean	46038	46545	48062	46868	58752	61730	114655	118844	118485	26367	42639
Std Dev	35719	28527	33015	28462	35055	42005	87990	120972	114831	42010	36562
Scale Reliability Statistics: Cronbach α										0.73	

Source: Tourism Department, Govt. of Sikkim

Table 1 portrays domestic tourists' arrival in Sikkim during 2011-21. Increasing mean and standard deviation till 2019 suggests a proportionate surge in the influx of domestic tourists for every successive year during the period under review, leaving Cronbach's α reliability score of 0.73. The extended lockdown for six months in succession and the subsequent reopening of tourism activities during 2020 and 2021 culminated in substantial downsizing of the mean and standard deviation for the periods. The mean score per month of domestic tourists for the nine years till 2019 ranges between 46038 to 118485, whereas the standard deviation for the same period ranges between 35719 and 114831. The post-pandemic scenario shows average mean domestic tourists per month ranges between 26367 and 42639 and a standard deviation of 42010 and 36562.

Table 2. Foreign Tourist Arrivals (FTAs) during 2011-21

Year	2011	2012	2013	2014	2015	2016	2017	2018	2019	2020	2021
Mean	1995	2207	2642	4098	3207	5501	4093	5391	11116	3323	959
Std D	1177	1132	1705	1946	1681	4193	2286	5909	12469	2013	874
Scale Reliability Statistics: Cronbach α										0.73	

Source: Tourism Department, Govt. of Sikkim

Foreign tourist arrivals (FTAs) during the same period (table 2) also registered a remarkable increase, especially from 2016 until 2019. The mean score for FTAs in 2011 was 1995, raised to 11116 in 2019. The standard deviation during the pre-Covid period of 2011-19 ranges between 1177 visitors to 12469 visitors and has increased by 10.69 times. The post-Covid period saw an overall decline in foreign tourist influx aggregating mean visitors of 3323 and 959 each for 2020 and 2021. The standard deviation also reduced during the same period leaving Cronbach's α reliability score of 0.73.

Figure 2 depicts month-wise average domestic tourists visiting Sikkim 2011-21. The range of tourists is measured along Y axis whereas X axis shows average monthly tourists across 11 years. Seasonal variability of domestic tourists till 2021 shows two convergences of curves, the first convergence visible in March, April, and May while the other in September, October, and November illustrates the prime tourist season for the destination. However, the distinct peak in April shows the top tourist month for domestic tourists visiting the state, and trend line slopes downward from left to right indicative of preferred month of domestic visitors during the period. In 2018, 1426127 domestic tourists traveled to the Himalayan State, followed by 1421823 tourists in 2019 (DoT, 2019-20), portraying the volume growth advocacy of the destination. Considering the fragility of the landscape, population size (6,10,577 persons as per the 2011 census), and geographical space (7096 square kilometers), these floating population from the domestic segment alone creates anthropogenic pressure upon land and masses much beyond generally acceptable limits.

Figure 2. Month-wise Average Domestic Tourists visiting Sikkim during 2011-21
Source: Department of Tourism, Govt. of Sikkim

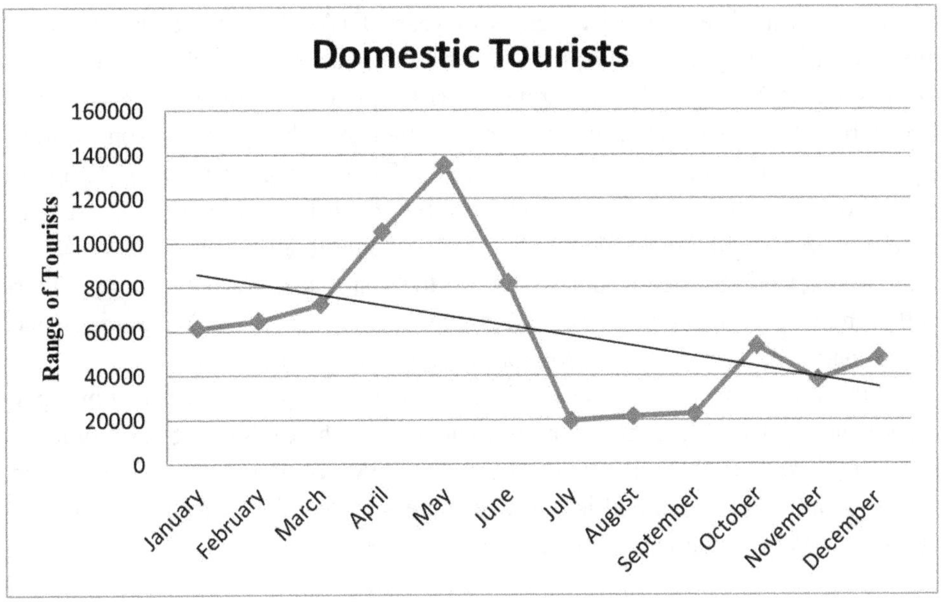

Figure 3. Month-wise Average Foreign Tourists visiting Sikkim during 2011-21
Source: Department of Tourism, Govt. of Sikkim

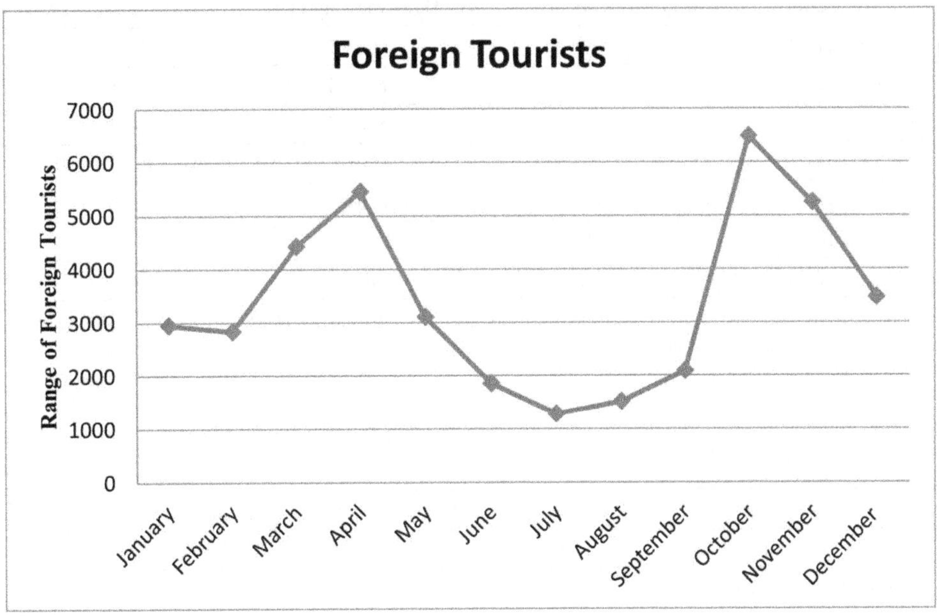

Figure 3 depicts month-wise average foreign tourists visiting Sikkim 2011-21. The range of foreign tourists is measured along Y axis whereas X axis shows average monthly foreign tourists across 11 years. FTAs show two convergence of months surrounding the peaks quite similar to the seasonality visible for domestic tourists during 2011-21; visiting foreign tourists opted for September, October, and November, the time of retreating monsoon, rather than the spring season (March, April, and May) preferred by domestic tourists. The Indian government gave up the restrictions for entry norms for visitors from Bangladesh to tour the state, resulting in hordes of Bangladeshi nationals visiting the state to enjoy the serenity of the destination. 133388 foreign tourists toured the state during 2019, exactly 62216 higher than the FTAs for 2018; a considerable portion of them was Bangladeshi and Nepali nationals over and above visitors from the rest of the world.

In the pre-Covid period, the tourism sector in Sikkim registered impressive growth rates of 10-12 percent per annum. The sector's share of Gross State Domestic Product (GSDP) for 2016-17 stands at Rs. 1,44,735 INR and tourism contribution is pegged at 7.68 percent and is likely to increase (DoT: 2020) in the post-Covid phases, given the volume growth advocacy.

Table 3. Relationship of Sikkim's Gross State Domestic Product (GSDP) at Current Price with Domestic and Foreign Tourist Arrivals during 2011-12 to 2017-18

Year	GSDP @	Domestic Tourist #	Foreign Tourist #
2011-12	1116510	500482	27766
2012-13	1233842	564777	26312
2013-14	1386190	540328	39540
2014-15	1540672	620277	49819
2015-16	1803394	815136	39972
2016-17	2002046	1073484	72912
2017-18	2224791	1277285	61898

Source: @ Directorate of Economics, Statistics, Monitoring and Evaluation (DESME), Govt. Of Sikkim. # Department of Tourism, Govt. of Sikkim.

Table 4. Model Summary

R	R^2	Adjusted R^2	Standard Error of the Estimate
0.970	0.941	0.912	142789.579

Source: Authors own calculation

Table 3 establishes the relationship between Sikkim's GSDP at its current price to domestic and foreign tourist arrivals during 2011-12 to 2017-18. GSDP and domestic tourist arrivals recorded consistent growth rates during the period under review. The FTAs also recorded periodical growth rates except for 2015-16 and 2017-18. Table 4 displays a correlation coefficient ® of 0.970, demonstrating GSDP's strong positive correlation with domestic and foreign tourist arrivals contributing to the state's Gross Domestic Product (GDP) growth. The value of adjusted R^2 stands at 91.2 percent signifies increased influx of tourists will magnify the state GSDP for the post-Covid phase.

Let us look at the tourism density (tourist arrival per square kilometer of area) and host-guest ratio in Sikkim, considering the total geographical space available for habitation and tourism-related activities (including a portion of protected areas). Of the total 7096 sq km area spread over four districts of the state, 81.28 percent (5767.63 sq km) of the spaces are covered with reserve forest (31.86%), Khasmal forest (4.01%), Goucharan forest (1.46%) are having land and tree cover partially useful for human habitation. Of the rest, alpine pastures and shrubs (14.44%) and perpetual snow-covered areas (29.50%) are inhabitable for humans (District Census Handbook: 2011). The protected area network, comprised of Khanchendzonga National Park (KNP), seven wildlife sanctuaries, and a conservation reserve), inside the state covers 30.77% of the geographical space (FSI:2019), thus limiting the space available for human habitation and adjoining activities.

Table 5. District-wise Forest Cover and Open Spaces in Sikkim (Area in sq km)

District	Area in sq km	As per 2019 Assessment (% of total area)				Open Spaces	
		Very dense forest	Moderate to Dense Forest	Open Forest	Total	Scenario I (5-1)	Scenario II (6+4)
	(1)	(2)	(3)	(4)	(5)	(6)	(7)
East	954	271.15 (3.82%)	313.61 (4.42%)	127.96 (1.80%)	712.72 (10.04%)	241.28 (3.40%)	369.24 (5.20%)
North	4,226	410.61 (5.79%)	586.18 (8.26%)	287.26 (4.01%)	1284.05 (18.10%)	2941.95 (41.46%)	3229.21 (45.51%)
South	750	173.35 (2.44%)	289.49 (4.08%)	108.98 (1.54%)	571.82 (8.06%)	178.18 (2.51%)	287.16 (4.05%)
West	1,166	246.85 (3.48%)	363.03 (5.12%)	164.02 (2.31%)	773.90 (10.91%)	392.10 (5.53%)	556.12 (7.84)
Total	7,096	1,101.96 (15.53%)	1552.31 (21.88%)	688.22 (9.70%)	3342.49 (47.11%)	3753.51 (52.90%)	4441.73 62.60%)

Source: Forest Survey in India (FSI), 2019, p-235

Table 5 shows district-wise forest cover and open spaces available for sectoral activities. The table depicts 47.11 percent of the total geographical area of 7096 sq km overlayed with very dense to open forest; the leftover 52.9 percent (or 3753.51 sq km) of the space is available for economic activities. Although open space availability was highest in the North District (41.46%), still rugged terrain made it the least densely populated area inside Sikkim. Of the rest, the West district offers 5.53 percent of the open space for sectoral performances, followed by East (3.4%) and South (2.52%) in that order.

In Scenario II, open forest locations appended to available open spaces broaden total geographical areas to 62.6 percent for sectoral activities. District-wise space availability also increased from Scenario I, in which overall space availability for North District rose to 45.51 percent, followed by West (7.84%), East (5.2%), and South (4.05%) in descending order. These two scenarios provide a fair idea about the total geographical space available for tourism and other economic activities inside the State, leaving streams and horas (rivers) outside the reserve forest areas.

Table 6. Tourism Density Ratio (TDR) vs Total Density (TD) in Sikkim during 2011-2017 (Area in sq km)

Year	Total Tourist Influx*	Total Area Scenario I @	TDR I (2/3)	Total Area Scenario II @	TDR II (2/5)	Year-wise Projected Populace #	Total projected & floating Populace (2+7)	TD I (8/3)	TD II (8/5)
(1)	(2)	(3)	(4)	(5)	(6)	(7)	(8)	(9)	(10)
2011	576398	3753.51	153.56	4441.73	129.77	610577	1186975	316.23	267.23
2012	585027	3753.51	155.86	4441.73	131.71	616079	1201106	319.99	270.41
2013	608447	3753.51	162.10	4441.73	136.98	621520	1229967	327.69	276.91
2014	611593	3753.51	162.94	4441.73	137.69	626899	1238492	329.96	278.83
2015	743502	3753.51	198.08	4441.73	167.39	632216	1375718	366.52	309.73
2016	806775	3753.51	214.94	4441.73	181.64	637471	1444246	384.77	325.15
2017	1424965	3753.51	379.64	4441.73	320.81	642664	2067629	550.85	465.50
	Scale Reliability Statistics: Cronbach α							0.760	

Source: *Tourism Annual Reports, Government of Sikkim. www.sikkimtourism.gov.in #https://www.census2011.co.in/census/state/sikkim.html @ as per table-5.

Tourist arrival per square kilometer area (TDR) provides a good measure of the tourism intensity of a destination over a while. Table 6 depicts a comprehensive review of TDR concerning scenarios I and II drawn from Table 5. The TDR I ascertains the ratio of tourist arrivals concerning the actual area available for tourists (Scenario I) inside the destination. The derived TDR I was 153.56 visitors during 2011, gone up by 2.47 times to 379.64 visitors during seven years. In the TDR II, although arrivals per square km area had declined due to the inclusion of open forest locations, it had recorded a similar increase by 2.47 times to 320.81 during seven years.

Based on scenarios I and II, total density (TD) (comprising native and floating populations) per square kilometer grew exponentially during seven years. Again, TD I per square km was 316.23 persons in 2011, incrementally rising to 550.85 persons in 2017, recording an increase of 1.74 times. Similarly, TD II also registered incremental growth, which amounted to 1.74 times (465.5/267.32) during the period under review. The Cronbach α reliability score of TD I and II stands at 0.760 for the period.

Table 7. Host-Guest (HG) Ratio in Sikkim during 2011-2021

Months (1)	Month-wise Average Influx of Tourists during 2011-21 (Nos)			Average Projected Population 2011-21 (5)	HG Ratio (%) (5/4) (6)
	Domestic (2)	Foreign (3)	Total (4)		
January	61368.71	2949.29	64318.00	626775	10.26
February	64705.57	2829.14	67534.71	626775	10.78
March	72216.71	4430.43	76674.14	626775	12.23
April	105040.86	5452.00	110492.86	626775	17.63
May	135335.43	3100.71	138436.14	626775	22.09
June	82090.14	1853.86	83944.00	626775	13.39
July	19486.00	1279.14	20765.14	626775	3.31
August	21481.57	1514.57	23023.14	626775	3.67
September	22722.14	2090.14	24812.28	626775	3.96
October	53385.00	6483.29	59868.29	626775	9.55
November	38053.00	5248.00	43301.00	626775	6.91
December	48156.00	3471.00	51627.00	626775	8.24

Source: same as table 6

To reinforce the subject matter discussed in the scenario analysis, Table 7 depicts the average host-guest (HG) ratio for 2011-21. The host-guest ratios for seven months, from January to June and again in October, show prime tourism months for the Himalayan State. These signify the concentration of tourists across fourteen popular locations viz., Gangtok (capital), Tsomgo Lake, Nathula Pass, Kanchenjunga Base Camp, Yumthang Valley, Pelling, Zuluk, Teesta Valley, Gurudongmar Lake, Goecha La, Ravangla, Yuksam, Namchi, and Temi Tea Garden, in particular, outnumber the local community at each location by several times much beyond the acceptable limits. The HG ratio ranges between 3.31 percent in July to 22.09 percent in May during the period under review.

Table 8 depicts per capita solid waste generation among eight North Eastern States (NESs) during 2009-12. Despite holding the second largest geographical area and the highest native population, Assam ranked first in terms of least per capita municipal solid waste generation kilograms per day, followed by Manipur, Sikkim, and Arunachal Pradesh in descending order. Mizoram is the poorest among the eight NESs in terms of per capita municipal solid waste generation per day for 2009-12. The magnitude of municipal solid waste generation causes irreparable damage to the Eastern Himalayan landscape created by native and floating populations of eight NESs. Sikkim, being the fewer in size and with several inhabitants, per capita municipal solid waste generation of 0.06 kilogram per day left a lasting impact on land and masses due to the volume growth advocacy model followed for its tourism development over the years.

Table 8. Per Capita Municipal Solid Waste Generation in the Eastern Himalayas during 2009-12

NE States	Geographical Area (in sq km)	Population (2011 Census)	Municipal Solid Waste (MT /day)	Per Capita Solid Waste Generation (gm/day)	Rank
Arunachal Pradesh	83,743	1382611	93.802	0.06	IV
Assam	78,438	31169272	1146.28	0.03	I
Manipur	22,327	2721756	112.9	0.04	II
Meghalaya	22,429	2964007	284.6	0.09	V
Mizoram	21,081	1091014	4742.0	4.34	VIII
Nagaland	16,579	1080602	187.6	0.09	VI
Sikkim	7096	607688	40.0	0.06	III
Tripura	10,492	3671032	360.0	0.09	VII

Source: ENVIS Centre, Sikkim

In accommodation, the demand for beds and rooms, especially during peak months, might propel stakeholders to make alternative infrastructural arrangements, given the destination carrying capacity for a fragile landscape like the Himalayan state. The projected demand for beds and rooms is arrived at as per Inskeep's (1991) formulae for estimating accommodation requirements as given below:

DEMAND FOR BEDS: Total number of tourists (period) X Average length of stay (nights)/ Number of nights (period) X Accommodation occupancy (nights).

Annual Projected bed requirement for 2024: 1200000 X 5 nights/ 365 X 70% equals 23,483 beds. Peak period projected bed requirements for 2024: 816000

As per figures 1 and 2, 68% of tourists traveled to Sikkim during March, April, May, September, October, and November. Thus, X 5 nights/ 180 X 90% equals 25,185 beds.

DEMAND FOR ROOMS: Number of beds demand/ Average room occupancy (per person)

Annual Projected room requirement for 2024: 23,483/ 1.8 equals 13,046 rooms.

Peak season projected requirement for 2024: 25,185/1.8 equals 13,992 rooms.

The projected beds and rooms requirement for the destination for 2024 presents a bleak picture as the pandemic saw the closure of small accommodation units across Sikkim. The annual projection for beds and rooms requirement, with a probable 1200000 tourist influx, arrived at 23,483 beds and 13,046 rooms, according to Inskeep's (1991) model. Peak season, composed of March, April, May, September, October, and November months', projected beds and rooms requirement stands at 25,185 beds and 13,992 rooms.

Table 9. Spot Specific Guidelines for Visitors in Sikkim

Spots	Limits	Spots	Limits
Forest Park	Upto 15 tourists	Fishing/Sailing	5-30 tourists
Suburban nature park	15-70 tourists	Speed boating	5-10 tourists
High density picnicking	300-600 tourists	Water Skiing	5-15 tourists
Low density picnicking	60-200 tourists	Skiing	100 tourists per hectare of trails
Golf	10-15 tourists	Nature trail hiking	40 tourists per km
Nature trail horse riding	25-80 tourists per km		

Source: Reproduced from the thesis titled 'Tourism Management and Environment: A study on the Carrying Capacity and Prospect of Ecotourism in Sikkim' by S.K. Banerjee, 2009.

The above statistics reproduced from Banerjee (2009) enlist spot-specific guidelines for visitors touring Sikkim and encompasses all segments of tourists. Depending on the carrying capacity of the landscape, the limit of tourists visiting the spots ranges between 5-600 tourists'. The forest park, golf course, and water skiing have a maximum limit of 15 tourists visiting the site. Suburban nature parks can sustain 15-70 visitors at a time, while nature trail hiking and horse riding limited to 40 and 25 to 80 tourists per kilometer. The magnitude of low and high-density picnicking lies between 60 -200 and 300-600 tourists each. The volume growth advocacy model pursued dispensed with spot-specific limits for embracing mass tourism beyond prescribed acceptable limits.

DISCUSSION

Tourists outnumbering locals in Sikkim are a common feature, and communities residing in the state play a catalytic role in hosting visitors. The perennial environment versus employment debates overpowers the former with a preference for volume growth advocacy to maximize revenues from tourism. Tourism emerged as a prime money spinner for the local populace vis-à-vis organic farming, and the latter's production pattern is suitably modified to meet the visitors' needs. Thus, the advancement of tourism persistently encroaches on the forest lands by downsizing the flora and fauna of the state.

Inskeep (1991, as cited in Vehbi 2012, p102) shed light on tourism-borne unsustainable practices impairing destination areas. These include: poorly designed accommodations and other tourist facilities inconsistent with local architectural designs drafted into the environment. Ill-suited building substances, improper planned arrangements for tourism facilities, and interference in scenic views through tourism development.

A traditional "Dokemoo Lee" Lepcha house signifies a 'sit on the top house' on hill slopes consistent with sustainable practices in mountainous terrains over the years. Of the 1228 homestays (Bhutia, 2022), bulk houses that follow Dokemoo Lee architecture in Sikkim while providing the best possible services to the guests were replaced with brick-and-mortar structures impairing the fragile landscapes. Mushrooming of concrete structures across urban centers exposes poorly planned arrangements for tourism facilities over sustainability considerations. Again, the collapse of a few concrete structures across urban settlements raises the question of ill-suited building substances. Notably, the construction of 'Chardham' within five kilometers of Namchi town draws several visitors of Hindu faiths to this Himalayan state, government initiatives of this kind encroach on open forest areas for tourism development beyond the acceptable limits. Similarly, vehicular movement, from inside and outside Sikkim, through narrow passages for connecting visitors to tourism sites takes a heavy toll on select tourism pockets, which enhances vehicular pollution as well.

In the premium database hotel sector, the number of active hotel rooms stands at 459 (CEIC, 2020) in Sikkim, leaving a sizeable segment for budget hotels. According to official estimates, 12,103 rooms accompanying 10,557 double beds and 5332 single beds can accommodate 21,785 visitors on a particular day for a stay-over in Sikkim. During 2018 and 2019, the state embraced 14,26,127 and 14,21,823 tourists, leaving 26,938 and 26,856 visitors per night, depending on the length of stay. The existing accommodation infrastructure and adjoining services are inadequate to meet these spiraling demands. Again, it is not feasible to invest in accommodation infrastructure primarily to meet peak season demand keeping the off-season occupancy factor in view.

Incidentally, twin cities', Gangtok (East) and Namchi (South), enlisted under National Smart City Mission projects pursued by the government of India. Smart tourism envisages stakeholders' application of technology in tourism which is also considered necessary in the post-pandemic scenarios. The use of technology in tourism, especially in travel bookings, getting tickets, scheduling destination trips, and tracking systems, will determine tourist activities in fragile landscapes.

CONCLUSION

Over-tourism and sustainability was the main issue for Sikkim tourism before the pandemic inroads. Despite negative tourism receipts, the pandemic outbreak provided a much-needed opportunity to address unsustainable tourism practices that cropped up. Zoning of landscape based on land use plans, keeping in mind stakeholder interests, is required for extending physical infrastructure in open forest areas, especially for the construction of hotels, restaurants, and road and rail networks (NA: 2018). Again, biodiversity conservation should accord priority, besides maintaining ecological balance for each developmental work.

According to Scenario I and II (table 5), the East district holds 3.4 percent and 5.2 percent of the habitable land area of the State. The available reports highlight 70 percent of tourists visiting Sikkim toured and stayed in the East district alone. The projected demand for beds and rooms suggests existing accommodation infrastructure and adjoining services are inadequate to meet these spiraling demands in the post-pandemic scenario. The concentration of this kind perpetually demeans destinations carrying capacity in favor of volume growth advocacy. Limiting the size of tourists (as per Table 9), dispersal to other destinations, overhauling accommodation and local transport, and revisiting pricing policies of services, health and safety measures, and adjoining services can address these impending issues. As the industry experts opined in favor of small but sustainable tourism, contrary to the tourism advocacy model, enforcing a regulatory mechanism coupled with an awareness campaign for the stakeholders is well on the cards. Stakeholders of the core group, including policymakers, have to shoulder broader responsibility concerning the planning and execution of tourism activities along the sustainability lines.

Homestay plays a catalytic role in dispersing tourists to satellite zones from overcrowded tourism sites, the best place to practice micro tourism away from the noisy surroundings of townships with Dokemoo Lee structures. Homestays provide a comprehensive package for the tourists on the one hand while generating revenues for the communities at large on the other. Incidentally, Sikkim's endowed with 1228 homestays across the territory, the bulk of them managed by village committees with eco-friendly practices.

Again, fostering pilgrimage tourism circuits can disperse tourists from urban centers. In this regard, vertical collaboration with the Ministry of Tourism, the Government of India, and tourism governing bodies from neighboring countries, like Nepal and Bhutan, seems a mutually beneficial exercise. Similarly, horizontal collaboration with the rest of the Indian states also facilitates a smoother transfer of information, besides ensuring safe passage for tourists.

The economic revitalization package of stakeholders adversely affected by the pandemic will refrain them from intensifying tourism activities. In this regard, provisions for enlisting tourism entrepreneurs in priority sector lending, tax rebates, a moratorium in existing and new loan installments, and provision for soft loans should accord due attention. The institution of travel insurance from the revenue generated from tourists can be an option to ponder over and exploit for those enterprises in distress. In this regard, re-purposing tourism as a supporter of community development is the need of the hour as the conducted tour operators from outside eat into tourism receipts generated.

Tourism's contribution to GSDP accentuated all sectors of this Himalayan economy. Apart from this, employing relative indicators like the Gross Happiness Index (GHI) of stakeholders bode well for the neighboring country Bhutan with similar topography. The thrust of GHI is on the well-being of those engaged in tourism for a considerable period. However, the approval of criteria for GHI is another issue that needs due cognizance of stakeholders.

Designing a joint code of conduct to address the post-pandemic situation and sustain resources for future generations has become a necessity. The role of tourism police, perhaps, extended to maintaining a code of conduct with strict adherence to a select number of tourists visiting per site. Stakeholders' involvement in devising spot-specific plans and their execution at the destination level help control the tourist movement along overcrowded zones. The strategical shift from budget-centric volume advocacy to high-value less-impact tourism can save the land from the recurrence of the Dehradun, Joshimath-like situation in the Northern Himalayas, known for unrestrained development works.

REFERENCES

Banerjee, S. K. (2009). Tourism Management and Environment: A study on the Carrying Capacity and Prospect of Ecotourism in Sikkim. PhD Thesis, North Bengal University, West Bengal. https://ir.nbu.ac.in/bitstream/123456789/1285/18/235578.pdf. Accessed on 25th March, 2023 at 10 am.

Bhattacharya, S. (2019). Environmental Crisis in the Eastern Himalayan Landscapes in India. *Consilience: Journal of Sustainable Development*, *21*(1), 65–85.

Bhutia, S. (2015). Spacio-Temporal Study on Urbanization in the Darjeeling Himalaya: A Demographic Perspective. *IOSR Journal of Humanities and Social Sciences*, *20*(4), 10–18.

Bhutia, S. G., Bhutia, J. W., & Mahapatra, S. S. (2022). Analysing the Issues and Challenges of Homestays for Sustainable Tourism Development in Sikkim and Darjeeling Hills. *Orissa Journal of Commerce*, *43*(4), 73–88. doi:10.54063/ojc.2022.v43i04.06

Boers, B., & Cottrell, S. (2007). Sustainable Tourism Infrastructure Planning: A GIS Supported Approach. *Tourism Geographies*, *9*(1), 1–21. doi:10.1080/14616680601092824

Brouder, P., Teoh, S., Salazar, N. B., Mostafanezhad, M., Pung, J. M., Lapointe, D., Higgins Desbiolles, F., Haywood, M., Hall, C. M., & Clausen, H. B. (2020). Reflections and discussions: Tourism matters in new normal post Covid 19. *Tourism Geographies*, *22*(3), 735–746. doi:10.1080/14616688.2020.1770325

Butowski, L., Makowska-Iskierka, M., & Pokojski, W. (2018). Tourism Space versus Tourism Destinations: Methodological Considerations and Empirical Testing of their Development. *Geografie (Utrecht)*, *121*(1), 85–106. doi:10.37040/geografie2018123010085

CEIC. (2020). Number of Hotel Rooms: Sikkim. https://www.ceicdata.com/en/india/number-of-hotel-rooms-by-states/number-of-hotel-rooms-sikkim. Accessed on 23rd January, 2023 at 9 pm.

Chand, D., Sati, J., Laxmi, V., & Nautiyal, M. C. (2016). Effects of Anthropogenic Activities in High Altitude Areas of Garhwal Himalayas. *International Journal for Environmental Rehabilitation and Conservation, II*(1), 78–87.

Department of Tourism (DoT). (2019). Tourism Annual Report 2018-29, Government of Sikkim.

Department of Tourism (DoT). (2020). Tourism Annual Report 2019-20, Government of Sikkim.

District Census Handbook (2011). Census of India, Sikkim, Series 12, Part XII B, Directorate of Census Operations, Sikkim.

Dwyer, L., Forsyth, P., & Spurr, R. (2006). Effects of SARS crisis on the economic contribution of tourism to Australia. *Tourism Review International, 10*(1), 47–55. doi:10.3727/154427206779307231

Economic Times (ET). **(2019). Where do tourists prefer to go in India?** https://economictimes.indiatimes.com/industry/services/travel/where-do-tourists-prefer-to-go-in-india/who-spends-the-most-on-tourism/slideshow/72516703.cms Accessed on 25th March, 2023 at 5.31 pm.

Economic Times (ET). (2020). Gross State Domestic Product to see sharp decline in FY21. https://economictimes.indiatimes.com/news/economy/indicators/state-gross-domestic-products-to-see-sharp-declines-this-fiscal/articleshow/76688108.cms?from=mdr. Accessed on 15th March, 2023 at 5.45 pm. e Silva, F.B., Herrera, M.A.M., Rosina, K., Barranco, R.R., Freire, S. and Schiavina, M. (2018). Analysing Spaciotemporal Pattern of Tourism in Europe at High-resolution with Conventional and Big Data sources. *Tourism Management*, 68, 101-115.

Federation of Indian Chamber of Commerce and Industry (FICCI) (2020). *Travel and Tourism Survive, Revive and Thrive in Times of Covid 19*. Grant Thornton, June 2020.

Forest Survey in India (FSI). (2019). *India State of Forest Report 2019, Sikkim*. Vol. II, Govt. of India. https://fsi.nic.in/isfr19/vol2/isfr-2019-vol-ii-sikkim.pdf. Accessed on 5th April, 2023 at 9 pm.

Gossling, S., Scott, D., & Hall, C. M. (2020). Pandemics, tourism and global change: A rapid assessment of COVID 19. *Journal of Sustainable Tourism, 29*(1), 1–21. doi:10.1080/09669582.2020.1758708

Government of Sikkim (GoS). (2011). *Sikkim Action Plan for Climate Change*. Draft Report, Water Security and Public Health Engineering Department.

Hall, C. M., Scott, D., & Gossling, S. (2020). Pandemic Transformations and Tourism: Be careful what you wish for. *Tourism Geographies*, *22*(3), 577–598. do i:10.1080/14616688.2020.1759131

Haywood, K. M. (2020). A Post Covid 19 future-tourism re-imagined and re-enabled. *Tourism Geographies*, *22*(3), 599–609. doi:10.1080/14616688.2020.1762120

Honeck, D. (2008). LDC Poverty Alleviation and the Doha Development Agenda: Is Tourism Being Neglected? Staff Working Paper ERSD 2008-03, Economic and Statistics Division, WTO. https://www.wto.org/English/res_e/reser_e/ersd200803_e.pdf. Accessed on 24th February, 2023 at 5 pm.

Hong, Y., Cai, G., Mo, Z., Gao, W., Xu, L., Jiang, Y., & Jiang, J. (2020). The Impact of Covid 19 on Tourists Satisfaction with B & B Zhejiang, China: An Importance-Performance Analysis. *International Journal of Environmental Research and Public Health*, *17*(10), 1–19. doi:10.3390/ijerph17103747

Hultsman, J., Cottrell, R. L., & Zales-Hultsman, W. (1999). *Planning Parks for People*. Venture Publishing Company.

IBEF. (2019). Sikkim State Presentation. https://www.ibef.org/states/sikkim-presentation. Accessed on 15th January, 2023 at 7 pm.

India Today (IT). (2018). Sikkim becomes the World's First Organic State wins Oscars for Best Policies by UN, October 16. https://www.indiatoday.in/education-today/gk-current-affairs/story/sikkim-becomes-world-s-first-organic-state-wins-oscar-for-best-policies-by-un-1369158-2018-10-16. Accessed on 25th March, 2023 at 9.30 pm.

India Tourism Statistics(ITS). (2022). *India Tourism Statistics 2022*. Market Research Division, Ministry of Tourism, Government of India.

Inskeep, E. (1991). *Tourism Planning: An Integrated and Sustainable Development Approach*. Van Nostrand Reinhold.

Ioannides, D., & Gyimóthy, S. (2020). The COVID-19 crisis as an opportunity for escaping the unsustainable global tourism path. *Tourism Geographies*, *22*(3), 624–632. doi:10.1080/14616688.2020.1763445

Johnston, D., & Braden, J. (2008). Green landscapes: Exogenous economic benefits of environmental improvement. In J. Wescoat & D. Johnston (Eds.), *Political Economies of Landscape Change* (pp. 131–144). Springer. doi:10.1007/978-1-4020-5849-3_7

Khazami, N., Lakner, Z., & Nefzi, A. (2020). Pandemic and Tourism: Re-preparation of Tourism Post Covid 19. *Journal of Hotel and Business Management*, *9*(2), 1–5.

Koscak, M., & O'Rourke, T. (2021). *Tourism Management: The New Reality of Managing Ethical and Responsible Tourism*. Routlegde.

Krugman, P. R., & Obstfeld, M. (2006). *International Economics, Theory and Policy* (7th ed.). Pearson Addison.

Kumari, S. Tiwari, H.R., and Inbakaran, Robert John (2006). Sustainability Criteria and Indicators as Tool to Evaluate the Visitors Impact on Fragile Ecosystem Destinations: A Case of Khanchendzonga Biosphere Reserve, Sikkim, India. https://www.researchgate.net/publication/237087340_Sustainability_Criteria_and_Indicators_as_Tools_to_Evaluate_the_Visitor_Impact_on_Fragile_Ecotourism_Destinations_A_Case_Study_of_Kanchandzonga_Biosphere_Reserve_Sikkim_India. Accessed on 10th January, 2023 at 7.30 pm.

Lama, R. (2018): Potential of Homestays as Tourism Products in Sikkim. https://www.researchgate.net/publication/325076920_Potential_of_Homestays_as_Tourism_Product_in_Sikkim Accessed on 25th March, 2023 at 10.02 pm.

Lisowski, A. (2003). Concept of Space in Geography. Studia Regionalne I Lokalne, UW, Warsaw.

Maan, J. S., & Chudhry, P. (2019). People and Protected Areas: Some Issues from India. *Animal Biodiversity and Conservation*, *42*(1), 79–90. doi:10.32800/abc.2019.42.0079

Malik, Z. A., Pandey, R., & Bhatt, A. B. (2016). Anthropogenic Disturbances and their Impact on Vegetation in Western Himalayas, India. *Journal of Mountain Science*, *13*(1), 69–82. doi:10.100711629-015-3533-7

Manera, C., & Valle, E. (2018). Tourist Intensity in the World 1995-2015: Two Measurement Proposals. *Sustainability (Basel)*, *10*(12), 1–20. doi:10.3390u10124546

Marques Santos, A., Madrid Gonzalez, C., Haegeman, K., & Rainoldi, A. (2020). *Behavioural Changes in tourism in times of Covid 19, Employment Scenarios and Policy Options*. Publication Office of European Union.

Mowforth, M., & Munt, I. (2003). *Tourism and Sustainability: Development and new tourism in the Third World* (2nd ed.). Routledge. doi:10.4324/9780203422779

Mowforth, M., & Munt, I. (2009). *Tourism and sustainability: Development and new tourism in the Third World*. Routledge.

Nepal, S. K. (2020). Adventure Tourism and Travel after Covid 19- business as usual or opportunity to reset? *Tourism Geographies*, *22*(3), 646–650. doi:10.1080/14616688.2020.1760926

Nientied, P., & Shutina, D. (2020). Tourism in Transition, the Post Covid 19 aftermath in the Western Balkans, Co-Plan Resilience Series 2. https://www.researchgate.net/publication/341607053_Tourism_in_Transition_The_post_COVID-19_Aftermath_in_the_Western_Balkans Accessed on 15th March, 2023 at 6.31 pm.

Niti Ayog (NA). (2018). Sustainable Tourism in the Indian Himalayan Region. Report of the Working Group II, Govt. of India, New Delhi.

Nyaupane, G. P., & Timothy, D. (2016). Bhutan's Low Volume, High Yield Tourism: The Influence of Power and Regionalism. https://www.semanticscholar.org/paper/Bhutan%E2%80%99s-Low-volume%2C-High-yield-Tourism%3A-The-of-and-Nyaupane-Timothy/4708bc5cea60e9d4215a394864e236134aec58c8. Accessed on 20th April, 2023 at 8.45 pm.

Organization for Economic Cooperation and Development (OECD). (2020). The Territorial Impact of Covid 19: Managing the Crisis across levels of Governments. https://read.oecd-ilibrary.org/view/?ref=128_128287-5agkkojaaa&title=The-territorial-impact-of-covid-19-managing-the-crisis-across-levels-of-government Accessed on 10th March, 2023 at 11.22 pm.

Orindaru, A., Popescu, M., Alexaoei, A. P., Caescu, S., Florescu, M., & Orzan, A. (2021). Tourism in a Post-Covid 19 Era: Sustainable Strategies for Industry's Recovery. *Sustainability (Basel)*, *13*(12), 1–22. doi:10.3390u13126781

Romagosa, F. (2020). The COVID-19 crisis: Opportunities for sustainable and proximity tourism. *Tourism Geographies*, *23*(3), 690–694. doi:10.1080/14616688.2020.1763447

Rubita, I. (2012). Sustainability Issues in Tourism: A Case study of Yuksam Village, Sikkim [AJTS]. *Atna Journal of Tourism Studies*, *7*(2), 117–129. doi:10.12727/ajts.8.7

Santos-Roldán, L., Canalejo, A. M. C., Berbel-Pineda, J. M., & Palacios-Florencio, B. (2020). Sustainable Tourism as a Source of Healthy Tourism. *International Journal of Environmental Research and Public Health*, *17*(15), 1–14. doi:10.3390/ijerph17155353 PMID:32722271

Seabra, C., & Bhatt, K. (2022). Tourism Sustainability and Covid 19 Pandemic: Is there a Positive Side? *Sustainability (Basel)*, *14*(14), 1–14. doi:10.3390u14148723

Sigala, M. (2020). Tourism and Covid 19: Impacts and Implications for Advancing and Resetting Industry and Research. *Journal of Business Research*, *117*, 312–321. doi:10.1016/j.jbusres.2020.06.015 PMID:32546875

Swarbrooke, J. (1999). *Sustainable Tourism Management*. CAB International. doi:10.1079/9780851993140.0000

Tamang, K., & Solanki, V. (2020). Tourist Perception and Satisfaction Level: A Study of East Sikkim Tourism Industry. *Journal of Xidian University*, *14*(4), 2559–2575.

Ulak, N. (2020). A Preliminary Study of Novel Corona Virus Disease (Covid 19) Outbreak: A Pandemic Leading Crisis in the Tourism Industry in Nepal. *Journal of Tourism and Hospitality Education*, *10*, 108–131. doi:10.3126/jthe.v10i0.28763

UNEP & UNWTO. (2005). *Making Tourism More Sustainable - A Guide for Policy Makers*. UNEP and UNWTO.

UNWTO. (2007). Tourism and Least Developed Countries: A Sustainable Opportunity to Reduce Poverty. https://www.unwto.org/sustainable/doc/tourism-and-ldc.pdf, Accessed on 10th March, 2023 at 11.32 pm.

UNWTO. (2022): Impact Assessment of the Covid-19 Outbreak on International Tourism, UNWTO. www.unwto.org. Accessed on 11th March, 2023 at 10.20 pm.

Vaisakh, J., Swarna Abirami, L., Krishnachandran, C., & Arun, K. (2020). Tourism Marketing Through Social Media After Covid 19 in Kerala. *International Journal of Advanced Science and Technology*, *29*(7), 11481–11486.

Vehbi, B. O. (2012). A Model for Assessing the Level of Tourism Impacts and Sustainability of Coastal Cities https://cdn.intechopen.com/pdfs/35710/InTech-A_model_for_assessing_the_level_of_tourism_impacts_and_sustainability_of_coastal_cities.pdf Accessed on 9th March, 2023 at 9.12 pm.

World Economic Forum (WEF). (2022). *Travel and Tourism Development Index 2021: Rebuilding for a Sustainable and Resilient Future*. Geneva, Switzerland.

Yassin, J., Aralas, S., & Basa, D. E. K. (2021). Tourism Density Effect on Environmental Performance Index: Evidence in ASEAN Countries, *Applied. Environmental Research*, *43*(1), 93–101.

Chapter 11
Study of Micro and Small Enterprises' Readiness in Implementing Industry 4.0:
A Study in Marathwada District of Maharashtra, India

Saibal Kumar Saha
https://orcid.org/0000-0002-7842-698X
Christ University, India

Janhavi Ravindra Deshmukh
Christ University, India

N. Ramakrishnan
https://orcid.org/0000-0002-2306-2218
Christ University, India

ABSTRACT

Industry 4.0 aims tp transform the development of global value chains and the development of a digital revolution, with intelligent machines capable of communicating via wireless connections and a connection thought system, resulting in autonomous decision-making. Although large sized firms are adopting Industry 4.0, the small and micro enterprises are facing great difficulties in adopting them. This study aims to identify the areas in which Enterprises need to focus for improving their level of readiness and develop strategies and plans to adopt Industry 4.0 technologies successfully. 219 samples were collected using snowball sampling from Marathwada District in Maharashtra, India. factor analysis was conducted using SPSS and different factors acting as barriers to implementation of Industry 4.0 technologies were identified.

DOI: 10.4018/978-1-6684-9089-1.ch011

INTRODUCTION

"Micro and Small-sized Enterprises" are found to be the emerging force of many industrial economies in India. Hence, they play an essential role across the sector (Pedraza, 2021). Recently, the Fourth Generation of the Industrial Revolution, dubbed "Industry 4.0," has had a significant influence by using technology such as artificial intelligence (A.I.), robotization, automation, and many more. Industry 4.0 was founded in Germany in 2011 by the "German Federal Government" to encourage the competitiveness of the "German manufacturing industry". Many technologies lie under Industry 4.0, including the "Internet of Services" (IoS), "Internet of Things" (IoT), "smart manufacturing", and "cyber-physical systems" (CPS) (Cardoso & Pereira, 2022; Brintha et al., 2019). Industry 4.0 will transform the development of global value chains and the development of a digital revolution, with intelligent machines capable of communicating via wireless connections and a connection thought system, resulting in autonomous decision-making by visualizing the manufacturing process. IoT is different from The Industrial Internet of Things (IIoT) as IOT stands for the Internet to a "thing," but IIoT is when data is collected from actuators, sensors, and various devices to be precise with the manufacturing environment of the industry. On the other hand, robots are used for similar tasks and automation for precision manufacturing operations. Robots are noted for performing tasks on schedule while maintaining flexibility, safety and collaboration with other humans and robots (Hassan et al., 2021).

The manufacturing industry, concerning "MICRO AND SMALL ENTERPRISEs", has limited Smart way of Manufacturing and Industry 4.0 roadmaps, maturity models, and frameworks (Grooss et al., 2022). The 4th industrial revolution has stressed micro and small enterprises to adopt the new technology. Few analyses specifically focus on supporting "micro and small enterprises" path, and "Micro and small enterprises" are not given priority while concentrating on "Advanced manufacturing technologies" (AMTs) hence micro and small enterprises do not retain a significant effect on intelligent factory adoption intention and readiness evaluations that are already available that represent the unique requirements and problems of "Micro and small enterprises" (Cotrino A et al. 2020)

MSMEs are the foundation of India's economic expansion (Kumar et al., 2009). Additionally, these MSMEs employ 120 million people, and 80% of them are located in rural areas (Shukla & Prajapti, 2019). The limited clarity of objectives in adopting new technological culture and priorities required to form strategies, and stakeholders in Micro and small enterprises need more resources and experience in engaging new technologies (Horváth & Szabó, 2019). Moreover, the Micro and small enterprises' Stakeholders will have financial constraints to upgrade and adopt advanced manufacturing technologies (AMTs). To realize the benefits of Industry 4.0, businesses must invest significantly (Ghobakhloo et al., 2022).

RESEARCH OBJECTIVES

The key objectives of this research are specified below:

1. To study the readiness of micro and small enterprises in the adoption of Industry 4.0
2. To identify the focus areas that the enterprises need to consider for enhancing their level of readiness towards Industry 4.0

LITERATURE REVIEW

Industry 4.0

Industry 4.0 is characterized as "a new degree of collaboration and command over the whole value chain of an item's life cycle; it is prepared for more individualized client needs" (Vaidya et al., 2018). Mechanical innovations have pushed and constrained every organization in the manufacturing industry to adapt to the new climate at some time in each modern upheaval that mankind has faced in the preceding few hundred years (Rüßmann et al., 2015). Nonetheless, Industry 4.0 stands out, especially in terms of the pace and extent of its evolution. This upheaval results in a considerable movement in responsibilities, administrations, and executives, as well as a transition from a concentration on large-scale manufacturing to recreated production. Industry 4.0 is unavoidable, owing to the opportunities for developing products, administrations, and activities to increase the intensity of any given substance (Ślusarczyk, 2018). Changes in the employable structural conditions were one of the primary goals of Industry 4.0, and several factors restricted this. The first is the more constrained development time frames, where there is a strong need for more imaginative solutions to boost production and shorten development cycles. The next is an increased interest in changed creation due to a move from an economically challenging market to a fast-moving commercial sector. Third, there is a more critical requirement for adaptation. Fourth, decentralization has become essential for enabling the treatment of the required speed of direction. Finally, asset effectiveness is critical due to the need for increased creation effectiveness (Lasi et al. 2014). The concept of Industry 4.0 is founded on several advancements. First and foremost, there is the Internet of Things, which refers to the infinitely extended network of everyday items that may communicate with one another over the Internet.

Furthermore, the "Industrial Internet" may be used to rebuild and further enhance an organization's business operations by breaking down the results of massive analyses from massive informational indexes. Smart Fabricating refers to the extensive and prevalent use of coordinated data-based innovations throughout a company operation's complete assembling and inventory network element. The next concept is "Cloud-based Manufacturing", which is a conglomeration of some newly developed innovations linked with the enlarged Advancements throughout the manufacturing process (Vaidya, et al., 2018). As Big Data is enabling these advances, it is vital for an organization to guarantee that it can cope with it (Coleman et al. 2016). Organizations will be able to gather and analyze data across and between machines as a result of Industry 4.0, resulting in quicker, better, and more adaptable cycles to manufacture higher-quality items at a cheaper cost than was previously the case. This will result in better fabrication usefulness, benefit current development, and a change in workforce profile, which will result in differences in seriousness across organizations within the company. Furthermore, Industry 4.0 advancements, communication and collaboration among people, machines, and parts will accelerate production to the point where it will be 30% faster and 25% more efficient, resulting in increased mass customization as this modern insurgency pushes toward customization and away from standardization.

Impact of Artificial Intelligence on Industry 4.0

"Artificial intelligence" is the most crucial component of Industry 4.0. A.I. works with big volumes of data, also known as "Big Data", which refers to the massive quantity of data collected over the life cycle of an object (Qi and Tao, 2018). Since it is practically impossible for any person to acquire and analyze such a massive quantity of information, "artificial intelligence" (A.I.) enables any organization or association to gather and analyze data using Machine Learning and Deep Learning (Tang, 2016). Industry 4.0, as previously said, moves the purpose of large-scale production assembly to modified manufacturing. As a result, A.I. may not enable organizations to save long-term expenditures through enhanced disappointment recognition throughout creation. Nonetheless, it enables organizations to adjust their output in response to specific client feedback (Vaidya et al., 2018).

Industry 4.0 In Micro and Small Enterprises

The ability to execute improvements for Industry 4.0 differs significantly between micro and small firms and large ventures since capabilities, assets, and tasks frequently vary considerably, rendering micro and small enterprises unsuitable for competing in Industry 4.0. Decentralization major component of Industry 4.0 for manufacturing sector. Since micro and small firms are less regulated, it may give them an advantage when carrying out Industry 4.0 advancements (Stentoft et al., 2021). Their motivations to improve would enable them to become more proficient and reduce their expenses are limitless. They fail to do so, however, due to a lack of specialized skills and financial resources.

However, many micro and small organizations are increasingly using cloud-based frameworks to socialize and provide rapid access to internal data. As a result, one approach for micro and small firms to implement at least some type of innovation fit for Industry 4.0 is to start with the least sophisticated or expensive technology (Rizos et al., 2016).

The world is changing because of new technology. The Industrial Revolution 4.0 necessitates using technology to manage operations across industries and organizations. Businesses must adapt to an ever-changing environment. Many current projects are based on automated procedures that employ Machine Learning approaches to execute projects with data as the key. Manufacturing and logistics industries have already embraced technological advancements. Organizational assessment models, often known as stages-of-growth models, are used to measure organizational maturity. Maturity models are a tool for assessing organizations and enterprises, exhibiting more paths and restricted ways of doing business (Amaral & Peças, 2021).

Readiness for Industry 4.0

With the implementation of new technologies, practices today are constantly evolving. The Industrial Revolution-4.0 necessitates using technology to manage operations across industries and organizations, including micro and small enterprises. According to Stentoft et al., (2021), the term "industry 4.0 readiness" might be substituted with "Information Technology readiness" to describe the extent to which organizations can use and benefit from Information technology. Industry 4.0 is more concerned with the implementation of technology; Information Technology readiness factors are also applicable in a digital technology context. That lack of readiness in micro and small enterprises has been recognized as one of the significant reasons for ERP implementation failures. Mittal et al., (2018) tried to assess micro and small enterprise preparation for industry 4.0 through well-known maturity models and determine whether each model would 'fit' with the micro and small enterprise standards. De Carolis' workshop series discovered that the essential pillars of intelligent manufacturing are technological maturity, information connection, organization, process, and personnel competency.

Industry 4.0 readiness is a way that can help businesses prepare for issues that may arise when transitioning to the new revolution (Grufman, 2020). The model of industry readiness evaluation happens before the start of the maturing process. To move forward with the model, the company must first evaluate if it is ready to embark on an Industry 4.0 transformation process. A maturity evaluation, on the other hand, establishes the company's maturity level.

The ubiquitous factory concept was conceived as a synonym for intelligent factories in terms of manufacturing technology integration and expansion, emphasizing the potential of ubiquitous computing technologies applied to the manufacturing sector (Yoon et al., 2012). As manufacturing enterprises in micro and small enterprises face significant challenges in disruptive concepts such as the Internet of Things, Cyber-Physical Systems, and so on, this has resulted in uncertainty in organizational, technological capabilities and complementary strategies to adopt (Schumacher, 2016). The model for assessing industry 4.0 maturity for industrial firms focuses on discrete production. The primary purpose is to expand the dominant technology by focusing on newly established models while considering organizational variables like Customers, Products, Technology, and Operations, which are discovered to be essential enablers.

The expansion of Industry 4.0 using technology implementation in micro and small businesses has been identified as a problem because it necessitates a multi-criteria strategy that includes quantitative and qualitative factors. Chang et al., (2021)'s study ties technology-organization-environment (TOE) to the technological implementation of Industry 4.0 is more multidimensional than previous methodologies, necessitating a full investigation. The study uses multiple-criteria decision-making (MCDM) to perform influence weight prioritization for the industry 4.0 technology implementation to micro and small enterprise aspects.

According to Brozzi et al., (2021), digital-level evaluation is critical for micro and small enterprises in the industrial sector since Industry 4.0 deployment is their digital strategy answer. Several assessment methods are available to test digital readiness and maturity; however, using such tools is highly successful, so firms reflect their digital level, shortcomings, strengths, and obstacles before Industry 4.0 deployment. The practical problems are more significant than large organizations, especially with the digital transformation of management, since expertise relies solely on their entire digital level owing to the restricted aspects of strategic decision-making. The study by Won & Park (2020) investigated the drivers and consequences of intelligent factory adoption in small and medium-sized enterprises (SMEs). The fourth industrial revolution's rise necessitated the use of information technology, emphasizing the company's experience in the manufacturing industry. Furthermore, it noticed that assistance from the government and academic research is apparent in several circumstances.

Various theories in existence could be linked to Industry 4.0 and its technologies. Some of the theories that are considered here include the following:

a) Institutional theory
b) Diffusion of Innovation theory
c) Social Cognitive theory
d) Strategic Management theories

Institutional theory proposed by John Meyer and Brian Rowan explains that the institutional environment could influence the creation of formal organizational structures. It encompasses processes through which rules, norms. structures etc. are formulated. The institutional theory suggests the tendency of an organization to shift towards conformance with norms and social influences in their operating environment. This is quite relevant in the adoption of Industry 4.0.

Diffusion of Innovation theory proposed by Rogers' Everett (1995) could be utilized to explain the adoption of new technologies by users. This theory specifies that the rate of diffusion could be impacted by various factors such as complexity, compatibility, innovation advantage, and observability. The adoption of various technologies while deploying Industry 4.0 practices could fit well with this theory.

Social Cognitive theory proposed by Albert Bandura could also be applied in the context of Industry 4.0. This theory has an emphasis on the dynamic interactions that occur between the individual factors of people, their behaviour, and their environment. This theory is relevant from Industry 4.0 perspective considering the interactions that happen between employees in an organization and the related ecosystem, the behaviour that they exhibit, and the operating cyber-physical environment.

Industry 4.0 practices and their adoption could also be associated with theories in the realm of strategic management. This is relevant considering the linkage between organization strategy, technology adoption strategy, and the ICT (Information and Communication Technologies) strategy.

METHODOLOGY

This research made use of a quantitative approach and data collection was done using a survey questionnaire. The questionnaire was developed with a 5-point Likert scale and sent to the target audience which comprised respondents in the Marathwada region. The questionnaire was shared through email, WhatsApp, telephone calls and face-to-face interview. Snowball sampling method was used for collecting the responses. 219 responses were collected from people working in micro and small enterprises from Marathwada District in Maharashtra in March 2023. The respondents belonged to different levels in the organization such as Manager, Supervisor, Owner, and other categories. Four categories were considered based on the number of employees in the organization, namely, less than 20 employees, number of employees between 21 and 40, and number of employees between 41 and 60. The survey responses were collected from respondents in various domains such as Agriculture, Manufacturing, Transportation, Textile and apparel, Retail, and Others.

SPSS was used as the data analysis tool. Correlation and Regression analysis, and Factor analysis were performed and the results were subjected to analysis. Table 1 provides the variables used for this study and Table 2 displays the sample characteristics.

The respondents were from four cities of Marathwada, indicated with different colors in a pie chart as shown in figure 1. The majority of respondents were from Aurangabad region 43%, followed by Nanded 23%, Parbhani 18% and Beed 16%.

Table 1. Description of variables

Sl. No.	Variables	Description
1	AII	Awareness about Internet of things
2	AAM	Awareness about additive manufacturing
3	AR	Awareness about Robotics
4	ICC	Implementation of cloud computing
5	ISC	Implementation of system integration
6	IAM	Implementation of additive manufacturing
7	IR	Implementation of robotics
8	PAM	Priority of adoption based on management
9	PAF	Priority of adoption based on funding
10	PAEA	Priority of adoption based on employee accessibility
11	PARI	Priority of adoption based on return of investment
12	BCE	Benefits from implementation based on cost effectiveness
13	BRI	Benefits from implementation based on easy return on investment
14	BIE	Benefits from implementation based on increased efficiency
15	BQA	Benefits from implementation based on quality assurance
16	BS	Benefits from implementation based on sustainability

Table 2. Sample characteristics

Designation of Respondents		No of employees		Sector	
a. Manager	27%	a. 1 - 20 employees	23%	a. Agriculture	19%
b. Supervisor	21%	b. 21 - 40 employees	40%	b. Manufacturing	31%
c. Owner.	41%	c. 41 - 60 employees	37%	c. Transportation.	18%
d. Other	11%			d. Textile and Apparel	20%
				e. Retail	11%
				f. Other	1%

Figure 1. Regions of survey

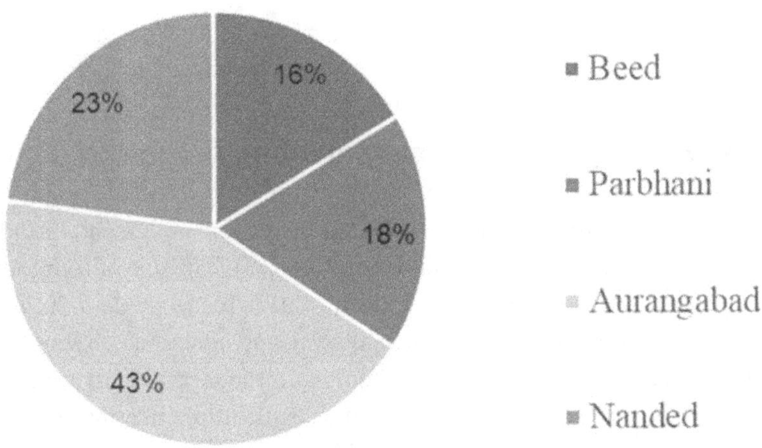

RESULTS AND DISCUSSION

Hurdles in Implementing Industry 4.0

According to the survey, the primary hurdle for the respondent to implement Industry 4.0 was unawareness. 35% of the respondents reported that they are not aware of the technological advancements that have been made in the sector of their work. 31% of respondents reported that they had financial constraints. 23% feared resistance from the employees while 10% said that they do not have skilled workforce to handle Industry 4.0.

Figure 2. Hurdles in Industry 4.0 implementation

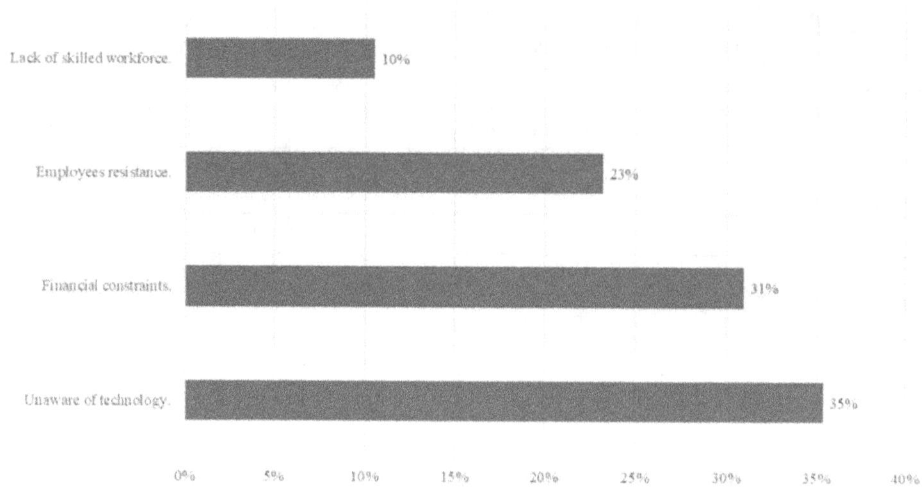

Most Influencing Factor in Implementing Industry 4.0

According to the survey, the most influential factor for implementing Industry 4.0 is Operational efficiency 34% (as shown in figure 3). The use of technologies like machine learning and use of data in Industry 4.0 helps to process the data in real time and increase operational efficiency, quality and adherence to standards (Javid et al., 2022). Return on investment in technology is the second most important factor that emerged from the survey which influences in implementing Industry 4.0 in micro and small enterprises. It is followed by commitment from top management and Government support.

Factors That Force to Adopt Industry 4.0

Figure 4 shows the factors that force a micro and small enterprise to adopt Industry 4.0. 37% of the respondents reported market competition to be the most compelling factor for adoption of Industry 4.0. It is followed by customer and self-driven, 26% each while only 12% was attributed to force from supplier.

Figure 3. Influencers in Industry 4.0 implementation

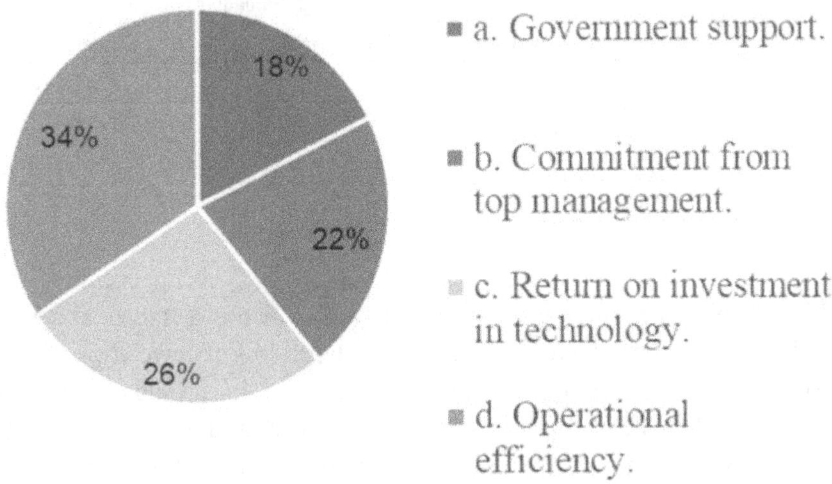

Figure 4. Factors forcing for adoption of Industry 4.0

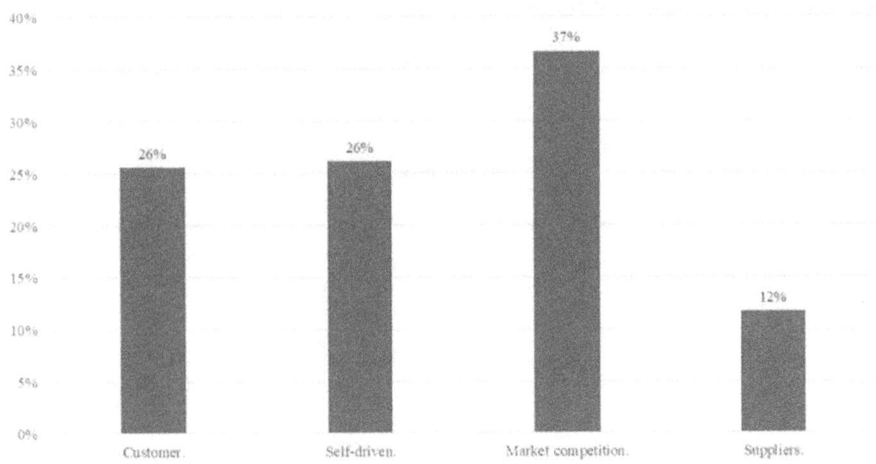

Factor Analysis

The "Kaiser-Meyer-Olkin" Measure of Sampling Adequacy is 0.876, which is greater than 0.5. Thus, the data is acceptable, and there is enough variance (Saha & Jha, 2023). Whereas the significance is 0.0001. $p<0.05$ implies that the data can be submitted to factor analysis. Hence Bartlett's Test is also acceptable (Saha & Jha, 2022).

Table 3. Total variance table

Component	Extraction S.S Loadings		
	Total	% of Variance	Cumulative %
1	7.651	47.820	47.820
2	1.814	11.338	59.158
3	1.256	7.850	67.007
4	1.178	7.361	74.368

Table 3 displays the findings of factor analysis, which is a statistical approach for identifying underlying factors that may explain patterns of correlations among a group of variables. As seen in the "Factor" column, the study yielded four factors. These factors explain the 74.3% of the total variance. The variation explained by each component after rotation is shown in the "Extraction Sums of Squared Loadings" column. The first component accounts for 47.82% of the variation, second component accounts for 11.33% of the variation, third component accounts for 7.85% of the variation whereas the second accounts for 7.36%. These four factors account for 74.36% of the total variation in the data.

Figure 5. Scree plot

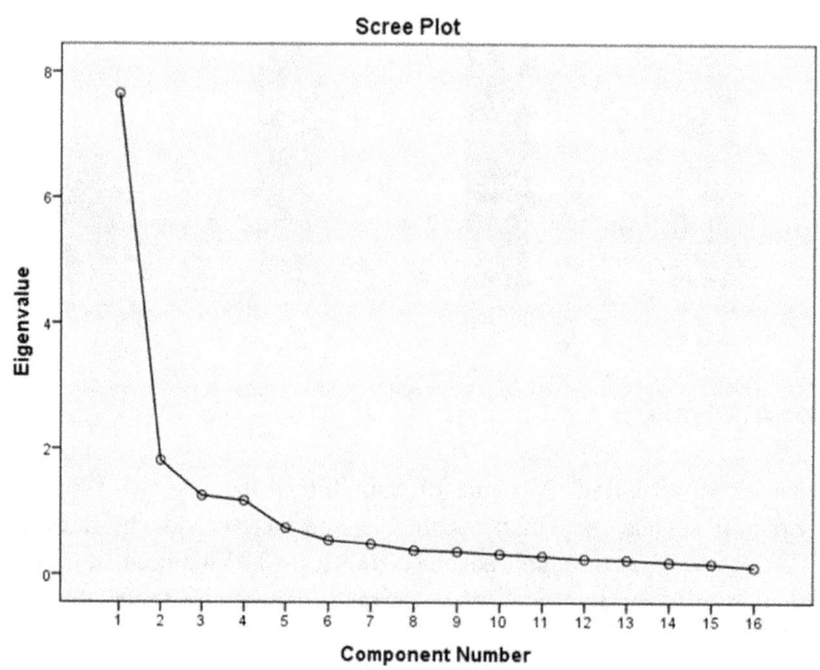

According to the scree plot as shown in figure 5, four factors are above eigenvalue 1. Hence, these four factors adequately represent the variance of the sample and hence 4 factors can be retained for further analysis (Saha et al., 2022).

Table 4. Pattern matrix

Variables	Component			
	1	2	3	4
AII			.972	
AAM			.818	
AR			.682	
ICC		.932		
ISC		.886		
IAM		.654		
IR		.722		
PAM				.585
PAF				.821
PAEA				.854
PARI				.776
BCE	.721			
BRI	.770			
BIE	.797			
BQA	.835			
BS	.809			

Table 4 shows the pattern matrix of different variables taken for the study. The variables AII (with factor loading 0.972), AAM (0.818), and AR (0.682) form factor 3. Since all these variables are about awareness, this factor can be termed as "awareness". The variables ICC (0.932), ISC (0.886), IAM (0.654), and IR (0.722) constitute factor 2. As these variables are related to implementation of different Industry 4.0 technologies, this factor is termed as "implementation." The variables PAM (0.585), PAF (0.821), PAEA (0.854), and PARI (0.776) form factor 4. As these variables are related to priority of adoption of different technologies, this factor is named "priority". The variables of factor 1 are BCE (0.721), BRI (0.770), BIE (0.797), BQA (0.835), and BS (0.809). These variables are related to benefits of implementation of different technologies. Hence, this factor is termed "benefits."

From the study it is evident that micro and small enterprises require support in defining individual and business-specific plans for Industry 4.0 adoption. These enterprises require skilled manpower to study and apply Industry 4.0 technologies and advanced business models. Solving these issues might be a game changer in micro and small enterprises ' smooth transition to Industry 4.0. The lack of awareness also acts as a major barrier to adoption to Industry 4.0. Micro and small enterprises should have a clear understanding of Industry 4.0 objectives and how they will be implemented in their current operations. These enterprises should educate themselves on disruptive technologies and how they might help their businesses. Technology should be the driving force behind enterprises. Thus, there is a need to improve their access to information, technology, skilled personnel, finance, and other resources. Since a company's geographical location has a significant impact on awareness, organizations must make the best use of government assistance and market trends. Considering human capital as the primary resource of any firm, as respondents perceived as a key barrier to the implementation of Industry 4.0, job enrichment is essential for all employees.

CONCLUSION

Industry 4.0 concept has many advantages under its umbrella such as increasing efficiency, productivity flexibility, and speed of operations. The manufacturing industry will be the forerunner and pilot industry in implementing Industry 4.0 principles on its shop floors. The Industry 4.0 idea has several benefits, including increased efficiency, productivity flexibility, and operational speed. "Cost" and "Smart factory" are two essential variables that Micro and small enterprises should begin to consider. As in Cost, this indicates that the organization should concentrate on financial strategy by providing the best financial assistance possible. In contrast, a Smart factory means the firm will establish the critical I.T. systems (Industry 4.0 technologies), and infrastructure, and guarantee that the valid and valuable data is used.

Furthermore, it is critical to understand that Industry 4.0 implementation is a continuous process of innovation and improvement rather than a one-time event. To be competitive in the long run, micro and small enterprises must be prepared to adapt and change in response to new technology and consumer expectations. Overall, researching enterprises' readiness to implement Industry 4.0 might give useful insights into the difficulties and possibilities that these businesses confront in the digital age. It can also aid in the formulation of policies to assist micro and small enterprises in embracing Industry 4.0 and contribute to the general economy's growth and development.

This study focused on assessing the readiness of micro and small enterprises for adopting Industry 4.0, by considering the Marathwada district of Maharashtra, India. Newly established models at the organizational level involving decision enablers such Customers, Products, Technology, and Operations constitute one of the focus areas that organizations need to consider for enhancing Industry 4.0 readiness. The formulation of a multi-criteria strategy with quantitative and qualitative dimensions is another focus area in this regard. Further, establishing appropriate mechanisms to address the issue of resistance to change due to deployment of Industry 4.0 and disruptive technologies such as IoT is also a focus area that is to be considered.

This study comprised respondents from selected cities of Marathwada district of Maharashtra, India. Further studies could be taken up for other cities and regions of India. However, there could be variations in outcome while doing the study for other cities / regions due to differences in culture and educational level of the population, and technological readiness of the geographical regions. This study considered sectors such as agriculture, textile, retail, and manufacturing. Future studies could also be done more elaborately for a specific sector.

REFERENCES

Amaral, A., & Peças, P. (2021). A framework for assessing manufacturing SMEs Industry 4.0 maturity. *Applied Sciences (Basel, Switzerland)*, *11*(13), 6127. doi:10.3390/app11136127

Brintha, N. C. JT, W. J., & Sukumaran, J. (2019). Integrating SMEs through cloud: an industrial revolution. In Organizational Transformation and Managing Innovation in the Fourth Industrial Revolution (pp. 143-164). IGI Global.

Brozzi, R., Riedl, M., & Matta, D. (2021). Key readiness indicators to assess the digital level of manufacturing SMEs. *Procedia CIRP*, *96*, 201–206. doi:10.1016/j.procir.2021.01.075

Cardoso, A. R., & Pereira, R. (2022). Relating Industry 4.0 and the SME Internationalization Process: A Case Study. In Implementing Automation Initiatives in Companies to Create Better-Connected Experiences (pp. 40-60). IGI Global.

Chang, S. C., Chang, H. H., & Lu, M. T. (2021). Evaluating industry 4.0 technology application in SMES: Using a Hybrid MCDM Approach. *Mathematics*, *9*(4), 414. doi:10.3390/math9040414

Coleman, S., Göb, R., Manco, G., Pievatolo, A., Tort-Martorell, X., & Reis, M. S. (2016). How can SMEs benefit from big data? Challenges and a path forward. *Quality and Reliability Engineering International*, *32*(6), 2151–2164. doi:10.1002/qre.2008

Cotrino, A., Sebastián, M. A., & González-Gaya, C. (2020). Industry 4.0 roadmap: Implementation for small and medium-sized enterprises. *Applied Sciences (Basel, Switzerland)*, *10*(23), 8566. doi:10.3390/app10238566

Ghobakhloo, M., Iranmanesh, M., Vilkas, M., Grybauskas, A., & Amran, A. (2022). Drivers and barriers of Industry 4.0 technology adoption among manufacturing SMEs: A systematic review and transformation roadmap. *Journal of Manufacturing Technology Management*, *33*(6), 1029–1058. doi:10.1108/JMTM-12-2021-0505

Grooss, O. F., Presser, M., & Tambo, T. (2022). Surround yourself with your betters: Recommendations for adopting Industry 4.0 technologies in SMEs. *Digital Business*, 100046.

Grufman, N., Lyons, S., & Sneiders, E. (2020). Exploring readiness of SMEs for industry 4.0. *Complex Systems Informatics and Modeling Quarterly*, (25), 54–86. doi:10.7250/csimq.2020-25.04

Hassan, M. B., Ahmed, E. S. A., & Saeed, R. A. (2021). Machine learning for industrial IoT systems. In Handbook of Research on Innovations and Applications of AI, IoT, and Cognitive Technologies (pp. 336-358). IGI global.

Horváth, D., & Szabó, R. Z. (2019). Driving forces and barriers of Industry 4.0: Do multinational and small and medium-sized companies have equal opportunities? *Technological Forecasting and Social Change*, *146*, 119–132. doi:10.1016/j.techfore.2019.05.021

Javaid, M., Haleem, A., Singh, R. P., Rab, S., & Suman, R. (2022). Exploring impact and features of machine vision for progressive industry 4.0 culture. *Sensors International*, *3*, 100132. doi:10.1016/j.sintl.2021.100132

Kumar, A., Batra, V., & Sharma, S. K. (2009). Micro, small and medium enterprises (MSMEs) in India: Challenges and issues in the current scenario. *Management Insight*, *5*(2), 75–87.

Lasi, H., Fettke, P., Kemper, H. G., Feld, T., & Hoffmann, M. (2014). Industry 4.0. *Business & Information Systems Engineering*, *6*(4), 239–242. doi:10.100712599-014-0334-4

Mittal, S., Khan, M. A., Romero, D., & Wuest, T. (2018). A critical review of smart manufacturing & Industry 4.0 maturity models: Implications for small and medium-sized enterprises (SMEs). *Journal of Manufacturing Systems*, *49*, 194–214. doi:10.1016/j.jmsy.2018.10.005

Pedraza, J. M. (2021). The micro, small, and medium-sized enterprises and its role in the economic development of a country. *Business and Management Research*, *10*(1), 33. doi:10.5430/bmr.v10n1p33

Qi, Q., & Tao, F. (2018). Digital twin and big data towards smart manufacturing and industry 4.0: 360 degree comparison. *IEEE Access : Practical Innovations, Open Solutions*, *6*, 3585–3593. doi:10.1109/ACCESS.2018.2793265

Rizos, V., Behrens, A., Van der Gaast, W., Hofman, E., Ioannou, A., Kafyeke, T., Flamos, A., Rinaldi, R., Papadelis, S., Hirschnitz-Garbers, M., & Topi, C. (2016). Implementation of circular economy business models by small and medium-sized enterprises (SMEs): Barriers and enablers. *Sustainability (Basel)*, *8*(11), 1212. doi:10.3390u8111212

Rüßmann, M., Lorenz, M., Gerbert, P., Waldner, M., Justus, J., Engel, P., & Harnisch, M. (2015). Industry 4.0: The future of productivity and growth in manufacturing

Saha, S., Saha, S. K., Pandey, J. R., & Jha, A. (2022). Employee motivation for training and development: A study of a pharmaceutical company in Sikkim. In *Handbook of research on developing circular, digital, and green economies in Asia* (pp. 370–386). IGI Global. doi:10.4018/978-1-7998-8678-5.ch018

Saha, S. K., & Jha, A. (2022). Impact of reminder on medication adherence: A structural equation model, based on study in Sikkim, India. *Bioscience Journal*, *38*(e38014), 1981–3163. doi:10.14393/BJ-v38n0a2022-59087

Saha, S. K., & Jha, A. (2023). Correlation Between Evaluative Beliefs of Patients, Reminder and Medication Adherence. [IJABIM]. *International Journal of Asian Business and Information Management*, *13*(1), 1–16. doi:10.4018/IJABIM.316174

Schumacher, A., Erol, S., & Sihn, W. (2016). A maturity model for assessing Industry 4.0 readiness and maturity of manufacturing enterprises. *Procedia CIRP*, *52*, 161–166. doi:10.1016/j.procir.2016.07.040

Shukla, S., & Prajapti, D. (2019). A Study on Various Schemes for Msmes: With Special Reference to SIDBI. *Adhyayan: A Journal of Management Sciences*, *9*(02), 27-30.

Ślusarczyk, B. (2018). Industry 4.0: Are we ready? *Polish Journal of Management Studies, 17*.

Stentoft, J., Adsbøll Wickstrøm, K., Philipsen, K., & Haug, A. (2021). Drivers and barriers for Industry 4.0 readiness and practice: Empirical evidence from small and medium-sized manufacturers. *Production Planning and Control*, *32*(10), 811–828. doi:10.1080/09537287.2020.1768318

Tang, T. A., Mhamdi, L., McLernon, D., Zaidi, S. A. R., & Ghogho, M. (2016, October). Deep learning approach for network intrusion detection in software defined networking. In 2016 international conference on wireless networks and mobile communications (WINCOM) (pp. 258-263). IEEE. doi:10.1109/WINCOM.2016.7777224

Vaidya, S., Ambad, P., & Bhosle, S. (2018). Industry 4.0–a glimpse. *Procedia Manufacturing*, *20*, 233–238. doi:10.1016/j.promfg.2018.02.034

Won, J. Y., & Park, M. J. (2020). Smart factory adoption in small and medium-sized enterprises: Empirical evidence of manufacturing industry in Korea. *Technological Forecasting and Social Change*, *157*, 120117. doi:10.1016/j.techfore.2020.120117

Yoon, J. S., Shin, S. J., & Suh, S. H. (2012). A conceptual framework for the ubiquitous factory. *International Journal of Production Research*, *50*(8), 2174–2189. doi:10.1080/00207543.2011.562563

Chapter 12
Application of Green Logistics in Supply Chain of Auto Parts: A South Indian Scenario

Saibal Kumar Saha
https://orcid.org/0000-0002-7842-698X
Christ University, India

Rashmi S
Christ University, India

Siva Kumar Pujari
Christ University, India

ABSTRACT

Green supply chain concept is used to reduce environmental degradation and emissions of air, water, and waste by incorporating green practices into business operations. Growing impacts of global warming, climate change, waste, and air pollution problems have prompted experts all over the world to think more environment friendly and find the best possible approach for "Green" solutions. green supply chain management is one of the factors that motivates organizations to be more sustainable. This study focuses on the green supply chain management in the auto parts industry in South Indian. Data from three green initiatives: recyclable packaging, green warehouse management and milk run approach for logistics is taken and compared with non-green approaches. It is found that there is significant reduction in costs by adopting the green approaches. With environmental issues growing all the time, green supply chain deserves to be a long-term community concern in developing countries.

DOI: 10.4018/978-1-6684-9089-1.ch012

INTRODUCTION

Global warming has garnered attention from around the world due to the fears and risks it brings with it as technology burdens the planet with expanding demands for energy production and consumption despite having finite resources (Hassan et al., 2022). However, Green Supply Chain Management (GSCM) is a viable option for resolving this issue. Researchers focused on the benefits: more environmentally friendly and sustainable supply chain practices offer competitive advantages, whereas less environmentally friendly practices pose a serious threat to the environment (Seman et al., 2019). Green logistics (GL) is a method of creating environmentally-friendly logistics and SCM (De Souza et al., 2022). Materials-related duties include acquiring, reducing, reusing, recycling, and replacing materials. Sustainable supply chain management solutions have become essential industrial success elements. Green procurement, manufacturing, materials management, distribution, marketing, and reverse logistics are all examples of green logistics (Sureeyatanapas & Yang, 2021).

Organizations all over the world have implemented green supply chain practices, yet Indian organizations continue to prioritize cost-benefit analysis. Though India has become one of the world's most significant industrial economies (Bosworth & Collins, 2008), but it still lags in the green supply chain concept, indicating that there is an opportunity for advancement from early to advanced. Green Manufacturing (GM) has permeated all industrial sectors, including automobiles, real estate, cement, steel, electronics, and construction, chemicals, water equipment and processes, and so on, as well as big energy users such as steel, cement, and paper goods. The publications "National Production Policy 8" prepared by the "Department of Industrial Policy and Promotion" of DIPP demonstrate the earnest attention of the Indian political leadership in this area. The adoption of more sustainable and financially viable corporate growth methods is leading the way in Indian industry. Real estate developer DLF has been concentrating on energy efficiency (Vijaya, 2011), "Hindustan Unilever" is aiming to reduce CO2 emissions by 22%, Kirloskar Brothers markets a range of ultra-energy-efficient pumps in Pune, and "Jaipur UltraTech Cement" has taken up the mission to burn 100 tonnes of household waste each day in its "waste plant", among other businesses that have chosen sustainable technology in their manufacturing (Dhunna, 2020). In India, there are already 48 GreenCo accredited businesses, and more than 140 other businesses are actively pursuing this certification (Ali et al., 2018).

Although there has been significant research on green supply chain management in recent years, there are still gaps in the literature. One of the main shortcomings of green supply chain research is the lack of standardization of definitions, metrics, and methods. This makes it difficult to compare and generalize the results of studies and limits the possibilities to compare and evaluate performance. Most research on green supply chain management has been conducted in developed countries, while research on developing countries is limited. As developing countries face unique challenges and opportunities in implementing green supply chain practices, more research is needed to understand their context and explore effective strategies. Although the circular economy is gaining attention as a strategy to promote sustainability, there is little research on the role of environmentally sustainable supply chain management in achieving a circular economy. Further research is needed to understand the potential of green supply chains in promoting the circular economy and to identify effective strategies for implementing circular supply chains. There is also limited research on milk run with respect of Indian SCM.

One of the major generators of industrial waste that harms the environment is the automobile sector (Zailani et al., 2015). As GSCM is challenging to adopt for automotive industries (Chandra Shukla et al., 2009), the auto parts industry is not ready to implement GSCM in their company as they find that GSCM initiative would make them lose profits and its difficulties in implementation. The auto parts industry mostly depends on storing, packaging, receiving, and transporting the parts through suppliers and customers (Boysen et al., 2015). Thus, the companies feel implementing green practices would be difficult. Hence, this study tries to understand the impact of GSCM implementation in auto parts industry and its prospects for cost savings.

LITERATURE REVIEW

Literature on green logistics and green supply chain management emerged as organizations and researchers are beginning to realize that management and environmental programs and activities do not stop organizational boundaries. (Jain & Sharma, 2012) conducted an empirical study to connect modern management strategies in the automotive and auto parts industries. The results showed that for GSCM to be successful the methods must begin with management process. In addition, they found that the automotive industry and auto parts Indian industry has potential to grow substantially in ecological management. The researchers also investigated the impact of green supply chain management practices on the organization's performance. Results show that green design, green production, vice versa logistics, environmental operations, economies in operation are closely related to the green corporation performance.

The research of Navavongsathian et al., (2020) identified four GSCM performance objectives:

1. Environmental performance: reduction of air emissions, reduction of wastewater, reduction of solid waste, reduction of consumption of hazardous/harmful/toxic

substances, reduction of the frequency of environmental accidents, improvement ecological situation of the company

2. Positive financial results: reduction of material acquisition costs, reduction of waste value and reduction of fines due to environmental accidents
3. Negative economic result: increase in investments, increase in operating costs, increase in acquisition costs of ecological materials, and increase in training costs and
4. Operational performance: increase the number of goods delivered on time, reduce inventory, reduce the amount of waste, promote product quality, increase product variety and increase capacity utilization.

Green supply chain management (GSCM) is an important concept that has received a lot of attention in recent years. This refers to incorporating environmental considerations into supply chain management practices to reduce negative environmental impacts of the supply chain (Al-Sheyadi et al., 2019). The GSCM literature is extensive and covers many topics.

Drivers of GSCM: The literature identifies several factors that motivate firms to adopt GSCM practices. These factors include regulatory pressure, customer demands, competitive advantage, cost savings and corporate social responsibility (de Sousa Jabbour et al., 2013).

GSCM Environmental Practices: GSCM practices include green procurement, green design, green manufacturing, green logistics and green marketing (Jassim et al., 2020). These practices aim to reduce the environmental impact of the entire supply chain, from raw material extraction to end-of-life processing.

GSCM Performance Outcomes: According to the literature, GSCM practices can lead to several outcomes such as cost savings, better environmental protection, increased customer loyalty, and reputation (Chavez et al., 2016).

Barriers to GSCM Adoption: Several barriers to the adoption of GSCM practices have been identified in the literature. These barriers include lack of awareness and knowledge, lack of funding, lack of government support and resistance from providers (Tumpa et al., 2019).

GSCM in specific industries: The literature also examines GSCM practices in specific industries such as manufacturing (Sheng et al., 2023a), retail (Sheng et al., 2023b), and construction (Badi & Murtagh, 2019). The results indicate that adoption of GSCM practices varies across industries due to differences in industry structure, market dynamics, and regulatory frameworks.

GSCM Tools and Frameworks: There are several tools and frameworks in the literature that help companies to adopt GSCM practices. These include life cycle assessment, green supply chain scorecards, ecological design principles and other tools to reduce carbon footprint (Gao et al., 2021; Gawusu et al., 2022).

GSCM In Automobile Industry

With a 7.1% share of the nation's gross domestic product (GDP), the Indian car sector is one of the major economic movers in the country (Sharma & Sharma, 2018). With a manufacturing volume of 23.37 million automobiles annually in 2014–2015, it is also among the biggest in the world, after the 8.68% rise in the prior year. According to the Auto Mission Plan (AMP) 2016-2026, the Indian government wants to make the auto sector a key driver of the "Make in India" effort since it anticipates that the market for passenger cars would quadruple to 9.4 million units by 2026. For the automobile sector, supply chain management is a crucial component of all business elements. To lead, manage, thrive, and generate a lasting competitive advantage, the whole process chain from procurement to delivery must be meticulously coordinated. In addition to making enormous profits, this industry is experiencing rapid growth and has a large negative influence on the environment. In order to reduce pollution in the capital, the prime minister removed the even/odd rule, and the Supreme Court banned a number of vehicle kinds (Garg & Singh, 2022). Environmental degradation calls for stringent action. The number of cars on the road has dramatically dropped, generating traffic congestion, notwithstanding the mixed benefits of this legislation.

For automakers, this poses a serious problem since the sector's constraints would lessen India's chances of becoming a worldwide leader in the production of compact cars. These issues underline the demand for environmentally friendly practices to lessen pollution, cut carbon dioxide emissions, and boost this industry's competitiveness. The sector is better able to achieve its environmental and sustainability goals and enhance financial performance, according to research released by the Centre for Science and the Environment.

Auto Parts Industry in India

An important segment of India's overall automotive industry is the car components sector. India is one of the biggest producers and exporters of car parts in the world, and it offers a wide variety of goods for both home and foreign markets (Miglani, 2019). Original equipment manufacturers (OEMs), aftermarket providers, and component makers are among the many big and minor participants in the sector. Bosch, Mahindra CIE, Motherson Sumi, Amtek car, and Minda Industries are important manufacturers of car components in India.

Companies actively engage in R&D to keep ahead of the curve in this highly competitive and continuously changing market. The National Automotive Testing and Rand Infrastructure Project (NATRIP) and the Automotive Mission Plan (AMP) 2026 are only two of the steps the Indian government has undertaken to encourage the expansion of the sector. The primary difficulties facing the Indian auto parts business are growing raw material costs, escalating international rivalry, and the requirement to adhere to ever-stricter safety and emission requirements (Patole, 2012). Due to strong local demand and rising exports to foreign markets, the Indian auto parts sector is anticipated to expand overall in the upcoming years.

Milk Run

Milk run is a logistics management strategy in which a vehicle follows a pre-planned route to collect goods from multiple suppliers or customers (Bocewicz et al., 2021). The aim is to maximize the use of vehicles and reduce transport costs. The literature on milk run logistics has shown the benefits of this strategy, such as reduced transport costs, improved supply reliability, and reduced carbon emissions (Minh et al., 2020). Factors influencing milk route planning include the number of suppliers and customers, the type of products being transported, the distance between suppliers and customers, the frequency of loading and delivery, and the capacity of the vehicle. Implementing milk run flow systems can be complex and coordination between suppliers and customers is critical to ensure efficient operation. Milk runs have been used in several industries such as automotive, retail and healthcare. Milk run systems can also be combined with other logistics strategies, such as cross-docking, to optimize the entire supply chain. Milk run literature can be found in various fields such as logistics, supply chain management and operations management.

Benefits of milk run: Several studies have shown the benefits of milk run systems (Minh et al., 2020). For example, during milk runs, transportation costs can be reduced by maximizing vehicle utilization and reducing the number of trips. In addition, milk runs can also reduce inventory by improving lead times and shortening lead times. Milk runs also have other benefits, such as reducing traffic congestion, improving supply reliability, and reducing carbon dioxide emissions (Bae et al., 2016).

Factors influencing the design of the milk run line: These include the number of suppliers and customers, the type of products transported, the distance between suppliers and customers, the frequency of pickups and deliveries, and the capacity of the vehicle. Researchers have developed mathematical models taking these factors into account to optimize the design of milk run flow systems (Grzegorz et al., 2021).

Challenges of implementing milk run handling systems: Although milk run systems offer many advantages, they can be difficult to implement. For example, coordination between suppliers and customers is crucial to ensure the proper functioning of the system. Changes in the production schedule, product mix or delivery locations can also affect the efficiency of the milk distribution system. Therefore, a strong information system is needed to ensure timely exchange of information between various stakeholders (Pawlewski, 2019). The milk run concept has been applied to several industries such as automotive, retail and healthcare. For example, in the automotive industry, a milk run is used to transport parts from suppliers to assembly plants. In retail, milk runs are used to collect goods from several suppliers and deliver them to stores. In healthcare, milk runs are used to collect medical waste from hospitals and deliver it to landfills.

Combination of milk runs with other logistics strategies: Milk-runs systems can also be combined with other logistics strategies to optimize the entire supply chain. For example, milk runs can be combined with cross-docking, where goods from multiple suppliers are combined in a distribution centre before being delivered to customers. This can further reduce transport costs and delivery times.

Packaging

Packaging is an integral part of any product and can also be a significant cost for businesses (Coelho et al., 2020). Therefore, package cost saving strategies have been an interesting topic in the literature. Following are some of the packaging costs saving strategies.

Choice of Material: Choosing the right packaging material can save a significant amount of money. Studies have shown that switching to lightweight materials such as thin-walled plastic can reduce packaging costs by up to 50%. In addition, the use of recycled materials or biodegradable materials can reduce disposal costs (Kozik, 2020).

Packaged design: Package design can also affect costs. For example, creating an easy-to-assemble model can reduce labour costs (Azzi et al., 2012). Similarly, designing packages that are easy to transport can reduce shipping costs. Studies have shown that using standard packaging can reduce packaging costs by up to 40%.

Lean production: Applying economical production principles to packaging can also save costs. Lean manufacturing focuses on reducing waste, including excess inventory and redundant processes. Applying these principles to packaging can reduce the amount of packaging material used, minimize transport costs, and reduce waste (Klimecka-Tatar, 2017).

Life cycle analysis: Life cycle analysis can help identify savings opportunities throughout the product's life cycle, including packaging. For example, identifying recyclable or reusable packaging materials can reduce disposal costs. In addition, analysing the environmental impact of packaging can help identify cost savings opportunities by reducing a company's environmental footprint.

Energy Consumption in Logistics

Route Optimization: Route optimization is a key strategy to reduce energy consumption in logistics. By optimizing delivery routes to reduce travel time and distance, companies can reduce fuel consumption and greenhouse gas emissions (Zhao et al., 2020).

Efficient vehicle technology: Efficient vehicle technology, such as electric and hybrid vehicles, can also help reduce energy consumption in logistics. Studies have shown that efficient vehicle technology can significantly reduce fuel consumption and greenhouse gas emissions.

Change the mode of transport: Switching to more energy-efficient modes of transport, such as rail or water transport, can also help reduce energy consumption in logistics (Mraihi et al., 2013). Studies have shown that switching from road to rail transport can reduce energy consumption by up to 75 percent.

Warehouse energy efficiency: Storage operations can also consume significant amounts of energy. By implementing energy-efficient lighting, heating and cooling systems, companies can reduce their energy consumption and promote sustainability (Minashkina & Happonen, 2020).

Cooperation in the supply chain: Supply chain collaboration can involve working with suppliers, customers, and other partners to optimize logistics operations and reduce energy consumption (Saha & Jha, 2018). In collaboration with supply chain partners, companies can identify opportunities to reduce energy consumption and promote sustainability (Saha et al., 2018). In short, energy saving in logistics is necessary to reduce energy consumption and greenhouse gas emissions. By optimizing delivery routes, using efficient vehicle technology, changing modes of transport, implementing energy-efficient storage operations, and collaborating with supply chain partners, companies can reduce their energy use and promote the sustainability of logistics (Saha, 2022).

METHODOLOGY

The study is quantitative in nature. Data from an auto parts company in Bengaluru, India, was used. A comparative analysis was done to find the impact of the green supply chain initiative in saving costs for the company. The company stored auto parts for different service stations in South India. The parts were sent from the Pune warehouse and stocked in the Bengaluru warehouse, which transported spare parts to all the service centres of South India. The comparative analysis of the company was done before and after the implementation of GSCM. Analysis was done for energy consumption in the warehouse, packaging material and milk run transportation of spare parts.

RESULTS AND DISCUSSION

Energy Consumption in Warehouse

The data is collected from 2015-2022. The company initially used Hybay bulbs in their warehouse. From 2019 LED lights were introduced. The LEDs significantly reduced the energy consumption in the warehouse.

Figure 1. Electricity consumption
Source: Generated by author from the company data

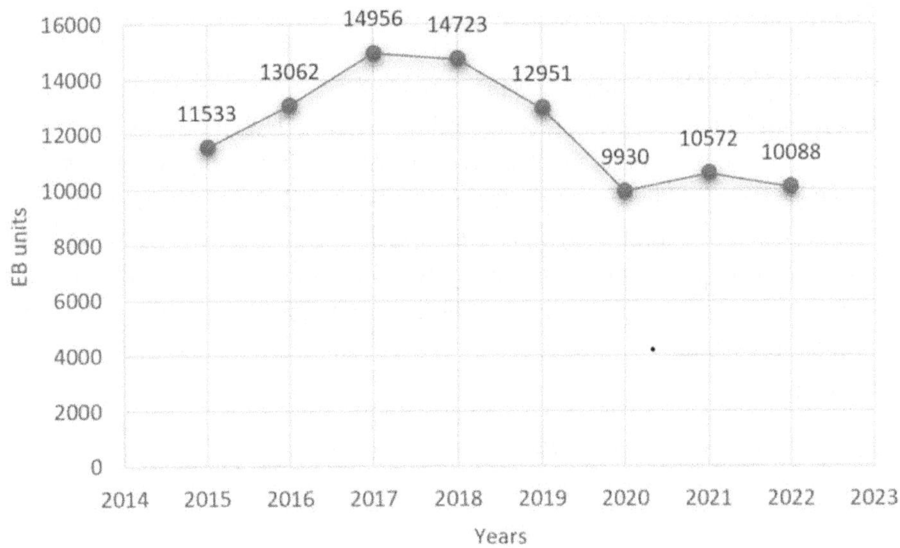

From the graph as shown in figure 1 we can see that, once the company installed LED bulbs in 2019, the energy consumption reduced significantly. Since energy consumption indirectly contributes to greenhouse emissions, the use of energy has a big environmental impact. The use of fossil fuels to generate energy, results in huge volumes of greenhouse gases (GHGs) being released into the atmosphere. These GHGs, especially CO2, trap heat in the Earth's atmosphere, contributing to climate change and global warming. The environmental impact of energy use is exacerbated by the extraction, transportation, and refining of fossil fuels. These processes can degrade and pollute the environment, wreaking havoc on ecosystems, human health, and biodiversity. Other dangerous pollutants released into the air by the combustion of fossil fuels include sulphur dioxide (SO2), nitrogen oxides (NOx), and particulate matter (PM) (Kampa & Castanas, 2008). These pollutants can induce asthma and bronchitis, as well as contribute to acid rain and smog. Furthermore, energy consumption is a major driver of water use, with various energy generating methods, such as fossil fuels and nuclear power, requiring large amounts of water. This water use has the potential to reduce the availability of freshwater for other uses, such as agriculture and human consumption. To lessen the environmental impact of energy consumption, it is critical to promote the use of renewable energy, which emit little to no greenhouse gases or toxic pollutants. Improving energy efficiency in buildings, transportation, and industrial processes can also reduce energy use and its environmental impact (Omer, 2009).

Packaging

The company had been using corrugated boxes for a long time. These boxes were used once and then dumped into waste bins, which generated lots of waste in landfills. The company implemented green logistics by replacing corrugated boxes with plastic boxes which was reused for all the distribution. These were reused multiple times in the supply chain before recycling. The company was able to save Rs. 14,99,303 on packaging material, by replacing one time use packaging with reusable plastic packaging and reusing the corrugated packaging along the supply chain more than once.

Figure 2. Packaging movement before green practice
Source: Author
WH: Warehouse

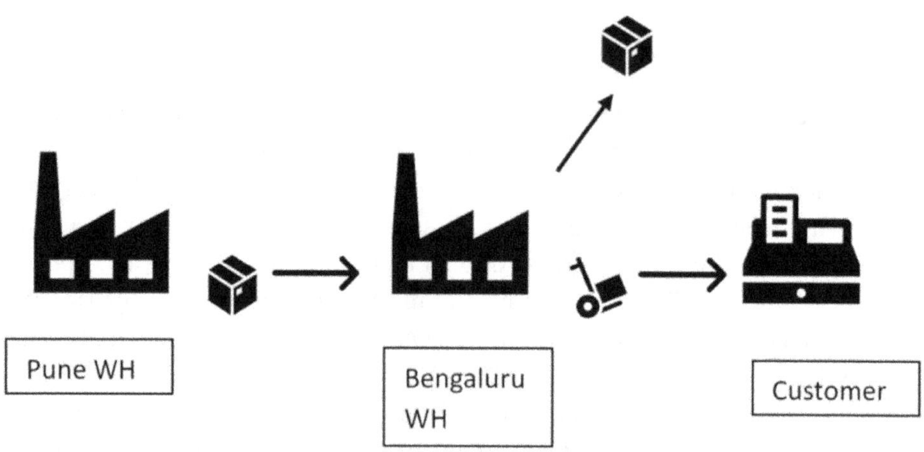

Figure 3. Packaging movement after green practice
Source: Author
WH: Warehouse

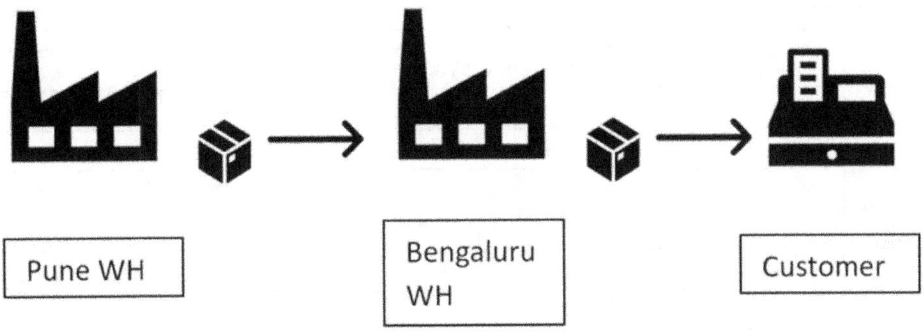

The above visual representation of figure 2 shows the packaging movement across the chain before implementation of reusable packaging. The packaging sent by Pune warehouse was discarded in Bengaluru warehouse, and new packaging boxes were used for supplying to customers. Thua, increasing the number of boxes used in the system. Figure 3 shows how the packaging material was stored in the Bengaluru warehouse and reused to send parts to various customer. Thus, reducing the number of boxes used in the system.

Table 1 shows the pervious expenditure and expenditure after the implementation of packaging returnable and re-usable project. The returnable boxes could go through 100 transactions before being discarded. The present returnable expenditure has 30% of previous expenditure (10% transportation and 20% investment) and 5% (manpower cost) of previous expenditure for re-use boxes.

Figure 4 shows the amount of cost savings that the company has been able to achieve by implementing recyclable packaging in its SCM. Packaging can have a substantial environmental impact throughout its lifecycle, from raw material extraction to disposal. Some of the most significant environmental effects of packaging include:

- Depletion of natural resources: Packaging frequently necessitates the use of considerable amounts of natural resources such as trees, petroleum, and water. These commodities' extraction and processing can have a variety of environmental consequences, including deforestation, habitat destruction, and water contamination.
- Energy consumption and greenhouse gas emissions: The production, delivery, and disposal of packaging materials all demand large quantities of energy, which can contribute to greenhouse gas emissions and climate change.

Table 1. Cost savings in packaging

SL.No	Months	Project Name	Previous expenditure	Present expenditure
1	June-22 To Dec-22	Returnable FLC Boxes	2,64,527	79,358
2	Jan-22 To Dec-22	Re-use Box-5V007 From IB	6,773	339
3	Jan-22 To Dec-22	Re-use Box From IB	5,29,267	26,463
4	Jan-22 To Dec-22	Returnable Small Blue Bin	23,182	6,955
5	March-22 To Dec-22	Returnable Big Blue Bin	55,751	16,725
6	Nov-22 To Dec-22	Returnable PP Box	1,21,180	36,354
7	Jan-22 To Dec-22	Re-use of Fender Box	6,99,807	34,990
			17,00,487	**2,01,184**

*1 Fender Box is equal to 2pc of Engine Box (7V002)
*1 Big Blue Bin Box is equal to 2pc of 5V002 Box
*1 Small Blue Bin Box is equal to 1 5V002 Box
*1 FLC Box is equal to 1 Engine Box 7V002
Source: Company data

Figure 4. Cost saving
Source: Generated by author from the company data

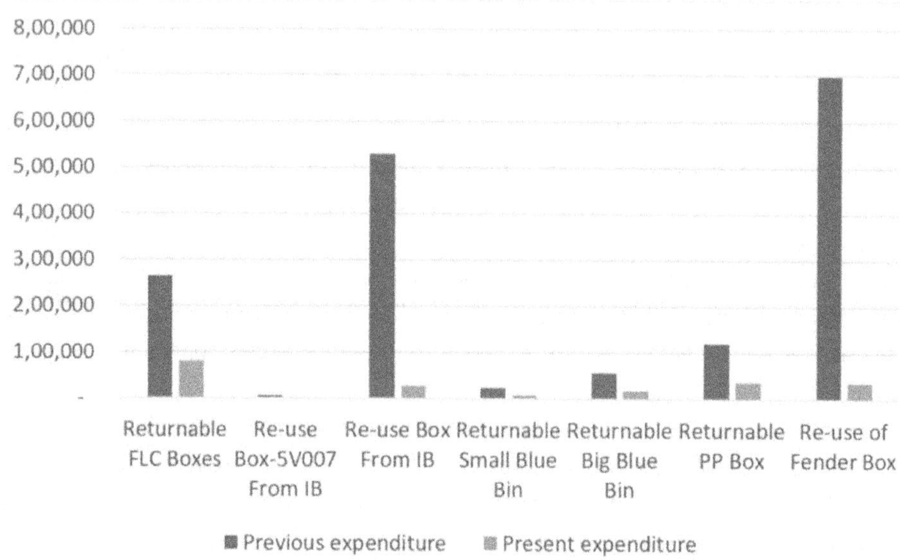

- Waste generation: Packaging accounts for a significant portion of municipal solid waste, which ends up in landfills. This can have several negative consequences, including soil and water pollution, wildlife injury, and greenhouse gas emissions from organic material decomposition.
- Toxic chemicals included in many packaging materials, such as phthalates, bisphenol A (BPA), and polyvinyl chloride (PVC), can seep into the environment and endanger wildlife and human health.

Milk Run- Transportation of Auto Spare Parts

The company supplied automotive spare parts throughout South India. For analysis we have considered only Bengaluru to all the customers in Tamil Nadu for duration of 5 months, as shown in table 2. It was found that milk run cost less than direct delivery.

Table 2. Comparison of Direct delivery cost and milk run cost

Month	Direct delivery cost	Milk run cost
July	₹ 44,49,847	₹ 22,28,844
August	₹ 34,34,274	₹ 21,55,044
September	₹ 27,60,102	₹ 22,50,601
October	₹ 22,71,987	₹ 19,37,515
November	₹ 23,14,872	₹ 19,22,986
December	₹ 24,34,629	₹ 20,51,086
	₹ 1,76,65,711	₹ 1,25,46,076

Source: Company data

The company used 17ft trucks, with utilisation of only 50-60%. Thus, in milk run they were able to use 32ft trucks to full utilisation. They had to pay the minimum base price for trucks, no matter how many parts they send in one trip. Figure 5 shows the amount the company paid in the direct delivery of parts.

Milk run approach to transport can benefit the environment in several ways:

- Reducing carbon dioxide emissions: By optimizing the route and combining multiple deliveries into a single journey, the milk run reduces the number of vehicles on the road and thus the carbon dioxide emissions they produce.

Figure 5. Delivery cost comparison
Source: Generated by author from the company data

- Reduce traffic congestion: The milk drive helps reduce traffic congestion by reducing the number of vehicles on the road. This can help reduce vehicle idling in traffic, which can also reduce emissions.
- Waste minimization: Milk run can help reduce waste by ensuring timely and efficient delivery of goods. This can help reduce the amount of inventory held at a destination, which in turn can reduce waste associated with excess inventory.
- Encourage sustainability: Adopting the milk run model in transport can demonstrate a company's commitment to sustainability, which in turn can encourage others to do the same. This could create a beneficial ripple effect for the environment as more companies adopt sustainable transport practices.

Sustainable materials: Many auto parts are made from materials that have a significant environmental impact, such as metals and plastics. Companies are increasingly looking to use sustainable materials such as recycled or bio-based plastics to reduce the environmental impact of their products. Manufacturing and transporting auto parts can be energy-intensive, increasing greenhouse gas emissions and other environmental impacts. To reduce the environmental impact of their operations, companies are increasingly adopting energy-efficient production methods and using more efficient modes of transport, such as rail or sea transport. Disposal of used or end-of-life car parts can also have a significant environmental impact. Companies implement remanufacturing and recycling programs to collect and reuse or recycle used parts, reducing waste and conserving resources. Companies are increasingly looking to work with suppliers who prioritize sustainability and environmental responsibility in their operations. This includes suppliers that use sustainable materials, implement energy efficient practices and have strong environmental systems.

CONCLUSION

Sustainability has emerged as a crucial issue in the auto parts sector, with corporations taking substantial steps to solve it. By cutting carbon emissions, minimizing waste output, and developing more efficient manufacturing processes, the sector is moving towards more environmentally friendly and sustainable practises. Furthermore, many businesses are undertaking sustainability efforts that emphasise the use of recyclable materials, the reduction of energy usage, and the promotion of sustainable mobility. The auto parts industry has made significant progress toward sustainability in recent years, but there is still much to do. Companies need to continue prioritizing sustainability in their operations and create a more sustainable future for the industry and the planet.

The study reveals that along with cost savings, environment can be protected by initiating GSCM and innovative technologies and frameworks. The adoption of GSCM practices can lead to several benefits, such as cost savings, improved environmental sustainability, and improved reputation. However, implementation of GSCM practices can be difficult due to various barriers such as lack of awareness and financial resources. It is also found that there is no one-size-fits-all approach to GSCM and firms must adapt their practices to their industry and context.

Milk run systems have emerged as a popular logistics management strategy to optimize transportation efficiency. The milk run systems offer several advantages, but their implementation can be difficult. Also, packaging cost-saving strategies can save significant costs while promoting sustainability. By choosing the right packaging materials, designing easy-to-assemble and transportable packages, automating the packaging process, applying low-pollution production principles, and performing life cycle analysis, companies can reduce packaging costs and improve their bottom line.

In this study only an auto parts company was studied with respect to its implementation of few GSCM initiatives. For further study, other initiatives can also be considered to improve cost savings and its impact on the environment.

REFERENCES

Al-Sheyadi, A., Muyldermans, L., & Kauppi, K. (2019). The complementarity of green supply chain management practices and the impact on environmental performance. *Journal of Environmental Management*, *242*, 186–198. doi:10.1016/j.jenvman.2019.04.078 PMID:31035181

Ali, S. S., Kaur, R., & Jaramillo, A. B. (2018). An assessment of green supply chain framework in Indian automobile industry using interpretive structural modelling and its validation using MICMAC analysis. *International Journal of Services and Operations Management*, *30*(3), 318–356. doi:10.1504/IJSOM.2018.092607

Azzi, A., Battini, D., Persona, A., & Sgarbossa, F. (2012). Packaging design: General framework and research agenda. *Packaging Technology & Science*, *25*(8), 435–456. doi:10.1002/pts.993

Badi, S., & Murtagh, N. (2019). Green supply chain management in construction: A systematic literature review and future research agenda. *Journal of Cleaner Production*, *223*, 312–322. doi:10.1016/j.jclepro.2019.03.132

Bae, K.-H. G., Evans, L. A., & Summers, A. (2016). Lean design and analysis of a milk-run delivery system: case study. *2016 Winter Simulation Conference (WSC)*, 2855–2866. 10.1109/WSC.2016.7822321

Bocewicz, G., Nielsen, P., & Banaszak, Z. (2021). Ordered–Fuzzy-Numbers-Driven Approach to Out-Plant Milk-Run Dynamic Routing and Scheduling. In Automation 2021: Recent Achievements in Automation, Robotics and Measurement Techniques (pp. 20–36). Springer. doi:10.1007/978-3-030-74893-7_3

Bosworth, B., & Collins, S. M. (2008). Accounting for growth: Comparing China and India. *The Journal of Economic Perspectives*, *22*(1), 45–66. doi:10.1257/jep.22.1.45

Boysen, N., Emde, S., Hoeck, M., & Kauderer, M. (2015). Part logistics in the automotive industry: Decision problems, literature review and research agenda. *European Journal of Operational Research*, *242*(1), 107–120. doi:10.1016/j.ejor.2014.09.065

Chandra Shukla, A., Deshmukh, S. G., & Kanda, A. (2009). Environmentally responsive supply chains: Learnings from the Indian auto sector. *Journal of Advances in Management Research*, *6*(2), 154–171. doi:10.1108/09727980911007181

Chavez, R., Yu, W., Feng, M., & Wiengarten, F. (2016). The effect of customer-centric green supply chain management on operational performance and customer satisfaction. *Business Strategy and the Environment*, *25*(3), 205–220. doi:10.1002/bse.1868

Coelho, P. M., Corona, B., ten Klooster, R., & Worrell, E. (2020). Sustainability of reusable packaging–Current situation and trends. *Resources, Conservation & Recycling: X*, *6*, 100037.

de Sousa Jabbour, A. B. L., Jabbour, C. J. C., Govindan, K., Kannan, D., Salgado, M. H., & Zanon, C. J. (2013). Factors affecting the adoption of green supply chain management practices in Brazil: Empirical evidence. *The International Journal of Environmental Studies*, *70*(2), 302–315. doi:10.1080/00207233.2013.774774

De Souza, E. D., Kerber, J. C., Bouzon, M., & Rodriguez, C. M. T. (2022). Performance evaluation of green logistics: Paving the way towards circular economy. *Cleaner Logistics and Supply Chain*, *3*, 100019. doi:10.1016/j.clscn.2021.100019

Dhunna, S. (2020). GREEN MARKETING: AN OVERVIEW. *New Horizons in Commerce, IT &. Social Sciences*, 106.

Gao, S., Qiao, R., Lim, M. K., Li, C., Qu, Y., & Xia, L. (2021). Integrating corporate website information into qualitative assessment for benchmarking green supply chain management practices for the chemical industry. *Journal of Cleaner Production*, *311*, 127590. doi:10.1016/j.jclepro.2021.127590

Garg, R. K., & Singh, D. (2022). Delhi Air Pollution: A Critical Analysis. *Issue 6 Int'l JL Mgmt. &. Human.*, *5*, 748.

Gawusu, S., Zhang, X., Jamatutu, S. A., Ahmed, A., Amadu, A. A., & Djam Miensah, E. (2022). The dynamics of green supply chain management within the framework of renewable energy. *International Journal of Energy Research*, *46*(2), 684–711. doi:10.1002/er.7278

Grzegorz, B., Izabela, N., Arkadiusz, G., & Zbigniew, B. (2021). Reference model of milk-run traffic systems prototyping. *International Journal of Production Research*, *59*(15), 4495–4512. doi:10.1080/00207543.2020.1766717

Hassan, T., Song, H., Khan, Y., & Kirikkaleli, D. (2022). Energy efficiency a source of low carbon energy sources? Evidence from 16 high-income OECD economies. *Energy*, *243*, 123063. doi:10.1016/j.energy.2021.123063

Jain, V. K., & Sharma, S. (2012). Green supply chain management practices in automobile industry: An empirical study. *Journal of Supply Chain Management Systems*, *1*(3), 20.

Jassim, S., Al-Mubarak, M., & Hamdan, A. (2020). The impact of green supply chain management on firm's performance. *Journal of Information & Knowledge Management*, *19*(01), 2040026. doi:10.1142/S0219649220400262

Kampa, M., & Castanas, E. (2008). Human health effects of air pollution. *Environmental Pollution*, *151*(2), 362–367. doi:10.1016/j.envpol.2007.06.012 PMID:17646040

Klimecka-Tatar, D. (2017). Value Stream Mapping as Lean Production tool to improve the production process organization–case study in packaging manufacturing. *Production Engineering Archives*, *17*(17), 40–44. doi:10.30657/pea.2017.17.09

Kozik, N. (2020). Sustainable packaging as a tool for global sustainable development. *SHS Web of Conferences*, *74*, 4012.

Miglani, S. (2019). The growth of the Indian automobile industry: Analysis of the roles of government policy and other enabling factors. *Innovation, Economic Development, and Intellectual Property in India and China: Comparing Six Economic Sectors*, 439–463.

Minashkina, D., & Happonen, A. (2020). Decarbonizing warehousing activities through digitalization and automatization with WMS integration for sustainability supporting operations. *E3S Web of Conferences*, *158*, 3002.

Minh, N., Kien, D. T., & Hau, P. K. (2020). Applying Milk-run Method to Optimize Cost of Transport: An Empirical Evidence. *International Journal of Supply and Operations Management*, *7*(2), 178–188.

Mraihi, R., ben Abdallah, K., & Abid, M. (2013). Road transport-related energy consumption: Analysis of driving factors in Tunisia. *Energy Policy*, *62*, 247–253. doi:10.1016/j.enpol.2013.07.007

Navavongsathian, A., Vuthipadadorn, D., Farangthong, S., Janthongpan, S., & Juntasart, R. (2020). Study of Green Logistics Managing Potential and the Preparedness of Auto Parts Industries in Thailand. *TEM Journal*, *9*(4), 1524–1534. doi:10.18421/TEM94-27

Omer, A. M. (2009). Energy use and environmental impacts: A general review. *Journal of Renewable and Sustainable Energy*, *1*(5), 53101. doi:10.1063/1.3220701

Patole, S. (2012). A study of auto components industry in India: Issues and challenges. *South Asian Journal of Marketing & Management Research*, *2*(7), 155–160.

Pawlewski, P. (2019). Interactive layout in the redesign of intralogistics systems. *Intelligent Systems in Production Engineering and Maintenance*, 462–473.

Saha, S. K. (2022). Bibliometric Analysis of Published Literature on the Pharmaceutical Supply Chain. [IJAL]. *International Journal of Applied Logistics*, *12*(1), 1–18. doi:10.4018/IJAL.309088

Saha, S. K., & Jha, A. (2018). Role of information technology on supply chain management of pharmaceutical industry. [IJAL]. *International Journal of Applied Logistics*, *8*(2), 39–68. doi:10.4018/IJAL.2018070103

Saha, S. K., Saha, S., & Jha, A. (2018). Use of Information Technology in the Supply Chain Management of the Pharmaceutical Industry: A Literature Review. *Supply Chain and Logistics Management: Concepts, Methodologies, Tools, and Applications*, *8*(2), 673–706.

Seman, N. A. A., Govindan, K., Mardani, A., Zakuan, N., Saman, M. Z. M., Hooker, R. E., & Ozkul, S. (2019). The mediating effect of green innovation on the relationship between green supply chain management and environmental performance. *Journal of Cleaner Production*, *229*, 115–127. doi:10.1016/j.jclepro.2019.03.211

Sharma, A., & Sharma, D. (2018). Impact of GST on Automobile Industry in India. *International Journal of Mnagement Science and Technology*, *9*(3), 145–154.

Sheng, X., Chen, L., Yuan, X., Tang, Y., Yuan, Q., Chen, R., Wang, Q., Ma, Q., Zuo, J., & Liu, H. (2023a). Green supply chain management for a more sustainable manufacturing industry in China: A critical review. *Environment, Development and Sustainability*, *25*(2), 1151–1183. doi:10.100710668-022-02109-9

Sheng, X., Chen, L., Yuan, X., Tang, Y., Yuan, Q., Chen, R., Wang, Q., Ma, Q., Zuo, J., & Liu, H. (2023b). Green supply chain management for a more sustainable manufacturing industry in China: A critical review. *Environment, Development and Sustainability*, *25*(2), 1151–1183. doi:10.100710668-022-02109-9

Sureeyatanapas, P., & Yang, J.-B. (2021). Sustainable manufacturing and technology: The development and evaluation. *Pursuing Sustainability: OR/MS Applications in Sustainable Design, Manufacturing, Logistics, and Resource Management*, 111–140.

Tumpa, T. J., Ali, S. M., Rahman, M. H., Paul, S. K., Chowdhury, P., & Khan, S. A. R. (2019). Barriers to green supply chain management: An emerging economy context. *Journal of Cleaner Production*, *236*, 117617. doi:10.1016/j.jclepro.2019.117617

Vijaya, P. (2011). Prospects of Clean Technologies for Sustainable Business. *Review of Management*, *1*(2).

Zailani, S., Govindan, K., Iranmanesh, M., Shaharudin, M. R., & Chong, Y. S. (2015). Green innovation adoption in automotive supply chain: The Malaysian case. *Journal of Cleaner Production*, *108*, 1115–1122. doi:10.1016/j.jclepro.2015.06.039

Zhao, Z., Li, X., & Zhou, X. (2020). Distribution route optimization for electric vehicles in urban cold chain logistics for fresh products under time-varying traffic conditions. *Mathematical Problems in Engineering*, *2020*, 1–17. doi:10.1155/2020/9864935

Chapter 13
Impact of the COVID-19 Pandemic on Industrial Recession in India and the Emerging Structural Breaks:
Testing Unit Root Hypothesis

Asim Kumar Karmakar
Netaji Subhas Open University, India

Tushar Das
Sadananda Mission High School, India

ABSTRACT

In this paper, an attempt has been made to look into the effects of external shock generated by Pandemic on the industrial output growth in India by analysing the macroeconomic time series data generating process. The authors have carried out a number of tests concerning the stochastic behaviour and nature of trend of the annual industrial output growth at national level as well as for some major states for the period 1989-90 to 2020-21 locating possible structural break in the series. The authors observe that average annual growth rate of industrial production has decreased significantly in every sector in the lockdown period as compared to pre lockdown period. Testing of unit root hypothesis confirms the presence of stochastic trend in the logarithmic values of industrial output series both at the national and at the state level. The total production declined sharply during the lockdown and the structural break towards the fall in industrial production appeared before the lockdown. Hence, it seems that the lockdown by the pandemic may not be the primary source of industrial recession in India.

DOI: 10.4018/978-1-6684-9089-1.ch013

I INTRODUCTION

The global economy experienced an unprecedented great shock leading to industrial recession due to the outbreak of Covid-19 and the resultant lockdown. The recessionary effects of this shock originated from both the demand and supply sides and naturally, the impact of this shock is expected to be more severe and long lasting. India is no exception to this.

Indian economy is, now, facing twin shocks of demand and supply simultaneously because of the pandemic. To control the pandemic situation, like most of the other countries, India locked down her economic activities completely or in some cases partially. Complete lockdown was imposed in India on March, 25 and continued till May 31, 2020. Thereafter, the unlocking process started in a phased manner and economic activities have been opening gradually, but not in full form. The supply shock (negative) caused by factory shut down is transmitted through supply chains disruption to downstream sectors around the world. In addition, the pandemic has caused income and demand to contract, which affects the upstream sectors everywhere. The resulting decline in income due to the sudden closures of production activities in almost every sector (excepting for essential services) of the economy can cause a downward spiral in demand for products and services. The market itself may not be able to solve this type of problem by its demand and supply mechanism. The role of the state once again becomes highly crucial as prescribed by Keynes in 1930.

Against this backdrop, an attempt has been made in this paper to look into the effects of external shock generated by Covid-19 Pandemic on the industrial output growth in India by analyzing the ' macroeconomic time series data generating process'. In order to do this, we have carried out a number of tests concerning the stochastic behavior and nature of trend of the annual industrial output growth at national level as well as for some major states for the period 1989-90 to 2020-21 locating possible structural break in the series. The location of break point in the time series of industrial output growth would be helpful in finding out whether the temporary shutdown is primarily responsible for industrial recession or not. Again, the presence of unit roots (if any) in such data series will have serious macroeconomic policy implication suggesting that any external shock could have a permanent effect (either positive or negative) on real output whereas in a stationary series, the effects of the shock will be temporary in nature.

Now, the paper has been organized as follows: In section I, after highlighting the background and purpose of the paper, section II reviews the relevant literature. Data base of the study and methodological framework have been presented in section III and section-IV respectively. The results and their interpretations are discussed in section V. Section VI deals with concluding remarks with some policy implication.

II LITERATURE REVIEW

We find several studies relating to the data generating process of macroeconomic time series for India and other countries. The most frequently cited study on the U.S. macroeconomic time series data are Nelson and Plosser (1982) and Perron (1989).

Nelson and Plosser(1982) investigates whether macroeconomic time series are better characterized as stationary fluctuations around a deterministic trend or as non-stationary processes that have no tendency to return to a deterministic path. Using long historical time series for the U.S, they are unable to reject the hypothesis that these series are non-stationary stochastic processes with no tendency to return to a trend line. Based on these findings and an unobserved components model for output that decomposes fluctuations into a secular or growth component and a cyclical component, they infer that shocks to the former, which they associate with real disturbances, contribute substantially to the variation in observed output.

Applying unit root test hypothesis on Nelson-Plosser(1982) data set and to the postwar quarterly real GNP series of US, Perron(1989) considered the null hypothesis that a time series has a unit root with possibly nonzero drift against the alternative that the process is trend-stationary and he showed how standard tests of the unit root hypothesis against trend stationary alternatives cannot reject the unit root hypothesis if the true data generating mechanism is that of stationary fluctuations around a trend function which contains a one-time break.

As far as Indian economy is concerned, the existing literature on the macroeconomic data generation process are quite illuminating and exciting too. Ghatak (1996) had carried out tests of unit root hypothesis on some macroeconomic data covering a period from 1900 to 1988 in the Indian Economy by following the methodologies developed by Nelson and Plosser (1982). In her exercise, she incorporated the possibility of structural break by choosing break points exogenously on the basis of some political and economical importance of the country and by applying the methodology popularized by Perron(1989).

Balakrishnan and Parameswaran (2007) had studied the recent development in the estimation and testing of multiple structural breaks in linear models to identify phases of growth in India since 1950. The noteworthy feature of the methodology is that it allows the data to parameterize the model, thus yielding results that are immune to the prior beliefs of the researcher. The resulting estimates reveal that there are two growth regimes in India since 1950. The authors then decompose by sector the contribution to the change in the growth rate across these regimes. Finally, by means of a simple econometric model, they test a hypothesis that emerges from their estimation and testing for structural break in the main sectors of the economy to provide an explanation of the growth transition in India.

III DATA BASE OF THE STUDY

The data base of our study have been official reports, journals and books which are mainly of secondary type. For our study, while the Central Statistical Office (CSO), Ministry of Statistics and Programme Implementation, Govt. of India provide us all the relevant information regarding the volume of output at constant prices of different sectors at national level, the Net State Domestic Product (NSDP) at constant prices for major Indian states are collected from the Hand Book of Statistics on Indian States, Reserve Bank of India. As different macroeconomic time series based on different base years(1978-79=100 and 1999-00=100) are not comparable, we have constructed a consistent chain linked time series data following the splicing method recommended by the CSO.

IV THE METHODOLOGICAL FRAMEWORK

Any macroeconomic time series may contain either deterministic trend or stochastic trend or both. Their Implications are qualitatively different. A time series with deterministic trend follows trend stationary process(TSP), while a non stationary time series showing stochastic trend is a difference stationary process(DSP). The issue whether a macroeconomic time series is of DSP or TSP is extremely important because the dynamic properties of the two processes are different (Nelson and Plosser, 1982) .While the former is predictable, the latter is completely unpredictable. In a series following TSP cyclical fluctuations are temporary around a stable trend, while for DSP any random shock to the series has a permanent effect. The cyclical components of a TSP originate from the residuals of a regression of the series on the variable time, and a DSP has a trend in the variance with or without trend in the mean. A random walk without drift has no trend in the mean values of the variable.

The most widely used model to take care of stochastic trend in time series is Vector Autoregression(VAR) of order p, [AR(p)] which is given by

$$Y_t = \alpha + \beta_1 Y_{t-1} + \beta_2 Y_{t-2} + \beta_3 Y_{t-3} + \ldots\ldots\ldots\ldots + \beta_p Y_{t-p} + \epsilon_t \ldots \quad (1)$$

Yt gives values in log form in time t and ϵ_t is a stationary series with mean zero and variance σ^2.

This model can generate the trend behavior of macroeconomic time series and the randomly fluctuating behavior of their growth rates, if, for example, Yt is generated by the model,

$$Y_t = \alpha + \beta_1 Y_{t-1} + \epsilon_t \ldots\ldots\ldots\ldots \quad (2)$$

Which is AR(1) with β_1 =1, accumulating Y_t starting with an initial value Y_0, we get

$$Y_t = Y_0 + \alpha t + å \epsilon_j \ldots \tag{3}$$

Which has the same form as the conventional log-linear trend equation, excepting for the fact that the disturbance is not stationary.

Now, in time series data, there may be the existence of correlation across observations. Income today, for example, is highly correlated with income of the last year. Thus Y_t tends to exhibit trend behavior and to be highly correlated over time. The non stationary time series containing a unit root will give a stochastic trend. If β_1 =1 for the AR(1) model, then Y_t has a unit root and exhibit trend behavior, especially when α ¹0 . Unit root series contain a so called stochastic trend.

The Augmented Dickey-Fuller (ADF) test is generally performed for unit root hypothesis. The more appropriate model for testing a unit root is the AR(p) with deterministic trend.

$$\Delta Y_t = \alpha + \rho Y_{t-1} + \eta_1 \Delta Y_{t-1\,+}\, \eta_2 \Delta\, Y_{t-2} + \ldots\ldots\ldots\ldots + \eta_{\rho-1} \Delta Y_{t-\rho+1} + \delta t + \epsilon_t \ldots \tag{4}$$

A series belongs to the class of Difference Stationary Process(DSP) exhibiting stochastic trend if ρ =0, δ =0 and it belongs to the class of Trend Stationary Process (TSP) if ρ <0, If ρ =0 . Then, Y_t contains a unit root. In this case we cannot perform hypothesis testing by utilizing the usual distributions appropriate for least square. Thus we have to follow ADF test. If the t statistics on ρ are less negative than the Dickey Fuller critical value, we conclude that the series Y_t has a unit root.

To test whether the series has a unit root, we have to choose lag length(p). Many sophisticated criteria and testing methods are available to determine the appropriate lag length in an AR(p) model. While determining lag length, econometricians either fix the lag length arbitrarily or choose it through some statistical procedure. For this study, we have used five lag order selection criterion such as Likelihood Ratio (LR), Final Prediction Error (FPE), Akaike Information Criterion (AIC), Schwarz Information Criterion (SC) and Hannan-Quinn Information Criterion (HQ) to select the optimum lag. The lowest value of each criterion is used to select the optimum lag. We found that the max lag length for our study is to be one.

Thus, our model becomes-

$$\Delta Y_t = \alpha + \rho Y_{t-1} + \eta_1 \Delta Y_{t-1\,+}\, \delta t + \epsilon_t \ldots \tag{5}$$

The Dicky-Fuller test sometimes gives a wrong signal particularly when the t-statistics is very close to its critical value. Again, a time series can also appear to exhibit unit root behavior owing to the presence of structural break.

The most commonly used test for structural break is attributed to Chow (1960). The conventional Chow test involves splitting the sample into two or more sub-periods, estimating the parameters for each of the sub-period and finally, testing for the equality of these sets parameters using F statistics. The basic weakness, however, of this test is its critical assumption that the break point is known a priori. CUSUM and CUSUM of squares tests on the residual may be used for testing structural break in mean and volatility of growth rates respectively with unknown break point. The CUSUM test does not require a prior specification of time period when the structural change takes place (Brown, Durbin and Evans, 1975). Bai and Perron(1998) have developed a methodology for testing multiple structural changes. But in Bai and Perron, break points are determined solely on the basis of deterministic trend. In this paper, we have computed the F statistics by allowing stochastic trend in the series.

V RESULTS AND INTERPRETATION

This section focuses on the results of our empirical study. Table- 1 presents the results of the stationary test of the time series employed in this investigation through the application of the Augmented Dickey-Fuller (ADF) stationary test. The hypothesis of a stationary series can be evaluated by testing whether the absolute value of ρ in equation – 5 is strictly less than zero. We have performed ADF test which takes the unit root as the null hypothesis Ho: $\rho = 0$, against the one sided alternative H_1: $\rho < 0$. Table-1 also shows the estimated test statistics corresponding to the logarithmic values and their first and second difference of the series. The results, as is observed from the table, indicate that the all the ADF - t values for the logarithmic series are statistically insignificant at 1%, 5% and even at 10% level indicating that the gross output of different sectors of Indian economy are non stationary at the use level with stochastic trend. But the estimated statistics for their first differences, implying annual growth rates are statistically significant, excepting for the 'construction' and service related sectors such as "Electricity, Gas, Water Supply and Other Utility Services", 'Trade, Repair, Hotels and Restaurants', Financial Services, Real Estate, 'Ownership of Dwelling and Professional Services' in the national economy. As far as their 2nd differences are concerned, the estimated results are statistically significant at all the levels (1%,5% and 10%).

Similarly, the ADF critical values presented in Table-2 suggest the presence of unit root or stochastic trend in the logarithmic series of NSDP for every state under consideration.

Table 1. Augmented Dickey-Fuller Tests for Unit Roots: National Level

Sl. No	Sectors	Log(y)	Δlog(y)	Δ^2log(y)
1	Agriculture, Forestry and Fishing	-2.0214	-5.2822	-7.3838
2	Mining and Quarrying	-3.0010	-4.2423	-4.3816
3	Manufacturing	-3.1221	-5.9254	-6.3021
4	Electricity, Gas, Water Supply and Other Utility Services	-2.6121	-0.3207	-4.8154
5	Construction	-1.6449	0.1277	-4.8058
6	Trade, Repair, Hotels and Restaurants	0.2294	2.6902	-5.5476
7	Transport, Storage, Communication and Services Related to Broadcasting	3.0401	1.1030	-5.7247
8	Financial Services	0.8233	-6.6404	-6.3122
9	Real Estate, Ownership of Dwelling and Professional Services	-1.0330	-2.5700	-7.3522
10	Public Administration and Defence	-2.1277	-6.8747	-8.4755
	Critical Values for Rejection of Hypothesis of a Unit Root			
	1% Critical Value	-4.184	-4.190	-4.154
	5% Critical Value	-3.516	-3.519	-3.503
	10% Critical Value	-3.188	-3.190	-3.180

Source: Authors' Own Calculation

Testing of unit root hypothesis shown above, confirms the presence of stochastic trend in the logarithmic values of gross output series both at the national as well as the state level. It is well known that attempt to find out any structural break in the growth path ignoring stochastic trend in the series may give misleading result. As our growth path bears stochastic trend, now we can safely apply Chow single break point test.

Table-3 shows the results of the test for sectoral structural break in growth paths at the national level . Our results suggest that different sectors of Indian economy experienced a structural break few years before the initiation of Covid-19 pandemic. As is observed from the table, while in Agriculture and Manufacture structural break occurred in the year-2017-18, and 2016-17 respectively, structural break for Financial Services started in 2013-14.

Table-4 shows the structural break points at the state level. It is observed from the table that almost all the states had experienced a structural break few years before the initiation of Covid-19 pandemic. As is highlighted from the table, while in Andhra Pradesh structural break occurred in the year- 2016-17, structural break in Karnataka started in 2015-16.

Table 2. Augmented Dickey-Fuller Tests for Unit Roots: State Level

Sl. No.	States	Log(y)	Δlog(y)	Δ^2log(y)
1	Andhra Pradesh	-1.218206	-2.183863	-5.319827
2	Bihar	-2.986761	-0.926906	0.217437
3	Chhattisgarh	-2.465921	-2.3301	-2.546591
4	Gujarat	0.15332	-0.754668	-2.753481
5	Haryana	-0.022078	-1.34115	-3.284065
6	Himachal Pradesh	2.709867	0.71704	-2.095282
7	Jharkhand	-2.772687	-3.515491	-5.444472
8	Karnataka	-0.012508	-0.99225	-18.68543
9	Madhya Pradesh	-1.519163	-1.813359	-3.394162
10	Odisha	-1.548372	-1.319293	-4.587141
11	Punjab	-0.214663	0.86705	-0.592804
12	Rajasthan	0.739507	-0.439517	-1.805183
13	Tamil Nadu	-0.92126	1.036696	-3.896838
14	Uttar Pradesh	-0.346913	-0.722991	-4.834294
15	West Bengal	-2.950656	-0.560259	-1.78728
	Critical Values for Rejection of Hypothesis of a Unit Root			
	1% Critical Value	-4.184	-4.190	-4.154
	5% Critical Value	-3.516	-3.519	-3.503
	10% Critical Value	-3.188	-3.190	-3.180

Source: Authors'Own Calculation

Table 3. Sectoral Structural Breaks: National Level

Sl.No.	Sectors	Year of Break	F-Statistics
1	Agriculture, Forestry and Fishing	2017-18	31.52994
2	Mining and Quarrying	2016-17	24.8137
3	Manufacturing	2016-17	24.8137
4	Electricity, Gas, Water Supply and Other Utility Services	2015-16	27.69586
5	Construction	2017-18	27.93288
6	Trade, Repair, Hotels and Restaurants	2016-17	22.55724
7	Transport, Storage, Communication and Services Related to Broadcasting	2016-17	12.65207
8	Financial Services	2013-14	10.42949
9	Real Estate, Ownership of Dwelling and Professional Services	2014-15	31.83806
10	Public Administration and Defence	2017-18	26.87616

Source: Authors' Own Calculation

Table 4. Structural Breaks: State Level

Sl.No.	Sectors	Year of Break	F-Statistics
1	Andhra Pradesh	2016-17	33.3655
2	Bihar	2019-20	23.46101
3	Chhattisgarh	2017-18	33.72105
4	Gujarat	2016-17	24.82601
5	Haryana	2016-17	33.75919
6	Himachal Pradesh	2015-16	31.59607
7	Jharkhand	2016-17	27.17504
8	Karnataka	2015-16	34.18648
9	Madhya Pradesh	2016-17	39.01869
10	Odisha	2018-19	50.85662
11	Punjab	2015-16	27.16724
12	Rajasthan	2015-16	35.38294
13	Tamil Nadu	2016-17	31.2386
14	Uttar Pradesh	2017-18	49.09449
15	West Bengal	2018-19	32.83871

Source: Authors' Own Calculation

Table-5 highlights the 'Average Annual Growth Rate' before and after the break point particularly at the national level. Almost all the sectors had suffered a lot after the break point as far as growth rate was concerned. In agriculture, forestry and fishing, the growth rate before the break point (2017-18) was 3.31 percent. It has marginally increased to 3.54 percent after the break had taken place. Manufacturing growth rate had declined from 5 percent to 2.5 percent after the break (2016-17). The growth rate of Trade, Repair, Hotels and Restaurants significantly decreased from roughly from 9 percent to 1 percent.

Table-6 presents the 'Average Annual Growth Rate' before and after the break point at the state level. Like national level, all the states had also experienced industrial recession after the break point. In Andhra Pradesh, the growth rate before the break point (2016-17) was 7.11 percent. It has decreased to 4.8 percent after the break had taken place. The growth rate of Gujarat had declined from 9.6 percent to7.2 percent after the break (2016-17). The growth rate of Uttar Pradesh significantly decreased from 6.21 percent to 1.02 percent.

Table 5. Average Annual Growth Rate Before and After Break Point

Sl.No	Sectors	Before Break Point	After Break Point
1	Agriculture, Forestry and Fishing	3.316423267	3.543264966
2	Mining and Quarrying	7.465884624	-5.868947423
3	Manufacturing	5.09925236	2.505244162
4	Electricity, Gas, Water Supply and Other Utility Services	6.011104961	4.542560574
5	Construction	3.624587929	0.452013978
6	Trade, Repair, Hotels and Restaurants	9.075895822	1.036306459
7	Transport, Storage, Communication and Services Related to Broadcasting	6.362320219	-1.707598192
8	Financial Services	8.847800698	5.675520307
9	Real Estate, Ownership of Dwelling and Professional Services	13.24477702	5.950013653
10	Public Administration and Defence	5.366968954	5.024215986

Source: Authors' Own Calculation

Table 6. Average Annual Growth Rate Before and After Break Point State Level

Sl.No.	Sectors	Before Break Point	After Break Point
1	Andhra Pradesh	7.113759882	4.800084935
2	Bihar	6.20453647	2.813618564
3	Chhattisgarh	5.377803843	3.587800972
4	Gujarat	9.644147574	7.232472802
5	Haryana	8.752281391	3.811317535
6	Himachal Pradesh	7.407669206	4.175239815
7	Jharkhand	5.408251891	3.944062275
8	Karnataka	7.853274457	6.013604829
9	Madhya Pradesh	8.277667765	4.163189678
10	Odisha	7.66565431	1.611310045
11	Punjab	5.372492763	3.330882044
12	Rajasthan	6.161699427	2.69171813
13	Tamil Nadu	6.462186365	5.740794815
14	Uttar Pradesh	6.213069993	1.026108205
15	West Bengal	4.867621291	3.553953285

Source: Authors' Own Calculation

VI CONCLUSION

In this paper, an attempt has been made to look into the effects of external shock generated by Covid-19 Pandemic on the industrial output growth in India by analyzing the ' macroeconomic time series data generating process'.

We observe that average annual growth rate of industrial production has decreased significantly in every sector in the lockdown period as compared to pre lockdown period. Testing of unit root hypothesis confirms the presence of stochastic trend in the logarithmic values of industrial output series both at the national and at the state level. We also observe that though the total production declined sharply during the lockdown, the structural break towards the fall in industrial production appeared before the lockdown. Hence, it seems that the lockdown induced by the Covid-19 pandemic may not be the primary source of industrial recession in India.

REFERENCES

Bai, J., & Perron, P. (1998). Estimating and testing linear models with multiple structural changes. *Econometrica*, *66*(1), 47–78. doi:10.2307/2998540

Balakrishnan, P., & Parameswaran, M. (2007). Understanding economic growth in India: A prerequisite. *Economic and Political Weekly*, 2915–2922.

Brown, R. L., Durbin, J., & Evans, J. M. (1975). Techniques for testing the constancy of regression relationships over time. *Journal of the Royal Statistical Society. Series B, Statistical Methodology*, *37*(2), 149–163.

Chow, G. (1960), Tests of Equality between Sets of Coefficients in Two Linear Regressions, *Econometrica*, 28, Pp-591-605.

Ghatak, A. (1996), Breaking Trend Functions in Macro Variables, The Case of India:1900-1988, *Sankhya,* Series B, Pp-274-87

Nelson, C. R., & Plosser, C. R. (1982). Trends and random walks in macroeconmic time series: Some evidence and implications. *Journal of Monetary Economics*, *10*(2), 139–162.

Perron P(1989), The great Crash, the oil price shock and the unit root hypothesis, *Econometrica*, Vol-57, No-2, PP-1361-1401

APPENDIX

Table 7. Sectoral Composition and Annual Growth Rate of GVA: 2019-20

Sectors	Percentage Share	Annual Growth Rate
Agriculture, Forestry and Fishing	14.6	4.05
Mining and Quarrying	2.7	3.08
Manufacturing	17.4	0.03
Electricity, Gas and Water Supply	2.3	4.14
Construction	7.8	1.27
Trade, Hotels, Transport and Communication	19.4	3.61
Financial, Real estate and Professional Services	21.9	4.62
Public Administration. Defence and Other Services	13.9	9.96
GVA at Basic prices	100	3.89

Source: National Statistical Office

Table 8. Quarterly Growth Rates of Real GDP at Market Price

	2018-19					2019-20		
Components of GDP	**Q1**	**Q2**	**Q3**	**Q4**	**Q1**	**Q2**	**Q3**	**Q4***
Private Final Consumption Expenditure	6.7	8.8	7	6.2	5	5.6	5.9	4.9
Govt. Final Consumption Expenditure	8.5	10.8	7	14.4	8.8	13.2	11.8	4.9
Grossed Fixed Capital Formation	12.9	11.5	11.4	4.4	4.3	-4.1	-5.2	2.5
Exports	9.5	12.5	15.8	11.6	3.2	-2.1	-5.5	-2.8
Imports	5.9	18.7	10.8	0.8	2.1	-9.3	-11.2	-3
GDP at Market Prices	7.1	6.2	5.6	5.7	5.6	5.1	4.7	4.7

Source: National Statistical Office. Note: * Projected Growth

Table 9. Quarterly Growth Rates of Real GDP at Basic Prices

	2018-19				2019-20			
Components of GVA	**Q1**	**Q2**	**Q3**	**Q4**	**Q1**	**Q2**	**Q3**	**Q4***
Agriculture, Forestry and Fishing	3.8	2.5	2	1.6	2.8	3.1	3.5	5
Mining and Quarrying	7.8	4.7	4.4	1.4	3.2	0.1	0.1	2.3
Manufacturing	-7.3	-7	-4.4	-4.8	4.7	0.2	3.2	2.6
Electricity, Gas and Water Supply	10.7	5.6	5.2	2.1	2.2	-0.4	-0.2	1.8
Construction	7.9	9.9	9.5	5.5	8.8	3.9	-0.7	6.5
Trade, Hotels, Transport and Communication	7.3	7.2	7.3	8.3	6.7	6.8	6.4	6.1
Financial, Real estate and Professional Services	6.4	5.2	6.6	6	5.5	2.9	0.3	3.2
Public Administration. Defence and Other Services	8.5	7.8	7.8	6.9	5.7	5.8	5.9	5.1
GVA at Basic prices	6	6.5	6.5	8.7	6.9	7.1	7.3	8
	8.8	8.9	8.1	11.6	8.7	10.1	9.7	6.7
	6.9	6.1	5.6	5.6	5.4	4.8	4.5	5

Source: National Statistical Office. Note: * Projected Growth

Compilation of References

Adams, E. C., Oduor, P., Wahome, A., & Tondapu, G. (2022). Reflections on two years teaching earth science at the Women in Science (WiSci) STEAM camp. *Journal of Women and Minorities in Science and Engineering*, *28*(1), 23–39. doi:10.1615/JWomenMinorScienEng.2021033536

Adelowo, C., Ilori, M. O., Siyanbola, W. O., & Oluwale, B. (2015). Technological learning mechanisms in Nigeria's technology incubation centre. *African Journal of Economic and Management Studies*, *6*(1), 72–89. doi:10.1108/AJEMS-10-2014-0071

Admiral, W., & Lockhorst, D. (2009). E-learning in small medium-sized enterprises across Europe. *International Small Business Journal*, *27*(6), 743–767. doi:10.1177/0266242609344244

Agarwal, N., & Ahuja, V. (2014). Creation of the Training-Chart: A Step Forward to Make the Training More Effective. [IJSSMET]. *International Journal of Service Science, Management, Engineering, and Technology*, *5*(2), 1–18. doi:10.4018/ijssmet.2014040101

Ahmed, S. F. A., Krishna, S. H., Ganeshkumar, K., Anthiyur, U., & Manivel, R. (2023). Explorıng the impact of artificial intelligence in business decision making. *Journal of Data Acquisition and Processing*, *38*(3), 686.

Akbari, M., Danesh, M., Rezvani, A., Javadi, N., Banihashem, S. K., & Noroozi, O. (2023). The role of students' relational identity and autotelic experience for their innovative and continuous use of e-learning. *Education and Information Technologies*, *28*(2), 1911–1934. doi:10.100710639-022-11272-5 PMID:35971410

Akman, I., & Turhan, C. (2018). Investigation of employers' performance expectations for new IT graduates in individual and team work settings for software development. *Information Technology & People*, *31*(1), 199–214. doi:10.1108/ITP-01-2017-0020

Al Karim, R. (2019). Impact of different training and development programs on employee performance in Bangladesh's perspective. *International Journal of Entrepreneurial Research*, *2*(1), 8–14. doi:10.31580/ijer.v1i2.497

Alarabiat, A., Hujran, O., Soares, D., & Tarhini, A. (2023). Examining students' continuous use of online learning in the post-COVID-19 era: An application of the process virtualization theory. *Information Technology & People*, *36*(1), 21–47. doi:10.1108/ITP-02-2021-0142

Alarcón, R., del Pilar-Jiménez, E., & de Vicente-Yagüe, M. I. (2020). Development and validation of the DIGIGLO, a tool for assessing the digital competence of educators. *British Journal of Educational Technology, 51*(6), 2407–2421. doi:10.1111/bjet.12919

Ali, S. S., Kaur, R., & Jaramillo, A. B. (2018). An assessment of green supply chain framework in Indian automobile industry using interpretive structural modelling and its validation using MICMAC analysis. *International Journal of Services and Operations Management, 30*(3), 318–356. doi:10.1504/IJSOM.2018.092607

Aliyu, M. M. (2017). Developing oral communication skills through collaborative learning: Implications for Nigerian teachers. *International Journal of English Literature and Social Sciences, 2*(5), 239225. doi:10.24001/ijels.2.5.15

Aljedaani, W., Krasniqi, R., Aljedaani, S., Mkaouer, M. W., Ludi, S., & Al-Raddah, K. (2022). If online learning works for you, what about deaf students? Emerging challenges of online learning for deaf and hearing-impaired students during COVID-19: A literature review. *Universal Access in the Information Society*, •••, 1–20. PMID:35910240

Aljohani, N. R., Aslam, M. A., Khadidos, A. O., & Hassan, S.-U. (2022). A Methodological Framework to Predict Future Market Needs for Sustainable Skills Management Using AI and Big Data Technologies. *Applied Sciences (Basel, Switzerland), 12*(14), 6898. doi:10.3390/app12146898

Allen, D., & Eve, A. (1968). Microteaching. *Theory into Practice, 7*(5), 81–185. doi:10.1080/00405846809542153

Almunawar, M. N., Islam, Z., & Ordóñez de Pablos, P. (2022). *Digital Transformation Management: Challenges and Futures in the Asian Digital Economy*. Routledge. doi:10.4324/9781003224532

Almunawar, N., Ordóñez de Pablos, P., & Anshari, M. (2023a). *Sustainable Development and the Digital Economy: Human-centricity, Sustainability and Resilience in Asia*. Routledge. doi:10.4324/9781003388753

Almunawar, N., Ordóñez de Pablos, P., & Anshari, M. (2023b). *Digital Transformation for Business and Society: Contemporary Issues and Applications in Asia*. Routledge.

Alnaim, N., & Alsarawi, A. (2023). Obstacles To Distance Education For Students With Learning Disabilities And Ways To Face Them: From The Point Of View Of Female Teachers. *Turkish Online Journal of Distance Education, 24*(1), 200–219. doi:10.17718/tojde.1076073

Alnasib, B. N. M. (2023). Digital competencies: Are pre-service teachers qualified for digital education? [IJEMST]. *International Journal of Education in Mathematics, Science, and Technology, 11*(1), 96–114. doi:10.46328/ijemst.2842

Al-Sheyadi, A., Muyldermans, L., & Kauppi, K. (2019). The complementarity of green supply chain management practices and the impact on environmental performance. *Journal of Environmental Management, 242*, 186–198. doi:10.1016/j.jenvman.2019.04.078 PMID:31035181

Amaral, A., & Peças, P. (2021). A framework for assessing manufacturing SMEs Industry 4.0 maturity. *Applied Sciences (Basel, Switzerland), 11*(13), 6127. doi:10.3390/app11136127

Ameliah, R., Negara, R.A. and Rahmawati, I. (2022). *Status Literasi Digital di Indonesia 2021*. Katadata Insight Center. Kementrian Komunikasi dan Informasi Indonesia.

Andrade, M. S. (2016). Curricular elements for learner success: 21st century skills. *Journal of Education and Training Studies*, *4*(8), 143–149. doi:10.11114/jets.v4i8.1743

Andreasen, J. K., Tømte, C. E., Bergan, I., & Kovac, V. B. (2022). Professional digital competence in initial teacher education: An examination of differences in two cohorts of pre-service teachers. *Nordic Journal of Digital Literacy*, *17*(1), 61–74. doi:10.18261/njdl.17.1.5

Anshari, M., Almunawar, M. N., & Ordóñez de Pablos, P. (2023). Disruptive Innovation and a Multi-Sided Digital Marketplace. In *Handbook of Research on Artificial Intelligence and Knowledge Management in Asia's Digital Economy* (pp. 366–377). IGI Global.

Anshari, M., Syafrudin, M., Tan, A., Fitriyani, N. L., & Alas, Y. (2023). Optimisation of Knowledge Management (KM) with Machine Learning (ML) Enabled. *Information (Basel)*, *14*(1), 35. Advance online publication. doi:10.3390/info14010035

APJII. (2020). Penetrasi dan Profil Perilaku Pengguna Internet Indonesia. Survei. [Online], Retrieved July 10, 2023, from.

Apunyo, R., White, H., Otike, C., Katairo, T., Puerto, S., Gardiner, D., Kinengyere, A. A., Eyers, J., Saran, A., & Obuku, E. A. (2022). Interventions to increase youth employment: An evidence and gap map. *Campbell Systematic Reviews*, *18*(1), e1216. doi:10.1002/cl2.1216 PMID:36913191

Ayuningtias, H. G., & Anggadwita, G. (2015). Human capital development of information and communication technology industry in Indonesia. *Journal of Research and Applications: Accounting and Management*, *1*(2), 157–164. doi:10.18382/jraam.v1i2.15

Ayuningtias, H. G., Anggadwita, G., & Prasetia, A. Y. (2018). Global talent program as determinants of employees' performance on telecommunication company in Indonesia. *International Journal of Learning and Intellectual Capital*, *15*(3), 242–257. doi:10.1504/IJLIC.2018.094721

Azzi, A., Battini, D., Persona, A., & Sgarbossa, F. (2012). Packaging design: General framework and research agenda. *Packaging Technology & Science*, *25*(8), 435–456. doi:10.1002/pts.993

Badi, S., & Murtagh, N. (2019). Green supply chain management in construction: A systematic literature review and future research agenda. *Journal of Cleaner Production*, *223*, 312–322. doi:10.1016/j.jclepro.2019.03.132

Bae, K.-H. G., Evans, L. A., & Summers, A. (2016). Lean design and analysis of a milk-run delivery system: case study. *2016 Winter Simulation Conference (WSC)*, 2855–2866. 10.1109/WSC.2016.7822321

Bahadorfar, M., & Omidvar, R. (2014). Technology in teaching speaking skill. *Acme International Journal of Multidisciplinary Research*, *2*(4), 9–13.

Bai, J., & Perron, P. (1998). Estimating and testing linear models with multiple structural changes. *Econometrica*, *66*(1), 47–78. doi:10.2307/2998540

Baird, A. M., & Parayitam, S. (2019). Employers' ratings of importance of skills and competencies college graduates need to get hired: Evidence from the New England region of USA. *Education + Training, 61*(5), 622–634. doi:10.1108/ET-12-2018-0250

Bakhshi, H., Downing, J., Osborne, M., & Schneider, P. (2017). *The Future of Skills: Employment in 2030.* Pearson and Nesta.

Banerjee, S. K. (2009). Tourism Management and Environment: A study on the Carrying Capacity and Prospect of Ecotourism in Sikkim. PhD Thesis, North Bengal University, West Bengal. https://ir.nbu.ac.in/bitstream/123456789/1285/18/235578.pdf. Accessed on 25th March, 2023 at 10 am.

Bang, H., Fuglesang, S. L., Ovesen, M. R., & Eilertsen, D. E. (2010). Effectiveness in Top Management Group Meetings: The Role of Goal Clarity, Focused Communication, and Learning Behavior. *Scandinavian Journal of Psychology, 51*(3), 253–261. doi:10.1111/j.1467-9450.2009.00769.x PMID:20132458

Bappenas. (2020). *Berdaulat, Maju, Adil dan Makmur.* [Online] Available at: https://www.bappenas.go.id/files/Visi Indonesia 2045/Ringkasan Eksekutif Visi Indonesia 2045_Final.pdf [Accessed: July 10, 2023]].

Basar, Z. M., Mansor, A. N., Jamaludin, K. A., & Alias, B. S. (2021). The Effectiveness and Challenges of Online Learning for Secondary School Students- A Case Study. *Asian Journal of University Education, 17*(3), 119–129. doi:10.24191/ajue.v17i3.14514

Basenko, G., & Baskakova, V. (2021). Podcasts in the teaching media space. In *E3S Web of Conferences* (pp. 1–8). EDP Sciences., doi:10.1051/e3sconf/202127312122

Bashir, S., & Miyamoto, K. (2020). Digital Skills: Frameworks and Programs. World Bank. [Online], Retrieved July 10, 2023, from. doi:10.1596/35080

Basten, D., & Haamann, T. (2018). Approaches for Organizational Learning: A Literature Review. *SAGE Open, 8*(3). doi:10.1177/2158244018794224

Beede, D. N., Julian, T. A., Langdon, D., McKittrick, G., Khan, B., & Doms, M. E. (2011). *Women in STEM: A gender gap to innovation.* Retrieved from https://files.eric.ed.gov/fulltext/ED523766.pdf

Beer, M., Voelpel, S., Leibold, M., & Tekie, E. (2005). Strategic management as organizational learning: Developing fit and alignment through a disciplined process. *Long Range Planning, 38*(5), 445–465. doi:10.1016/j.lrp.2005.04.008

Belgrade Open School. (2017). *The Regional Youth Employability Skills Portfolio*, Belgrade, Serbia. https://bos.rs/rs/uploaded/Portfolio%2025_7.pdf

Bell, B., Tannenbaum, S., Ford, K., Noe, R., & Kraiger, K. (2017). 100 Years of training and development research: What we know and where we should go. *The Journal of Applied Psychology, 102*(3), 305–323. doi:10.1037/apl0000142 PMID:28125262

Bennett, S., Maton, K., & Kervin, L. (2008). The digital natives. *British Journal of Educational Technology*, *39*(5), 775–786. doi:10.1111/j.1467-8535.2007.00793.x

Bervell, B., & Umar, I. N. (2018). Utilization decision towards LMS for blended learning in distance education: Modeling the effects of personality factors in exclusivity. *Knowledge Management & E-Learning*, *10*(3), 309–333.

Beyene, W. M., Mekonnen, A. T., & Giannoumis, G. A. (2023). Inclusion, access, and accessibility of educational resources in higher education institutions: Exploring the Ethiopian context. *International Journal of Inclusive Education*, *27*(1), 18–34. doi:10.1080/13603116.2020.1817580

Bhasin, B. (2012). Integration of information and communication technology in enhancing teaching and learning. *Contemporary Educational Technology*, *3*(2), 130–140. doi:10.30935/cedtech/6073

Bhatia, M. (2022). *Women Empowerment with the Power of STEAM*. Retrieved from http://library.ediindia.ac.in:8181/jspui/handle/123456789/13212

Bhattacharya, S. (2019). Environmental Crisis in the Eastern Himalayan Landscapes in India. *Consilience: Journal of Sustainable Development*, *21*(1), 65–85.

Bhatt, G., & Zaveri, J. (2002). The enabling role of decision support systems in organizational learning. *Decision Support Systems*, *32*(3), 297–309. doi:10.1016/S0167-9236(01)00120-8

Bhutia, S. (2015). Spacio-Temporal Study on Urbanization in the Darjeeling Himalaya: A Demographic Perspective. *IOSR Journal of Humanities and Social Sciences*, *20*(4), 10–18.

Bhutia, S. G., Bhutia, J. W., & Mahapatra, S. S. (2022). Analysing the Issues and Challenges of Homestays for Sustainable Tourism Development in Sikkim and Darjeeling Hills. *Orissa Journal of Commerce*, *43*(4), 73–88. doi:10.54063/ojc.2022.v43i04.06

Bianchi, G., Pisiotis, U., & Cabrera Giraldez, M. (2022). *GreenComp-The European sustainability competence framework*. Publication Office of the European Union.

Bijelovic Bosanac, B., Pavlovic, Lj., & Martinic, M. (2017). *Stanje ducausegve politike zapošljavanja mladih u Republici Srbiji*, Beograd: Udruženje građana: Inicijativa za razvoj i saradnju. http://www.poslodavci.rs/wp-content/uploads/2017/09/stanje-i-perspektive-politike-zapoljavanja-mladih.pdf

Blackburn, H. (2017). The status of women in STEM in higher education: A review of the literature 2007–2017. *Science & Technology Libraries*, *36*(3), 235–273. doi:10.1080/0194262X.2017.1371658

Blaique, L., Ismail, H. N., & Aldabbas, H. (2023). Organizational learning, resilience and psychological empowerment as antecedents of work engagement during COVID-19. *International Journal of Productivity and Performance Management*, *72*(6), 1584–1607. doi:10.1108/IJPPM-04-2021-0197

Blickstein, S. (1996). Does training pay off? *Across the Board*, *33*(6), 16–20.

Bloodhart, B., Balgopal, M. M., Casper, A. M., Sample McMeeking, L. B., & Fischer, E. V. (2020). Outperforming yet undervalued: Undergraduate women in STEM. *PLoS One*, *15*(6), e0234685. Advance online publication. doi:10.1371/journal.pone.0234685 PMID:32584838

Bocewicz, G., Nielsen, P., & Banaszak, Z. (2021). Ordered–Fuzzy-Numbers-Driven Approach to Out-Plant Milk-Run Dynamic Routing and Scheduling. In Automation 2021: Recent Achievements in Automation, Robotics and Measurement Techniques (pp. 20–36). Springer. doi:10.1007/978-3-030-74893-7_3

Boers, B., & Cottrell, S. (2007). Sustainable Tourism Infrastructure Planning: A GIS Supported Approach. *Tourism Geographies*, *9*(1), 1–21. doi:10.1080/14616680601092824

Boettcher, J. V., & Conrad, R. M. (2021). *The online teaching survival guide: Simple and practical pedagogical tips*. John Wiley & Sons.

Bolliger, D. U., Supanakorn, S., & Boggs, C. (2010). Impact of podcasting on student motivation in the online learning environment. *Computers & Education*, *55*(2), 714–722. doi:10.1016/j.compedu.2010.03.004

Borsheim, C., Merritt, K., & Reed, D. (2008). Beyond technology for technology's sake: Advancing multiliteracies in the twenty-first century. *The Clearing House: A Journal of Educational Strategies, Issues and Ideas*, *82*(2), 87–90. doi:10.3200/TCHS.82.2.87-90

Bosworth, B., & Collins, S. M. (2008). Accounting for growth: Comparing China and India. *The Journal of Economic Perspectives*, *22*(1), 45–66. doi:10.1257/jep.22.1.45

Boysen, N., Emde, S., Hoeck, M., & Kauderer, M. (2015). Part logistics in the automotive industry: Decision problems, literature review and research agenda. *European Journal of Operational Research*, *242*(1), 107–120. doi:10.1016/j.ejor.2014.09.065

BPS. (2018). *Survey Angkatan Kerja Nasional 2018*.

BPS. (2020). Survey Angkatan Kerja Nasional 2020.

Bradic-Martinovic, A., Pavlovic, D., & Zdravkovic, A. (2019). *Digitalne veštine studenata u Srbiji. Naucni skup "Trendovi razvoja: Kvalitet visokog obrazovan"a"*, Fakultet tehničkih nauka, Novi Sad, Srbija, 210-213. http://www.trend.uns.ac.rs/stskup/trend_2019/radovi/T4.1/T4.1-1.pdf

Brandenburg, D., & Ellinger, A. (2003). The future: Just-in-time learning expectations and potential implications for human resource development. *Advances in Developing Human Resources*, *5*(3), 308–320. doi:10.1177/1523422303254629

Brantlinger, E., Jimenez, R., Klingner, J., Pugach, M., & Richardson, V. (2005). Qualitative studies in special education. *Exceptional Children*, *71*(2), 195–207. doi:10.1177/001440290507100205 PMID:15757375

Brevik, L. M., Gudmundsdottir, G. B., Lund, A., & Strømme, T. A. (2019). Transformative agency in teacher education: Fostering professional digital competence. *Teaching and Teacher Education*, *86*, 102875. doi:10.1016/j.tate.2019.07.005

Brewer, L. (2013). Enhancing youth employability: what? Why? and how? Guide to core work skills, published by International Labor Organization-ILO, Geneva.

Brintha, N. C. JT, W. J., & Sukumaran, J. (2019). Integrating SMEs through cloud: an industrial revolution. In Organizational Transformation and Managing Innovation in the Fourth Industrial Revolution (pp. 143-164). IGI Global.

Brouder, P., Teoh, S., Salazar, N. B., Mostafanezhad, M., Pung, J. M., Lapointe, D., Higgins Desbiolles, F., Haywood, M., Hall, C. M., & Clausen, H. B. (2020). Reflections and discussions: Tourism matters in new normal post Covid 19. *Tourism Geographies*, *22*(3), 735–746. doi:10.1080/14616688.2020.1770325

Brown, R. L., Durbin, J., & Evans, J. M. (1975). Techniques for testing the constancy of regression relationships over time. *Journal of the Royal Statistical Society. Series B, Statistical Methodology*, *37*(2), 149–163.

Brown, T. (2000). Lifelong learning: An organizing principle for reform. *Australian Journal of Adult Learning*, *58*(3), 312. https://www.researchgate.net/publication/329574090_Lifelong_Learning_An_organising_principle_for_reform#fullTextFileContent

Brozzi, R., Riedl, M., & Matta, D. (2021). Key readiness indicators to assess the digital level of manufacturing SMEs. *Procedia CIRP*, *96*, 201–206. doi:10.1016/j.procir.2021.01.075

Bughin, J., Hazan, E., Ramaswamy, S., Chui, M., Allas, T., Dahlström, P., Henke, N., & Trench, M. (2017). Artificial intelligence the next digital frontier? ***McKinsey and Company Global Institute***, Discussion Paper.

Bundsgaard, J., & Hansen, T. I. (2011). Evaluation of learning materials: A holistic framework. *Journal of Learning Design*, *4*(4), 31–44. doi:10.5204/jld.v4i4.87

Buranova, D., & Latipova, D. (2023). Involvement of a medical student in the process of learning foreign languages. ***British medical, 3(3)***.

Burgess, J., & Russell, J. (2003). The effectiveness of distance learning initiatives in organizations. *Journal of Vocational Behavior*, *63*(2), 289–303. doi:10.1016/S0001-8791(03)00045-9

Burke, B. (2020). Gamification 2020: What is the future of gamification. **Gartner Research,** https://www.gartner.com/en/documents/2226015 Accessed on: 11.01.2023.

Bustillo, E., & Aguilos, M. (2022). The Challenges of Modular Learning in the wake of COVID-19: A Digital Divide in the Philippine Countryside Revealed. *Education Sciences*, *12*(7), 449. doi:10.3390/educsci12070449

Butowski, L., Makowska-Iskierka, M., & Pokojski, W. (2018). Tourism Space versus Tourism Destinations: Methodological Considerations and Empirical Testing of their Development. *Geografie (Utrecht)*, *121*(1), 85–106. doi:10.37040/geografie2018123010085

Caena, F., & Redecker, C. (2019). Aligning teacher competence frameworks to 21st century challenges: The case for the European Digital Competence Framework for Educators (Digcompedu). *European Journal of Education, 54*(3), 356–369. doi:10.1111/ejed.12345

Cardoso, A. R., & Pereira, R. (2022). Relating Industry 4.0 and the SME Internationalization Process: A Case Study. In Implementing Automation Initiatives in Companies to Create Better-Connected Experiences (pp. 40-60). IGI Global.

Cavanaugh, C., Repetto, J., Wayer, N., & Spitler, C. (2013). Online learning for students with disabilities: A framework for success. *Journal of Special Education Technology, 28*(1), 1–8. doi:10.1177/016264341302800101

Çebi, A., Özdemir, T. B., Reisoğlu, İ., & Çolak, C. (2022). From digital competences to technology integration: Re-formation of pre-service teachers' knowledge and understanding. *International Journal of Educational Research, 113*, 101965. doi:10.1016/j.ijer.2022.101965

CEIC. (2020). Number of Hotel Rooms: Sikkim. https://www.ceicdata.com/en/india/number-of-hotel-rooms-by-states/number-of-hotel-rooms-sikkim. Accessed on 23rd January, 2023 at 9 pm.

Cekic, J., & Markovic, J. (2019). *Work-based learning in Serbia. Turin: European Training Foundation. (Report)*, ETF Europe. https://www.etf.europa.eu/sites/default/files/2019-07/Work-based%20learning%20Serbia.pdf

Chaiyaso, P., Piyaphimolsit, C., Ekwarangkun, P., Sritanyu, E., Chantra, C., Wanichwasin, P., Taksino, P., & Gamket, W. (2013). *Management innovation development. Teaching and learning to enhance the learning assessment competencies of student teachers*. Research and Development Institute of Kasetsart University.

Chamchantharawong, C., Chanjarain, I., Suwanjinda, C., & Siririrth, W. (2017). Factors affecting the functional competency of teacher of graduate diploma students in teaching profession at Chiangrai College. Veridian E-Journal. *Silpakorn University, 10*(1), 269–280.

Chand, D., Sati, J., Laxmi, V., & Nautiyal, M. C. (2016). Effects of Anthropogenic Activities in High Altitude Areas of Garhwal Himalayas. *International Journal for Environmental Rehabilitation and Conservation, II*(1), 78–87.

Chandra Shukla, A., Deshmukh, S. G., & Kanda, A. (2009). Environmentally responsive supply chains: Learnings from the Indian auto sector. *Journal of Advances in Management Research, 6*(2), 154–171. doi:10.1108/09727980911007181

Chang, S. C., Chang, H. H., & Lu, M. T. (2021). Evaluating industry 4.0 technology application in SMES: Using a Hybrid MCDM Approach. *Mathematics, 9*(4), 414. doi:10.3390/math9040414

Chan, N. N., Walker, C., & Gleaves, A. (2015). An exploration of students' lived experiences of using smartphones in diverse learning contexts using a hermeneutic phenomenological approach. *Computers & Education, 82*, 96–106. doi:10.1016/j.compedu.2014.11.001

Chavez, R., Yu, W., Feng, M., & Wiengarten, F. (2016). The effect of customer-centric green supply chain management on operational performance and customer satisfaction. *Business Strategy and the Environment, 25*(3), 205–220. doi:10.1002/bse.1868

Chen, K. S., & Liao, H. T. (2020, December). Mapping the Field of Educational Assessment and Professional Accreditation in Higher Education: Bibliographic Coupling and Co-Citation Analyses for Green and Digital Transformation. *Management Science Informatization and Economic Innovation Development Conference (MSIEID)* (pp. 281-284). IEEE. 10.1109/MSIEID52046.2020.00058

Chiaburu, D. S., & Marinova, S. V. (2005). What predicts skill transfer? An exploratory study of goal orientation, training self-efficacy and organizational supports. *International Journal of Training and Development, 9*(2), 110–123. doi:10.1111/j.1468-2419.2005.00225.x

Chow, G. (1960), Tests of Equality between Sets of Coefficients in Two Linear Regressions, *Econometrica*, 28, Pp-591-605.

Cinquin, P. A., Guitton, P., & Sauzéon, H. (2019). Online e-learning and cognitive disabilities: A systematic review. *Computers & Education, 130*, 152–167. doi:10.1016/j.compedu.2018.12.004

Clark, A. (2011). Embedding transferable skills and enhancing student learning in a political science research methods module. Evidence from the United Kingdom. *PS, Political Science & Politics, 44*(1), 135–139. doi:10.1017/S1049096510002039

Cloete, N. (2014). The South African higher education system: Performance and policy. *Studies in Higher Education, 39*(8), 1355–1368. doi:10.1080/03075079.2014.949533

Coelho, P. M., Corona, B., ten Klooster, R., & Worrell, E. (2020). Sustainability of reusable packaging–Current situation and trends. *Resources, Conservation & Recycling: X, 6*, 100037.

Cohen, L., Manion, L., & Morrison, K. (2017). *Research methods in education*. Routledge. doi:10.4324/9781315456539

Coleman, S., Göb, R., Manco, G., Pievatolo, A., Tort-Martorell, X., & Reis, M. S. (2016). How can SMEs benefit from big data? Challenges and a path forward. *Quality and Reliability Engineering International, 32*(6), 2151–2164. doi:10.1002/qre.2008

Collins, C. (2008). Looking to the future: Higher education in the Metaverse. *EDUCAUSE Review, 43*(5), 50–52. https://www.learntechlib.org/p/102249/

Collis, B., & Margaryan, A. (2004). Applying activity theory to CSCL and work-based activities in corporate settings. *Educational Technology Research and Development, 52*(4), 38–52. doi:10.1007/BF02504717

Colliver, J. A. (2000). Effectiveness of problem-based learning curricula: Research and theory. *Academic Medicine, 75*(3), 259–266. doi:10.1097/00001888-200003000-00017 PMID:10724315

ColomboE.StancaL. (2008). The impact of training on productivity: evidence from a large panel of firms. *Available at* SSRN.

Comey, W. (2009). Blended learning and the classroom environment: A comparative analysis of students' perception of the classroom environment across community college courses taught in traditional face-to-face, online and blended methods. (Unpublished Doctoral Disertation), The George Washington University, The Faculty of The Graduate School of Education and Human Development.

Contini, D., Filandri, M., & Pacelli, L. (2019). Persistency in the NEET state: A longitudinal analysis. *Journal of Youth Studies*, *22*(7), 1–22. doi:10.1080/13676261.2018.1562161

Contreras, G. S., González, A. H., Fernández, M. I., Martínez, C. B., Cepa, J., & Escobar, Z. (2022). The importance of the application of the metaverse in education. *Modern Applied Science*, *16*(3), 1–34. doi:10.5539/mas.v16n3p34

Cooper, M. (2006). Making online learning accessible to disabled students: An institutional case study. *Research in Learning Technology*, *14*(1), 103–115. doi:10.3402/rlt.v14i1.10936

Cope, B., & Kalantzis, M. (Eds.). (2000). *Multiliteracies: Literacy learning and the design of social futures*. Psychology Press.

Cotrino, A., Sebastián, M. A., & González-Gaya, C. (2020). Industry 4.0 roadmap: Implementation for small and medium-sized enterprises. *Applied Sciences (Basel, Switzerland)*, *10*(23), 8566. doi:10.3390/app10238566

Cramer, J., & Crocco, M. S. (2010). Women and technology, upon reflection: Linking global women's issues to the digital gender divide in urban social studies education. In Technology Implementation and Teacher Education: Reflective Models (pp. 184-202). IGI Global.

Crosling, G., & Ward, I. (2002). Oral communication: The workplace needs and uses of business graduate employees. *English for Specific Purposes*, *21*(1), 41–57. doi:10.1016/S0889-4906(00)00031-4

Crow, Kevin L. "Four types of disabilities: Their impact on online learning." TechTrends 52.1 (2008): 51.

Damşa, C., Langford, M., Uehara, D., & Scherer, R. (2021). Teachers' agency and online education in times of crisis. *Computers in Human Behavior*, *121*, 106793. doi:10.1016/j.chb.2021.106793 PMID:36568039

Damuluri, S., Islam, K., Ahmadi, P. and Qureshi, N. S. (2020). Analyzing Navigational Data and Predicting Student Grades Using Support Vector Machine. *Emerging Science Journal*, 4(4), pp,243–252. doi:10.28991/esj-2020-01227

Das, K., Wibowo, P., Chui, M., Agarwal, V., & Lath, V. (2019). *Automation and the future of work in Indonesia*. McKinsey & Company.

de Sousa Jabbour, A. B. L., Jabbour, C. J. C., Govindan, K., Kannan, D., Salgado, M. H., & Zanon, C. J. (2013). Factors affecting the adoption of green supply chain management practices in Brazil: Empirical evidence. *The International Journal of Environmental Studies*, *70*(2), 302–315. doi:10.1080/00207233.2013.774774

De Souza, E. D., Kerber, J. C., Bouzon, M., & Rodriguez, C. M. T. (2022). Performance evaluation of green logistics: Paving the way towards circular economy. *Cleaner Logistics and Supply Chain*, *3*, 100019. doi:10.1016/j.clscn.2021.100019

Deloitte. (2017). *What Key Competencies Are Needed in the Digital Age? The Impact of Automation on Employees, Companies, and Education.* [Online] Available at: https://www2.deloitte.com/content/dam/Deloitte/ch/Documents/innovation/ch-en-innovation-automation-competencies.pdf [Accessed: July 10, 2023]].

Department of Education. (2004, September 2). White Paper on e-Education: Transforming Learning and Teaching through Information and Communication Technologies (ICTs). *Government Gazette* (26734), pp. 3 - 46.

Department of Tourism (DoT). (2019). Tourism Annual Report 2018-29, Government of Sikkim.

Department of Tourism (DoT). (2020). Tourism Annual Report 2019-20, Government of Sikkim.

Dhunna, S. (2020). GREEN MARKETING: AN OVERVIEW. *New Horizons in Commerce, IT &. Social Sciences*, 106.

Diningrat, R. A., Rizku, M., Oley, J. D. B., Natih, P., & Astini, F. N. (2022). *Digital Skills Landscape in Indonesia*. The SMERU Research Institute.

District Census Handbook (2011). Census of India, Sikkim, Series 12, Part XII B, Directorate of Census Operations, Sikkim.

Djukic, M., & Pavlovic, D. (2020). *Istrazivanje o potrebama anketara za sprovodjenje integrisanih usluga za zaposljivost NEET mladih sa fokusom na 4.0 industrijsku revoluciju*, Youth Work Centre, Serbia. https://socijalnoukljucivanje.gov.rs/wp-content/uploads/2021/12/Istrazivanje_o_potrebama_aktera_za_sprovodjenje_integrisanih_usluga_za_zaposljivost_NEET_mladih.pdf

Dodgson, M. (2021). The strategic management of technology and innovation. In Oxford Research Encyclopedia of Business and Management. doi:10.1093/acrefore/9780190224851.013.145

Do, H., Budhwar, P., Shipton, H., Nguyen, H.-D., & Nguyen, B. (2022). Building organizational resilience, innovation through resource-based management initiatives, organizational learning and environmental dynamism. *Journal of Business Research*, *141*, 808–821. doi:10.1016/j.jbusres.2021.11.090

Doornbos, A., Simons, R.-J., & Denessen, E. (2008). Relations between characteristics of workplace practices and types of informal work-related learning: A survey study among Dutch police. *Human Resource Development Quarterly*, *19*(2), 129–151. doi:10.1002/hrdq.1231

Do, T. T., & Mai, N. K. (2022). Organizational Learning and firms performance: A systematic review. *International Journal of Productivity and Performance Management*, *71*(4), 1230–1253. doi:10.1108/IJPPM-02-2020-0051

Drotner, K. (2020). Children's digital content creation: Towards a processual understanding of media production among Danish children. *Journal of Children and Media, 14*(2), 221–236. do i:10.1080/17482798.2019.1701056

Duffey, D., & Fox, C. (2012). National educational technology trends: 2012. State leadership empowers educators, transforms teaching and learning. ***Glen Burnie, MD: State Educational Technology Directors Association (SETDA).*** https://files.eric.ed.gov/fulltext/ED536746.pdf Access date: 25.02.2023.

Dulskaia, I., Bellini, F., Zabal, R., Garcia Pañella, O., & D'ascenzo, F. (2017). Gamification as an instrument for organizational behaviour change during the meeting: case study «ROBATIEMPOS». *Proceedings of the International Conference on Business Excellence,* 11(1). DOI: 10.1515/picbe-2017-0039

Dwyer, L., Forsyth, P., & Spurr, R. (2006). Effects of SARS crisis on the economic contribution of tourism to Australia. *Tourism Review International, 10*(1), 47–55. doi:10.3727/154427206779307231

Easterby-Smith, M., Crossan, M., & Nicolini, D. (2000). Organizational Learning: Debates Past, Present and Future. *Journal of Management Studies, 37*(6), 783–796. doi:10.1111/1467-6486.00203

Economic Times (ET). **(2019). Where do tourists prefer to go in India?** https://economictimes.indiatimes.com/industry/services/travel/where-do-tourists-prefer-to-go-in-india/who-spends-the-most-on-tourism/slideshow/72516703.cms Accessed on 25th March, 2023 at 5.31 pm.

Economic Times (ET). (2020). Gross State Domestic Product to see sharp decline in FY21. https://economictimes.indiatimes.com/news/economy/indicators/state-gross-domestic-products-to-see-sharp-declines-this-fiscal/articleshow/76688108.cms?from=mdr. Accessed on 15th March, 2023 at 5.45 pm. e Silva, F.B., Herrera, M.A.M., Rosina, K., Barranco, R.R., Freire, S. and Schiavina, M. (2018). Analysing Spaciotemporal Pattern of Tourism in Europe at High-resolution with Conventional and Big Data sources. *Tourism Management*, 68, 101-115.

Ellinger, A. (2005). Contextual factors influencing informal learning in a workplace setting: The case of "Reinventing Itself Company. *Human Resource Development Quarterly, 16*(3), 389–415. doi:10.1002/hrdq.1145

Elo, S., & Kyngäs, H. (2008). The qualitative content analysis process. *Journal of Advanced Nursing, 62*(1), 107–115. doi:10.1111/j.1365-2648.2007.04569.x PMID:18352969

Escudeiro, N., Cruz, M., & Escudeiro, P. (2021, April). ATHENA: A Novel Higher Education Approach to Advance a Green Digital Europe. *Open Conference of Electrical, Electronic and Information Sciences (eStream)* (pp. 1-6). IEEE.

Eshet-Alkalai, Y., & Amichai-Hamburger, Y. (2004). Experiments in digital literacy. *Cyberpsychology & Behavior, 7*(4), 421–429. doi:10.1089/cpb.2004.7.421 PMID:15331029

Esteve-Mon, F. M., Llopis, M. Á., & Adell-Segura, J. (2020). Digital competence and computational thinking of student teachers. [iJET]. *International Journal of Emerging Technologies in Learning, 15*(2), 29–41. doi:10.3991/ijet.v15i02.11588

Europe, I. A. B. (2019). *Human Capital in the Digital Environment. Report.* Retrieved from https://iabeurope.eu/wp-content/uploads/2019/09/HCitDE-Report-2019-FINAL-1.pdf

European Commission. (2019). *Digital Opportunity traineeships: boosting digital skills on the job.* https://ec.europa.eu/digital-single-market/en/digital-opportunity-traineeships-boosting-digital-skills-job

European Commission. (2020). *Digital Education Action Plan (2021-2027); Resetting education and training for the digital age.* https://ec.europa.eu/education/education-in-the-eu/digital-education-action-plan_en

European Commission. (2020). *Digital education action plan 2021–2027. Resetting education and training for the digital age.* https://ec.europa.eu/education/sites/default/files/document-library-docs/deap-communication-sept2020_en.pdf

European Commission. (2023a). Green Transition. Accessed on July 27, 2023 https://reform-support.ec.europa.eu/what-we-do/green-transition_en

European Commission. (2023b). A European Green Deal Striving to be the first climate-neutral continent. Accessed on July 27, 2023 https://commission.europa.eu/strategy-and-policy/priorities-2019-2024/european-green-deal_en

European Commission. (2023c). Digital Education Action Plan (2021-2027). Accessed on July 27, 2023 https://education.ec.europa.eu/focus-topics/digital-education/action-plan

European Commission. (2023d). Communication from the Commission to the European Parliament, The Council, The European Economic And Social Committee And The Committee Of The Regions Digital Education Action Plan 2021-2027 Resetting Education And Training For The Digital Age. Accessed on July 27, 2023 https://eur-lex.europa.eu/legal-content/EN/TXT/?uri=CELEX%3A52020DC0624

European Foundation for the Improvement of Living and Working Conditions. (2013). *Working conditions of young entrants to the labor market.* ETF Report. Retrieved from https://www.eurofound.europa.eu/sites/default/files/ef_files/docs/ewco/tn1306013s/tn1306013s.pdf

Eurostat (2021). *Database Labor Market Survey*, Retrieved from https://ec.europa.educauseat/data/database

Eurostat. (2021). *Youth unemployment rate by sex, age and country of birth.* Retrieved from: https://appsso.eurostat.ec.europa.eu/nui/show.do?dataset=yth_empl_100&lang=en

Evans, C. (2008). The effectiveness of m-learning in the form of podcast revision lectures in higher education. *Computers & Education, 50*(2), 491–498. doi:10.1016/j.compedu.2007.09.016

Eyibe. (2009). *Modern curriculum development* (second edition). Adson Educational Publisher.

Facione, P. A. (2011). *Measured reasons and critical thinking*. The California Academic Press.

Farikayi, B., & Goosen, L. (2022). Designing Effective E-Learning Environments in Gweru, Midlands Province, Zimbabwe: An E-School's Community Engagement Case Study. In T. Driscoll (Ed.), *Designing Effective Distance and Blended Learning Environments in K-12* (pp. 258–281). IGI Global., doi:10.4018/978-1-7998-6829-3.ch015

Federation of Indian Chamber of Commerce and Industry (FICCI) (2020). *Travel and Tourism Survive, Revive and Thrive in Times of Covid 19*. Grant Thornton, June 2020.

Fernando, Y., & Wulansari, P. (2021). Perceived understanding of supply-chain integration, communication, and teamwork competency in the global manufacturing companies. European. *Journal of Management and Business Economics*, *30*(2), 191–210.

Ferrari, A. (2012). DigComp: *A framework for developing and understanding digital competence in Europe*. Luxembourg: Publications Office of the European Union. Retrieved from https://ec.europa.eu/jrc/sites/default/files/lb-na-26035-enn.pdf

Field, A. (2009). Discovering Statistics Using SPSS (and Sex and Drugs and Roc' 'n' Roll), Los Angeles [i.e. Thousand Oaks, Calif.], SAGE Publications.

Fiol, M., & Lyles, M. (1985). Organizational learning. *Academy of Management Review*, *10*(4), 803–813. doi:10.2307/258048

Fitria, T. N. (2022). Microlearning in Teaching and Learning Process: A Review. ***CENDEKIA: Jurnal Ilmu Sosial, Bahasa dan Pendidikan, 2(4).***

Fitria, T. N., Afdaleni, Sibolon, N. E., & Suamba, I. B. P. (2022). Online Learning Implementation and Challenges during Covid-19 Pandemic: English Lecturers' Perspective in Indonesia. ***SOSHUM : Jurnal Sosial Dan Humaniora, 12(2)***, 171-183. https://doi.org/ doi:10.31940/soshum.v12i2.171-183

Flores, M. A., & Swennen, A. (2020). The COVID-19 pandemic and its effects on teacher education. *European Journal of Teacher Education*, *43*(4), 453–456. doi:10.1080/02619768.2020.1824253

Foreman-Peck, L., & Winch, C. (2010). *Using educational research to inform practice: A practical guide to practitioner research in universities and colleges*. Routledge. doi:10.4324/9780203847619

Forest Survey in India (FSI). (2019). *India State of Forest Report 2019, Sikkim*. Vol.II, Govt. of India. https://fsi.nic.in/isfr19/vol2/isfr-2019-vol-ii-sikkim.pdf. Accessed on 5th April, 2023 at 9 pm.

Fraillon, J., Ainley, J., Schulz, W., Friedman, T., & Gebhardt, E. (2014). *Preparing for life in a digital age. The IEA International Computer and Information Literacy Study International Report*. Springer Open. doi:10.1007/978-3-319-14222-7

Frand, J. (2000). The information-age mindset. *International Journal of Higher Education*, *35*(2), 14–24.

Fraser, J., Atkins, L., & Richard, H. (2013). *DigiLit leicester. Supporting teachers, promoting digital literacy, transforming learning*. Leicester City Council.

Galindo-Domínguez, H., & Bezanilla, M. J. (2021). Digital competence in the training of pre-service teachers: Perceptions of students in the degrees of early childhood education and primary education. *Journal of Digital Learning in Teacher Education, 37*(4), 262–278. doi:10.1080/21532974.2021.1934757

Gao, S., Qiao, R., Lim, M. K., Li, C., Qu, Y., & Xia, L. (2021). Integrating corporate website information into qualitative assessment for benchmarking green supply chain management practices for the chemical industry. *Journal of Cleaner Production, 311*, 127590. doi:10.1016/j.jclepro.2021.127590

Garg, R. K., & Singh, D. (2022). Delhi Air Pollution: A Critical Analysis. *Issue 6 Int'l JL Mgmt. &. Human., 5*, 748.

Gartner. (2012). Gamification: engagement strategies for business and IT. ***Gartner,*** https://www.gartner.com/en/documents/2246217 Accessed on: 22.01.2023.

Garvin, D. (1993). Building a learning organization. *Harvard Business Review, 71*(4), 78–92. PMID:10127041

Gawusu, S., Zhang, X., Jamatutu, S. A., Ahmed, A., Amadu, A. A., & Djam Miensah, E. (2022). The dynamics of green supply chain management within the framework of renewable energy. *International Journal of Energy Research, 46*(2), 684–711. doi:10.1002/er.7278

Ghatak, A. (1996), Breaking Trend Functions in Macro Variables, The Case of India:1900-1988, *Sankhya,* Series B, Pp-274-87

Ghazinoory, S., Ali, A. A., Hassanzadeh, A. R., & Majidpour, M. (2019). Examining systematic technological learning of Syrian textile industry. *Journal of Science and Technology Policy Management, 10*(1), 116–142. doi:10.1108/JSTPM-10-2017-0049

Ghobakhloo, M., Iranmanesh, M., Vilkas, M., Grybauskas, A., & Amran, A. (2022). Drivers and barriers of Industry 4.0 technology adoption among manufacturing SMEs: A systematic review and transformation roadmap. *Journal of Manufacturing Technology Management, 33*(6), 1029–1058. doi:10.1108/JMTM-12-2021-0505

Gill, L., Dalgarno, B., & Carlson, L. (2015). How does pre-service teacher preparedness to use ICTs for learning and teaching develop through their degree program? *The Australian Journal of Teacher Education, 40*(1), 36–59. doi:10.14221/ajte.2015v40n1.3

Giousmpasoglou, C., & Ikonen, E. (2019). The importance of genuine service and human interaction for millennials' guest experience: the case of hotel reception and front line employees.

Gligorov, V., Ognjenovic, K., & Vidovic, H. (2011). *Assessment of the labour market in Serbia* (No. 371). WIIW Research Report.

Gómez-Pablos, V. B., & Muñoz-Repıso, A. G.-V. (2019). Opinión del profesorado hacia proyectos colaborativos con Tecnologías de la Información y Comunicación: Un estudio psicométrico. *Educação e Pesquisa*, *45*, e213768. doi:10.15901678-4634201945213768

González-Pérez, L. I., & Ramírez-Montoya, M. S. (2022). Components of Education 4.0 in 21st century skills frameworks: Systematic review. *Sustainability (Basel)*, *14*(3), 1493. doi:10.3390u14031493

Goosen, L. (2015). Educational Technologies for an ICT4D MOOC in the 21st Century. In D. Nwaozuzu, & S. Mnisi (Ed.), *Proceedings of the South Africa International Conference on Educational Technologies* (pp. 37 - 48). Pretoria: African Academic Research Forum.

Goosen, L. (2008). A Brief History of Choosing First Programming Languages. In J. Impagliazzo (Ed.), *History of Computing and Education 3* (Vol. 269, pp. 167–170). Springer. doi:10.1007/978-0-387-09657-5_11

Goosen, L. (2018a). Ethical Data Management and Research Integrity in the Context of e-Schools and Community Engagement. In C. Sibinga (Ed.), *Ensuring Research Integrity and the Ethical Management of Data* (pp. 14–45). IGI Global., doi:10.4018/978-1-5225-2730-5.ch002

Goosen, L. (2018b). Ethical Information and Communication Technologies for Development Solutions: Research Integrity for Massive Open Online Courses. In C. Sibinga (Ed.), *Ensuring Research Integrity and the Ethical Management of Data* (pp. 155–173). IGI Global., doi:10.4018/978-1-5225-2730-5.ch009

Goosen, L. (2019). Research on Technology-Supported Teaching and Learning for Autism. In L. Makewa, B. Ngussa, & J. Kuboja (Eds.), *Technology-Supported Teaching and Research Methods for Educators* (pp. 88–110). IGI Global., doi:10.4018/978-1-5225-5915-3.ch005

Goosen, L., & Naidoo, L. (2014). Computer Lecturers Using Their Institutional LMS for ICT Education in the Cyber World. In C. Burger, & K. Naudé (Ed.), *Proceedings of the 43rd Conference of the Southern African Computer Lecturers' Association (SACLA)* (pp. 99-108). Port Elizabeth: Nelson Mandela Metropolitan University.

Gossling, S., Scott, D., & Hall, C. M. (2020). Pandemics, tourism and global change: A rapid assessment of COVID 19. *Journal of Sustainable Tourism*, *29*(1), 1–21. doi:10.1080/09669582.2020.1758708

Government of Indonesia. (2020). Rencana Pembangunan Jangka Menengah Nasional (RPJMN) Tahun 2020–2024 [2020–2024 National Medium-Term Development Plan (RPJMN)]. Government of Indonesia. [Online], Retrieved July 10, 2023, from.

Government of Sikkim (GoS). (2011). *Sikkim Action Plan for Climate Change*. Draft Report, Water Security and Public Health Engineering Department.

Griful-Freixenet, J., Struyven, K., Verstichele, M., & Andries, C. (2017). Higher education students with disabilities speaking out: Perceived barriers and opportunities of the universal design for learning framework. *Disability & Society*, *32*(10), 1627–1649. doi:10.1080/09687599.2017.1365695

Grooss, O. F., Presser, M., & Tambo, T. (2022). Surround yourself with your betters: Recommendations for adopting Industry 4.0 technologies in SMEs. *Digital Business*, 100046.

Grufman, N., Lyons, S., & Sneiders, E. (2020). Exploring readiness of SMEs for industry 4.0. *Complex Systems Informatics and Modeling Quarterly*, (25), 54–86. doi:10.7250/csimq.2020-25.04

Grzegorz, B., Izabela, N., Arkadiusz, G., & Zbigniew, B. (2021). Reference model of milk-run traffic systems prototyping. *International Journal of Production Research*, *59*(15), 4495–4512. doi:10.1080/00207543.2020.1766717

Gudmundsdottir, G. B., Gassó, H. H., Rubio, J. C. C., & Hatlevik, O. E. (2020). Student teachers' responsible use of ICT: Examining two samples in Spain and Norway. *Computers & Education*, *152*, 103877. doi:10.1016/j.compedu.2020.103877

Gudmundsdottir, G. B., & Hatlevik, O. E. (2018). Newly qualified teachers' professional digital competence: Implications for teacher education. *European Journal of Teacher Education*, *41*(2), 214–231. doi:10.1080/02619768.2017.1416085

Guglielman, E. (2010, September). E-learning and Disability: Accessibility as a Contribute to Inclusion. In EC-TEL Doctoral Consortium (pp. 31-36).

Guillén-Gámez, F. D., Mayorga-Fernández, M. J., Bravo-Agapito, J., & Escribano-Ortiz, D. (2021). Analysis of teachers' pedagogical digital competence: Identification of factors predicting their acquisition. *Technology. Knowledge and Learning*, *26*(3), 481–498. doi:10.100710758-019-09432-7

Gutiérrez-Porlán, J., & Serrano-Sánchez, J. L. (2016). Evaluation and development of digital competence in future primary school teachers at the University of Murcia. *Journal of New Approaches in Educational Research*, *5*(1), 51–56. doi:10.7821/naer.2016.1.152

Haas, R., Sayer, M., Ajanovic, A., & Auer, H. (2023). Technological learning: Lessons learned on energy technologies. *Wiley Interdisciplinary Reviews. Energy and Environment*, *12*(2), e463. doi:10.1002/wene.463

Haenlein, M., & Kaplan, A. (2019). A brief history of artificial intelligence: On the past, present, and future of artificial intelligence. *California Management Review*, *61*(4), 5–14. doi:10.1177/0008125619864925

Haleem, A., Javaid, M., Qadri, M. A., & Suman, R. (2022). Understanding the role of digital technologies in education: A Review. *Sustainable Operations and Computers*, *3*, 275–285. doi:10.1016/j.susoc.2022.05.004

Hall, C. M., Scott, D., & Gossling, S. (2020). Pandemic Transformations and Tourism: Be careful what you wish for. *Tourism Geographies*, *22*(3), 577–598. doi:10.1080/14616688.2020.1759131

Hall, R., Atkins, L., & Fraser, J. (2014). Defining a self-evaluation digital literacy framework for secondary educators: The DigiLit Leicester project. *Research in Learning Technology, 22*. Advance online publication. doi:10.3402/rlt.v22.21440

Hammer, P. E. (2003). The last decade. *History Today, 53*(5), 53.

Hanson, J., Niqab, M., & Bangert, A. (2021). Determination and Validation of the 'Organizational Learning Capabilities' Scale in a New Context of Educational Instituties in Pakistan. *Journal of Education and Educaitonal Development, 8*(2), 270–295. doi:10.22555/joeed.v8i2.408

Hartley, K., & Andújar, A. (2022). Smartphones and learning: An extension of M-learning or a distinct area of inquiry. *Education Sciences, 12*(1), 50. doi:10.3390/educsci12010050

Hassan, M. B., Ahmed, E. S. A., & Saeed, R. A. (2021). Machine learning for industrial IoT systems. In Handbook of Research on Innovations and Applications of AI, IoT, and Cognitive Technologies (pp. 336-358). IGI global.

Hassan, T., Song, H., Khan, Y., & Kirikkaleli, D. (2022). Energy efficiency a source of low carbon energy sources? Evidence from 16 high-income OECD economies. *Energy, 243*, 123063. doi:10.1016/j.energy.2021.123063

Hay Group. (2003). *Using competencies to identify high performers: An overview of the basics*. Hay Group, Inc.

Haywood, K. M. (2020). A Post Covid 19 future-tourism re-imagined and re-enabled. *Tourism Geographies, 22*(3), 599–609. doi:10.1080/14616688.2020.1762120

Helsper, E. J. (2016). *Slipping through the net report: Are disadvantaged young people being further left behind in the digital era?* in Socio-digital Skills and Wellbeing of Disadvantaged Young People. Retrieved from https://www.lse.ac.uk/media@lse/research/DiSTO/DiSTO-NEETs.aspx

Hinojo-Lucena, F. J., Aznar-Díaz, I., Cáceres-Reche, M. P., Trujillo-Torres, J. M., & Romero-Rodriguez, J.-M. (2019). Factors influencing the development of digital competence in teachers: Analysis of the teaching staff of permanent education centres. *IEEE Access : Practical Innovations, Open Solutions, 7*, 178744–178752. doi:10.1109/ACCESS.2019.2957438

Honeck, D. (2008). LDC Poverty Alleviation and the Doha Development Agenda: Is Tourism Being Neglected? Staff Working Paper ERSD 2008-03, Economic and Statistics Division, WTO. https://www.wto.org/English/res_e/reser_e/ersd200803_e.pdf. Accessed on 24th February, 2023 at 5 pm.

Hong, Y., Cai, G., Mo, Z., Gao, W., Xu, L., Jiang, Y., & Jiang, J. (2020). The Impact of Covid 19 on Tourists Satisfaction with B & B Zhejiang, China: An Importance- Performance Analysis. *International Journal of Environmental Research and Public Health, 17*(10), 1–19. doi:10.3390/ijerph17103747

Horváth, D., & Szabó, R. Z. (2019). Driving forces and barriers of Industry 4.0: Do multinational and small and medium-sized companies have equal opportunities? *Technological Forecasting and Social Change, 146*, 119–132. doi:10.1016/j.techfore.2019.05.021

Howard, Y. (2018). *Developing 3R8C skills to become innovative leaders for Thailand 4.0.* Tribune.

Huang, H-C. (2013). Technological learning and technological innovation creation: An Empirical analysis of biotechnology R&D teams. ***Journal of Engineering Science & Technology Review, 6(2)***, 120-124. https://doi.org/. 25 doi:10.25103/jestr.062

Huber, G. (1991). Organizational learning: The contributing processes and the literatures. *Organization Science*, *2*(1), 88–115. doi:10.1287/orsc.2.1.88

Hultsman, J., Cottrell, R. L., & Zales-Hultsman, W. (1999). *Planning Parks for People*. Venture Publishing Company.

Hwang, G. J., & Chien, S. Y. (2022). Definition, roles, and potential research issues of the metaverse in education: An artificial intelligence perspective. *Computers and Education: Artificial Intelligence*, *3*, 100082. Advance online publication. doi:10.1016/j.caeai.2022.100082

IBEF. (2019). Sikkim State Presentation. https://www.ibef.org/states/sikkim-presentation. Accessed on 15th January, 2023 at 7 pm.

Ignatius, J., Leen, J. Y. A., Ramayah, T., Kah-Hin, C., & Jantan, M. (2012). The impact of technological learning on NPD outcomes: The moderating effect of project complexity. ***Technovation, 32(7-8)***, 452-463. https://doi.org/. technovation.2012.03.003 doi:10.1016/j

İhtiyaroğlu, N., Karakuş Yıldız, B., & Yılmaz, F. (2022). Okullarda Pozitif Psikolojik Sermaye ve Örgütsel Öğrenme İlişkisinin İncelenmesi. *Sinop Üniversitesi Sosyal Bilimler Dergisi*, *6*(2), 315–352. doi:10.30561inopusd.1139670

Ilomäki, L., Kantosalo, A., & Lakkala, M. (2011). What is digital competence? In *Linkedportal*. European Schoolnet., http://linked.eun.org/web/guest/in-depth3

Ilomäki, L., Paavola, S., Lakkala, M., & Kantosalo, A. (2016). Digital competence–an emergent boundary concept for policy and educational research. *Education and Information Technologies*, *21*(3), 655–679. doi:10.100710639-014-9346-4

IMD World Competitiveness Center. (2022). *World Digital Competitiveness Rankings 2020*. [Online]. Available at: https://www.imd.org/centers/wcc/world-competitiveness-center/rankings/world-digital-competitiveness-ranking// [Accessed: July 10, 2023].

Inceoglu, M. M., & Ciloglugil, B. (2022, July). Use of Metaverse in education. *Computational Science and Its Applications (ICCSA) Workshops: Malaga, Spain, Proceedings, Part I* (pp. 171-184). Cham: Springer International Publishing. Retrieved from https://link.springer.com/chapter/10.1007/978-3-031-10536-4_12

India Today (IT). (2018). Sikkim becomes the World's First Organic State wins Oscars for Best Policies by UN, October 16. https://www.indiatoday.in/education-today/gk-current-affairs/story/sikkim-becomes-world-s-first-organic-state-wins-oscar-for-best-policies-by-un-1369158-2018-10-16. Accessed on 25th March, 2023 at 9.30 pm.

India Tourism Statistics(ITS). (2022). *India Tourism Statistics 2022*. Market Research Division, Ministry of Tourism, Government of India.

Indonesian Information Portal. (2023). Indonesia Profile. [Online]. Available at: https://indonesia.go.id/profil//] Accessed: July 10, 2023]

Inskeep, E. (1991). *Tourism Planning: An Integrated and Sustainable Development Approach*. Van Nostrand Reinhold.

Instefjord, E. J., & Munthe, E. (2016). Preparing pre-service teachers to integrate technology: An analysis of the emphasis on digital competence in teacher education curricula. *European Journal of Teacher Education*, *39*(1), 77–93. doi:10.1080/02619768.2015.1100602

INTEF. (2017). *Marco común de competencia digital docente*. Retrieved from http://educalab.es/documents/10180/12809/MarcoComunCompeDigiDoceV2.pdf

International Labor Organization. (2020). *Global Employment Trends for Youth 2020: Technology and the future of jobs*. Retrieved from: https://www.ilo.org/wcmsp5/groups/public/---dgreports/---dcomm/---publ/documents/publication/wcms_737648.pdf

International Labor Organization. (2023). Youth Unemployment Database. Available at: https://www.ilo.org/shinyapps/bulkexplorer2/?lang=en&id=UNE_3UNE_SEX_AGE_EDU_NB_A, accessed at July, 22nd, 2023

Ioannides, D., & Gyimóthy, S. (2020). The COVID-19 crisis as an opportunity for escaping the unsustainable global tourism path. *Tourism Geographies*, *22*(3), 624–632. doi:10.1080/14616688.2020.1763445

Islam, M. A. (2022). Industry 4.0: Skill set for employability. *Social Sciences & Humanities Open*, *6*(1), 100280. doi:10.1016/j.ssaho.2022.100280

Ivic, I., & Pesikan, A. (2012). Education system reforms in an unstable political situation: The case of Serbia in the first decade of the 21st century. *CEPS Journal*, *2*(2), 31–53. doi:10.26529/cepsj.383

Jain, V. K., & Sharma, S. (2012). Green supply chain management practices in automobile industry: An empirical study. *Journal of Supply Chain Management Systems*, *1*(3), 20.

Janssen, J., Stoyanov, S., Ferrari, A., Punie, Y., Pannekeet, K., & Sloep, P. (2013). Experts' views on digital competence: Commonalities and differences. *Computers & Education*, *68*, 473–481. doi:10.1016/j.compedu.2013.06.008

Jassim, S., Al-Mubarak, M., & Hamdan, A. (2020). The impact of green supply chain management on firm's performance. *Journal of Information & Knowledge Management*, *19*(01), 2040026. doi:10.1142/S0219649220400262

Javaid, M., Haleem, A., Singh, R. P., Rab, S., & Suman, R. (2022). Exploring impact and features of machine vision for progressive industry 4.0 culture. *Sensors International*, *3*, 100132. doi:10.1016/j.sintl.2021.100132

Jimoyiannis, A., & Komis, V. (2007). Examining teachers' beliefs about ICT in education: Implications of a teacher preparation programme. *Teacher Development*, *11*(2), 149–173. doi:10.1080/13664530701414779

Johnston, D., & Braden, J. (2008). Green landscapes: Exogenous economic benefits of environmental improvement. In J. Wescoat & D. Johnston (Eds.), *Political Economies of Landscape Change* (pp. 131–144). Springer. doi:10.1007/978-1-4020-5849-3_7

Jojkic, I., & Mitrovic, P. (2017). *Da znanje ostane u Srbiji: Studija o položaju mladih osoba na tržištu rada u Republici Srbiji.* Beograd: Friedrich-Ebert-Stiftung. https://library.fes.de/pdf-files/bueros/belgrad/13784.pdf

Julio, C. A., Guillén-Gámez, F. D., Ruiz-Palmero, J., & Palacios-Rodríguez, A. (2021). Digital competence of higher education professor according to DigCompEdu. Statistical research methods with ANOVA between fields of knowledge in different age ranges. *Education and Information Technologies*, *26*(4), 4691–4708. doi:10.100710639-021-10476-5 PMID:33758572

Kahn, S., & Ginther, D. (2017). *Women and STEM.* National Bureau of Economic Research. Retrieved from https://www.nber.org/system/files/working_papers/w23525/w23525.pdf

Kampa, M., & Castanas, E. (2008). Human health effects of air pollution. *Environmental Pollution*, *151*(2), 362–367. doi:10.1016/j.envpol.2007.06.012 PMID:17646040

Kang, Y. D. (2021). Determinants of youth unemployment: Empirical analysis of OECD and EU member countries in 2000-2017. *Journal of Economic Development*, *46*(3), 109–131.

Kaya, N., & Taşgit, Y. E. (2013). Teknolojik Öğrenme, Örgütsel Öğrenme ve Ar-Ge Yeteneklerinin Entellektüel Sermaye Üzerine Etkisi. *Düzce Üniversitesi Sosyal Bilimler Dergisi*, *3*(2), 1–12.

Kaykı, İ. E. (2008). ***Modüler Öğretim Yöntemi ve Uygulamalı Dersler.*** (Unpublished Master Thesis), Beykent Üniversitesi, Sosyal Bilimler Enstitüsü, İstanbul.

Kem, D. (2022). Personalised and adaptive learning: Emerging learning platforms in the era of digital and smart learning. *International Journal of Social Science and Human Research*, *5*(2), 385–391. doi:10.47191/ijsshr/v5-i2-02

Khammanee, T. (2014). *Teaching science: Body of knowledge for an effective learning process.* Publishing House of Chulalongkorn University.

Khan, M. S., Hasan, M., & Clement, C. K. (2012). Barriers to the introduction of ICT into education in developing countries: The example of Bangladesh. *International Journal of Instruction*, *5*(2), 61–80.

Khazami, N., Lakner, Z., & Nefzi, A. (2020). Pandemic and Tourism: Re-preparation of Tourism Post Covid 19. *Journal of Hotel and Business Management*, *9*(2), 1–5.

Khribi, M. K., Othman, A., & Al-Sinani, A. (2022, July). Toward Closing the Training and Knowledge Gap in ICT Accessibility and Inclusive Design Harnessing Open Educational Resources. In *2022 International Conference on Advanced Learning Technologies (ICALT)* (pp. 289-291). IEEE. 10.1109/ICALT55010.2022.00093

Kim, D. (1993). The link between individual and organizational learning. *Sloan Management Review, 35*(1), 37–51.

Kim, J. Y., & Fienup, D. M. (2022). Increasing access to online learning for students with disabilities during the COVID-19 pandemic. *The Journal of Special Education, 55*(4), 213–221. doi:10.1177/0022466921998067

Kim, K.-J., Bonk, C., & Oh, E. G. (2008). The present and future state of blended learning in workplace learning settings in the United States. *Performance Improvement, 47*(8), 5–16. doi:10.1002/pfi.20018

Kivunja, C. (2015). Teaching students to learn and to work well with 21st century skills: Unpacking the career and life skills domain of the new learning paradigm. *International Journal of Higher Education, 4*(1), 1–11. doi:10.5430/ijhe.v4n1p1

Klimecka-Tatar, D. (2017). Value Stream Mapping as Lean Production tool to improve the production process organization–case study in packaging manufacturing. *Production Engineering Archives, 17*(17), 40–44. doi:10.30657/pea.2017.17.09

Kluge, A., & Schilling, J. (2003). Organizational learning and learning organizations. *The Psychologist Manager Journal, 6*(1), 31–50. doi:10.1037/h0095917

Koscak, M., & O'Rourke, T. (2021). *Tourism Management: The New Reality of Managing Ethical and Responsible Tourism.* Routlegde.

Kowal, B., Włodarz, D., Brzychczy, E., & Klepka, A. (2022). Analysis of Employees' Competencies in the Context of Industry 4.0. *Energies, 15*(19), 7142. doi:10.3390/en15197142

Kozik, N. (2020). Sustainable packaging as a tool for global sustainable development. *SHS Web of Conferences, 74*, 4012.

Krugman, P. R., & Obstfeld, M. (2006). *International Economics, Theory and Policy* (7th ed.). Pearson Addison.

Krumsvik, R. J. (2008). Situated learning and teachers' digital competence. *Education and Information Technologies, 13*(4), 279–290. doi:10.100710639-008-9069-5

Krumsvik, R. J. (2014). Teacher educators' digital competence. *Scandinavian Journal of Educational Research, 58*(3), 269–280. doi:10.1080/00313831.2012.726273

Kumar, A., Batra, V., & Sharma, S. K. (2009). Micro, small and medium enterprises (MSMEs) in India: Challenges and issues in the current scenario. *Management Insight, 5*(2), 75–87.

Kumari, S. Tiwari, H.R., and Inbakaran, Robert John (2006). Sustainability Criteria and Indicators as Tool to Evaluate the Visitors Impact on Fragile Ecosystem Destinations: A Case of Khanchendzonga Biosphere Reserve, Sikkim, India. https://www.researchgate.net/publication/237087340_Sustainability_Criteria_and_Indicators_as_Tools_to_Evaluate_the_Visitor_Impact_on_Fragile_Ecotourism_Destinations_A_Case_Study_of_Kanchandzonga_Biosphere_Reserve_Sikkim_India. Accessed on 10th January, 2023 at 7.30 pm.

Kuzey, C., Uyar, A., & Delen, D. (2014). The impact of multinationality on firm value: A comparative analysis of machine learning techniques. *Decision Support Systems*, *59*, 127–142. doi:10.1016/j.dss.2013.11.001

Kuzmanovic, D. (2017). *Empirical validation of digital literacy construct and analysis of predictors of achievement* (doctoral dissertation). National Repository of Dissertations in Serbia (1234567899324).

Lakic, D., Parojcic, J., Kovacevic, N., & Antonijevic, M. (2012). Digital literacy of employers and students in Serbia. Conference Employer Engagement in a Digital Age. University of Greenwich, UK.

Lalima, & Dangwal, K. L. (2017). Blended Learning: An Innovative Approach. ***Universal Journal of Educational Research, 5(1)***, 129-136. DOI: . doi:10.13189/ujer.2017.050116

Lama, R. (2018): Potential of Homestays as Tourism Products in Sikkim. https://www.researchgate.net/publication/325076920_Potential_of_Homestays_as_Tourism_Product_in_Sikkim Accessed on 25th March, 2023 at 10.02 pm.

Lampropoulos, G., Keramopoulos, E., Diamantaras, K., & Evangelidis, G. (2022). Augmented reality and gamification in education: A systematic literature review of research, ***Applications, and empirical studies. applied sciences, 12(13)***, 6809.

Lasi, H., Fettke, P., Kemper, H. G., Feld, T., & Hoffmann, M. (2014). Industry 4.0. *Business & Information Systems Engineering*, *6*(4), 239–242. doi:10.100712599-014-0334-4

Lawn, S., Battersby, M., Lindner, H., Mathews, R., Morris, S., Wells, L., & Reed, R. (2009). What skills do primary health care professionals need to provide effective self-management support? Seeking consumer perspectives. *Australian Journal of Primary Health*, *15*(1), 37–44. doi:10.1071/PY08053

Lebedeva, M., Taranova, M., & Beketov, V. (2022). Assessment of academic achievements in m-learning. *Education and Information Technologies*, *1*, 21. PMID:36406788

Lebedinski, L., & Vladisavljević, M. (2022). Parenthood and labor market outcomes in Serbia. *Ekonomski anali*, 67(232), 7-47.

LeCompte, M. D. (2000). Analyzing qualitative data. *Theory into Practice*, *39*(3), 146–154. doi:10.120715430421tip3903_5

Ledward, B. C., & Hirata, D. (2011). *An overview of 21st century skills: Summary of 21st century skills for students and teachers, by Pacific Policy Research Center.* Kamehameha Schools–Research & Evaluation.

Lee, J. (2013). Employment and earnings in high tech ethnic niches. *Social Forces, 91*(3), 747–784. doi:10.1093fos199

Lee, J., Davari, H., Singh, J., & Pandhare, V. (2018). Industrial Artificial Intelligence for Industry 4.0-based manufacturing systems. *Manufacturing Letters, 18*, 20–23. doi:10.1016/j.mfglet.2018.09.002

Lei, D., Slocum, J., & Pitts, R. (1999). Designing organizations for competitive advantage: The power of unlearning and learning. *Organizational Dynamics, 27*(3), 24–38. doi:10.1016/S0090-2616(99)90019-0

S. Lin, & X. Huang (Eds.). (2011, June). Preface. *Advanced Research on Computer Education, Simulation and Modeling (CESM): International Conference, Proceedings, Part I.* 175. Wuhan, China: Springer. 10.1007/978-3-642-21783-8

Lin, W. C., Tsai, C. F., & Zhong, J. R. (2022). Deep learning for missing value imputation of continuous data and the effect of data discretization. *Knowledge-Based Systems, 239*, 108079. doi:10.1016/j.knosys.2021.108079

Liotti, G. (2022). Labour market regulation and youth unemployment in the EU-28. *Italian Economic Journal, 8*(1), 77–103. doi:10.100740797-021-00154-3

Lisowski, A. (2003). Concept of Space in Geography. Studia Regionalne I Lokalne, UW, Warsaw.

List, A., Brante, E. W., & Klee, H. L. (2020). A framework of pre-service teachers' conceptions about digital literacy: Comparing the United States and Sweden. *Computers & Education, 148*, 103788. doi:10.1016/j.compedu.2019.103788

Li, X., Xu, H., Zhang, J., & Chang, H.-H. (2022). Deep Reinforcement Learning for Adaptive Learning Systems. *Journal of Educational and Behavioral Statistics, 48*(2), 220–243. Advance online publication. doi:10.3102/10769986221129847

Llurda, E. (2000). On competence, proficiency, and communicative language ability. *International Journal of Applied Linguistics, 10*(1), 85–96. doi:10.1111/j.1473-4192.2000.tb00141.x

López-Belmonte, J., Pozo-Sánchez, S., Fuentes-Cabrera, A., & López-Núñez, J. A. (2019). Content creation and flipped learning: A necessary pairing for education in the new millennium. *Revista Española de Pedagogía, 77*(274), 535–555. doi:10.22550/REP77-3-2019-07

Louviere, J. J., Flynn, T. N., & Marley, A. A. J. (2015). *Best-worst scaling: Theory, methods, and applications*. Cambridge University Press. doi:10.1017/CBO9781107337855

Lu, L., Chen, Q., Huang, R., & Usman, A. (2023). Education and its impact on renewable energy demand, carbon intensity, and green growth: Do digital financial inclusion and environmental policy stringency matter in China. *Environmental Science and Pollution Research International, 30*(5), 12020–12028. doi:10.100711356-022-22759-6 PMID:36103074

Lund, A., Furberg, A., Bakken, J., & Engelien, K. L. (2014). What does professional digital competence mean in teacher education?. *Nordic journal of digital literacy, 9*(4), 280-298. https:// doi.org/ doi:10.18261/ISSN1891-943X-2014-04-04

Maan, J. S., & Chudhry, P. (2019). People and Protected Areas: Some Issues from India. *Animal Biodiversity and Conservation, 42*(1), 79–90. doi:10.32800/abc.2019.42.0079

Mahawichit, P. (2017). Innovation in learning from Finland. *IPST Magazine, 46*(209), 40–45.

Maimunah Yohanes, A. K. Y., & Puspitasari, N. (2012). Konsep dan penerapan cloud computing untuk meningkatkan mutu pembelajaran. *CSRID Journal, 4*(3), 220–230.

Malik, Z. A., Pandey, R., & Bhatt, A. B. (2016). Anthropogenic Disturbances and their Impact on Vegetation in Western Himalayas, India. *Journal of Mountain Science, 13*(1), 69–82. doi:10.100711629-015-3533-7

Manera, C., & Valle, E. (2018). Tourist Intensity in the World 1995-2015: Two Measurement Proposals. *Sustainability (Basel), 10*(12), 1–20. doi:10.3390u10124546

Marimuthu, M., Arokiasamy, L., & Ismail, M. (2009). Human capital development and its impact on firm performance: Evidence from developmental economics. *The journal of international social research, 2* (8), 265-272.

Marino, L., & Capone, V. (2021). Smart Working and Well-Being before and during the COVID-19 Pandemic: A Scoping Review. *European Journal of Investigation in Health, Psychology and Education, 11*(4), 1516–1536. doi:10.3390/ejihpe11040108 PMID:34940386

Marques Santos, A., Madrid Gonzalez, C., Haegeman, K., & Rainoldi, A. (2020). *Behavioural Changes in tourism in times of Covid 19, Employment Scenarios and Policy Options*. Publication Office of European Union.

Masalimova, A. R., Khvatova, M. A., Chikileva, L. S., Zvyagintseva, E. P., Stepanova, V. V., & Melnik, M. V. (2022). Distance learning in higher education during COVID-19. *Frontiers in Education, 7*, 822958. doi:10.3389/feduc.2022.822958

Matijevic, M., & Solaja, M. (2020). *ICT in Serbi– - At a Glance*, 2020. Novi Sad: Vojvodina ICT Cluster & Deutsche Gesellschaft für Internationale Zusammenarbeit (GIZ) GmbH. https://www.ict-cs.org/fileadmin/user_upload/ICT-in-Serbia-At-a-Glance-2020.pdf

Mautso, K., & Goosen, L. (2023). Using Applicable E-Learning Principles to Promote Effectiveness and Quality in the Digital Era: Toward a More Inclusive and Resilient E-School. In P. Ordóñez de Pablos, X. Zhang, & M. Almunawar (Eds.), *Handbook of Research on Education Institutions, Skills, and Jobs in the Digital Era* (pp. 359–381). IGI Global., doi:10.4018/978-1-6684-5914-0.ch021

McGarr, O., & McDonagh, A. (2021). Exploring the digital competence of pre-service teachers on entry onto an initial teacher education programme in Ireland. *Irish Educational Studies*, *40*(1), 115–128. doi:10.1080/03323315.2020.1800501

McMillan, J. H., & Schumacher, S. S. (2014). *Research in education: Evidence–based inquiry* (7th ed.). Pearson.

Menggo, S., Suastra, I., Budiarsa, M., & Padmadewi, N. N. (2019). Needs analysis of academic-English speaking material in promoting 21st century skills. *International Journal of Instruction*, *12*(2), 739–754. doi:10.29333/iji.2019.12247a

Merriam, S. (2009). *Qualitative research: A guide to design and implementation.* Jossey-Bass.

Miglani, S. (2019). The growth of the Indian automobile industry: Analysis of the roles of government policy and other enabling factors. *Innovation, Economic Development, and Intellectual Property in India and China: Comparing Six Economic Sectors*, 439–463.

Minashkina, D., & Happonen, A. (2020). Decarbonizing warehousing activities through digitalization and automatization with WMS integration for sustainability supporting operations. *E3S Web of Conferences, 158*, 3002.

Minh, N., Kien, D. T., & Hau, P. K. (2020). Applying Milk-run Method to Optimize Cost of Transport: An Empirical Evidence. *International Journal of Supply and Operations Management*, *7*(2), 178–188.

Ministry of Communication and Informatics. (2020). *Roadmap Literasi Digital 2020-2024.*

Ministry of Education, Science and Technological Development of the Republic of Serbia. (2021). *The Law on the Dual Model of Studies in Higher Education.* Laws. https://www.mpn.gov.rs/wp-content/uploads/2020/01/ZAKON.pdf

Ministry of Education. Culture, Research, and Technology (Kemendikbudristek). (2021). *The Merdeka Campus Program Invites Indonesian Students to Become Creative and Adaptive Human Resources*. [Online] Available at: https://dikti.kemdikbud.go.id/kabar-dikti/kabar/program-kampus-merdeka-ajak-mahasiswa-indonesia-menjadi-sdm-kreatif-dan-adaptif/[Accessed: July 10, 2023].

Ministry of Education. Singapore. (2010). *Learning of 21st-century competency.* https://www.moe.gov.sg/media/press/2010/03/moe-to-enhance-learning-of-21s.php

Mishra, P., & Koehler, M. (2009). Too cool for school? No way! Using the TPACK framework: You can have your hot tools and teach with them, too. *Learning and Leading with Technology*, *36*(7), 14–18.

Mittal, S., Khan, M. A., Romero, D., & Wuest, T. (2018). A critical review of smart manufacturing & Industry 4.0 maturity models: Implications for small and medium-sized enterprises (SMEs). *Journal of Manufacturing Systems*, *49*, 194–214. doi:10.1016/j.jmsy.2018.10.005

Mohanty, S. B. (1988). *Lifelong and adult education.* Ashish Publishing House.

Montenegro-Rueda, M., & Fernández-Batanero, J. (2022). Digital competence of special education teachers: Impact, challenges and opportunities. *Australasian Journal of Special and Inclusive Education*, *46*(2), 178–192. doi:10.1017/jsi.2022.8

Mora, M., O'Connor, R., Raisinghani, M., Macías-Luévano, J., & Gelman, O. (2011). An IT service engineering and management framework (ITS-EMF). *International Journal of Service Science, Management, Engineering, and Technology*, *2*(2), 1–15. doi:10.4018/jssmet.2011040101

Morgan, G. A., Barrett, K. C., & Leech, N. L. (2011). IBM SPSS For Intermediate Statistics: Use and Interpretation, 4 ed. Routledge, Taylor & Francis group, New York.

Motteram, G., & Sharma, P. (2009). Blended learning in a Web 2.0 world. *International Journal of Emerging Technologies and Society*, *7*, 83–96.

Mowforth, M., & Munt, I. (2003). *Tourism and Sustainability: Development and new tourism in the Third World* (2nd ed.). Routledge. doi:10.4324/9780203422779

Mowforth, M., & Munt, I. (2009). *Tourism and sustainability: Development and new tourism in the Third World.* Routledge.

Mqina, B. P., & Goosen, L. (2021). Investigating the Roles of School Management Teams in Curriculum Delivery Through ICTs in Education: e-Schools' Community Engagement. In A. Mawela, M. van Wyk, L. Lebeloane, & A. Mudau (Eds.), *Investigating the Roles of School Management Teams in Curriculum Delivery* (pp. 158–178). IGI Global., doi:10.4018/978-1-7998-7168-2.ch011

Mraihi, R., ben Abdallah, K., & Abid, M. (2013). Road transport-related energy consumption: Analysis of driving factors in Tunisia. *Energy Policy*, *62*, 247–253. doi:10.1016/j.enpol.2013.07.007

Mustafa, B. (2022). Analyzing education based on metaverse technology. *Technium Social Sciences Journal*, *32*, 278–295. doi:10.47577/tssj.v32i1.6742

Muzambi, A., & Goosen, L. (2022). Factors Affecting the Successful Implementation of an E-Education Policy and Community Engagement: Acquiring 21st Century Skills Through E-Learning. In C. Lane (Ed.), *Handbook of Research on Acquiring 21st Century Literacy Skills Through Game-Based Learning* (pp. 580–605). IGI Global., doi:10.4018/978-1-7998-7271-9.ch030

Namhiran, U. (2016). A development of framework of competencies assessment for Thai student teachers' learning management and specific content knowledge. *Rommayasan: Journal of Humanities and Social Sciences*, 14(2), 207-—214.

Napal-Fraile, M., Peñalva-Vélez, A., & Mendióroz-Lacambra, A. M. (2018). Development of digital competence in secondary education teachers' training. *Education Sciences*, *8*(3), 104. doi:10.3390/educsci8030104

National Education Association. (2010). *Preparing 21st century students for a global society: An educator's guide to the "four Cs."* https://www.nea.org/assets/docs/A-Guideto-Four-Cs.pdf

National Innovation Agency. (2010). *Innovation management for executives (IMEs)* (3rd ed.). Innovation Culture Promotion Department National Innovation Agency.

Navavongsathian, A., Vuthipadadorn, D., Farangthong, S., Janthongpan, S., & Juntasart, R. (2020). Study of Green Logistics Managing Potential and the Preparedness of Auto Parts Industries in Thailand. *TEM Journal*, *9*(4), 1524–1534. doi:10.18421/TEM94-27

Nelson, M. J., Voithofer, R., & Cheng, S. L. (2019). Mediating factors that influence the technology integration practices of teacher educators. *Computers & Education*, *128*, 330–344. doi:10.1016/j.compedu.2018.09.023

Nepal, S. K. (2020). Adventure Tourism and Travel after Covid 19- business as usual or opportunity to reset? *Tourism Geographies*, *22*(3), 646–650. doi:10.1080/14616688.2020.1760926

Nientied, P., & Shutina, D. (2020). Tourism in Transition, the Post Covid 19 aftermath in the Western Balkans, Co-Plan Resilience Series 2. https://www.researchgate.net/publication/341607053_Tourism_in_Transition_The_post_COVID-19_Aftermath_in_the_Western_Balkans Accessed on 15th March, 2023 at 6.31 pm.

Niti Ayog (NA). (2018). Sustainable Tourism in the Indian Himalayan Region. Report of the Working Group II, Govt. of India, New Delhi.

Nyaupane, G. P., & Timothy, D. (2016). Bhutan's Low Volume, High Yield Tourism: The Influence of Power and Regionalism. https://www.semanticscholar.org/paper/Bhutan%E2%80%99s-Low-volume%2C-High-yield-Tourism%3A-The-of-and-Nyaupane-Timothy/4708bc5cea60e9d4215a394864e236134aec58c8. Accessed on 20th April, 2023 at 8.45 pm.

Obisi, C. (2001). Employee development, Issues and dimensions. *Unical Journal of public Administrator Sept, 1*.

Obisi, C. (2011). Employee Training and Development in Nigerian Organizations: Some Observations and Agenda for Research. *Australian Journal of Business and Management Research*, *1*(9), 82–91. doi:10.52283/NSWRCA.AJBMR.20110109A09

Oblinger, D. (2003). *Boomers, Gen-xers, and Millennials: Understanding th' 'new students'*. Educause, 4:38-47. https://er.educause.edu/-/media/files/article-downloads/erm0342.pdf

OECD. (2015). *OECD Skills Outlook 201– - Youth, Skills and Employability*. https://www.oecd-ilibrary.org/education/oecd-skills-outlook-2015_9789264234178-en

Office of the Education Council, Education Council. (2018). *Report on the results of research and development of the competency framework for students at the lower elementary school for basic education courses*. NESDB.

Office of the Education Council, Ministry of Education. (2009). *Educational reform proposals in the second decade (2009-2018)*. Office of the Council of Education Secretariat.

Office of the Education Council, Ministry of Education. (2017). *National Education Plan 2017 –2036*. Office. Secretary of the Education Council.

Office of the Education Council, Ministry of Education. (2019). *Framework for learner competency at the basic education level and Lower primary education (Grade 1-Grade 3)*. NESDB.

Office of the National Economic and Social Development Council. (2020). *National strategy 2018–2037*. Office of the Prime Minister.

Office of the National Education Commission, Office of the Prime Minister. (2002). National Education Act 1999, amended (2nd edition) 2002. Office of the Education Commission, National Office of the Prime Minister.

Ognjenovic, K. (2018). Training Intentions and Skills Needs in the Private Sector Companies in Serbia. *Journal of Wom'n's Entrepreneurship and Education*, *1-2*(1-2), 68–85. doi:10.28934/jwee18.12.pp68-85

Ognjenovic, K., & Djukic, M. (2023). *Response of the Economic System to COVID-19: A Review of Evaluation Methods Applied in Assessing Economic Policies' Effects in Serbia. Reimagining Systems Thinking in a Post-Pandemic World*. IGI Global. doi:10.4018/978-1-6684-7285-9.ch006

Okoli, A. C. (2017). Relating communication competence to teaching effectiveness: Implication for teacher education. *Journal of Education and Practice*, *8*(3), 150–154.

Oliveira, W., Hamari, J., Shi, L., Toda, A. M., Rodrigues, L., Palomino, P. T., & Isotani, S. (2023). Tailored gamification in education: A literature review and future agenda. *Education and Information Technologies*, *28*(1), 373–406. doi:10.100710639-022-11122-4

Olofsson, A. D., Lindqvist, M. H., Lindberg, O. J., From, J., Stödberg, U., & Holmgren, T. (2017). Pedagogical digital competence for police teachers in relation to distance-based police education. In *INTED2017 Proceedings* (pp. 4219-4227). IATED. 10.21125/inted.2017.1010

Omer, A. M. (2009). Energy use and environmental impacts: A general review. *Journal of Renewable and Sustainable Energy*, *1*(5), 53101. doi:10.1063/1.3220701

Ong, E. C., & Tan, C. L. (2022). Soft TQM, agility, and knowledge management deliver organizational performance: A study of Malaysian manufacturing organizations in the electrical and electronics sector. *Global Business and Organizational Excellence*, *41*(4), 28–47. doi:10.1002/joe.22155

Ordóñez de Pablos, P. (2022). Unlocking the potential of intellectual capital, innovation and higher education: Some keys for competitiveness in the post-pandemic era. *International Journal of Learning and Intellectual Capital*, *19*(5), 391–394.

Ordóñez de Pablos, P. (2023). Digital economy, intellectual capital and competitiveness. Some lessons for the post-COVID-19 era. *International Journal of Learning and Intellectual Capital*, *20*(1), 1–5.

Ordóñez de Pablos, P., Lytras, M. D., & Zhang, X. (Eds.). (2020). *IT and the Development of Digital Skills and Competences in Education*. IGI Global.

Ordóñez de Pablos, P., Lytras, M. D., Zhang, X., & Chui, K. T. (Eds.). (2019). *Opening Up Education for Inclusivity across Digital Economies and Societies*. IGI Global. doi:10.4018/978-1-5225-7473-6

Ordóñez de Pablos, P., Zhang, X., & Almunawar, M. (Eds.). (2023). *Handbook of Research on Education Institutions, Skills, and Jobs in the Digital Era*. IGI Global.

Ordóñez de Pablos, P., Zhang, X., & Almunawar, M. N. (Eds.). (2022). *Handbook of Research on Green, Circular, and Digital Economies as Tools for Recovery and Sustainability*. IGI Global., Retrieved from https://www.igi-global.com/book/handbook-research-green-circular-digital/280314 doi:10.4018/978-1-7998-9664-7

Ordóñez de Pablos, P., Zhang, X., & Almunawar, M. N. (Eds.). (2023). *Handbook of Research on Artificial Intelligence and Knowledge Management in Asia's Digital Economy*. IGI Global., Retrieved from https://www.igi-global.com/book/handbook-research-artificial-intelligence-knowledge/297912

Organisation of Economic Cooperation and Development [OECD] (2005). *Definition and selection of key competencies. executive summary (online resource)*. OECD Publishing.

Organization for Economic Cooperation and Development & International Labor Organization-OECD & ILO. (2014). *Promoting better labor market outcomes for youth. Report on youth employment and apprenticeships prepared for the G20 Labor and Employment Ministerial Meeting*. https://www.oecd.org/g20/topics/employment-and-social-policy/OECD-ILO-Youth-Apprenticeships-G20.pdf

Organization for Economic Cooperation and Development (OECD). (2020). The Territorial Impact of Covid 19: Managing the Crisis across levels of Governments. https://read.oecd-ilibrary.org/view/?ref=128_128287-5agkkojaaa&title=The-territorial-impact-of-covid-19-managing-the-crisis-across-levels-of-government Accessed on 10th March, 2023 at 11.22 pm.

Organization for Economic Co-operation and Development. (2015). *OECD Skills Outlook 201– - Youth, Skills and Employability*. https://www.oecd-ilibrary.org/education/oecd-skills-outlook-2015_9789264234178-en

Orindaru, A., Popescu, M., Alexaoei, A. P., Caescu, S., Florescu, M., & Orzan, A. (2021). Tourism in a Post-Covid 19 Era: Sustainable Strategies for Industry's Recovery. *Sustainability (Basel)*, *13*(12), 1–22. doi:10.3390u13126781

Ottenbreit-Leftwich, A., Liao, J. Y. C., Sadik, O., & Ertmer, P. (2018). Evolution of teachers' technology integration knowledge, beliefs, and practices: How can we support beginning teachers use of technology? *Journal of Research on Technology in Education*, *50*(4), 282–304. doi:10.1080/15391523.2018.1487350

Ottestad, G., Kelentrić, M., & Gudmundsdottir, G. B. (2014). Professional digital competence in teacher education. *Nordic Journal of Digital Literacy*, *9*(4), 243–249. doi:10.18261/ISSN1891-943X-2014-04-02

Özcan, H., & Emiroğlu, B. G. (2020). Bulut Tabanlı Öğrenme Yönetim Sistemi Seçiminde Bulanık Çok Kriterli Karar Analizi Yaklaşımı. *Bilişim Teknolojileri Dergisi, 13*(1), 97–111. doi:10.17671/gazibtd.589757

Ozdamar-Keskin, N., Ozata, F. Z., Banar, K., & Royle, K. (2015). Examining digital literacy competences and learning habits of open and distance learners. *Contemporary Educational Technology, 6*(1), 74–90. doi:10.30935/cedtech/6140

Özkan, H. H. (2005). Öğrenme Öğretme Modelleri Açısından Modüler Öğretim. *Atatürk Üniversitesi Sosyal Bilimler Enstitüsü Dergisi, 6*(2), 117–128.

Palmer, E. (2015a). Introduction: The sustainable development goals forum. *Journal of Global Ethics, 11*(1), 3–9. doi:10.1080/17449626.2015.1021091

Palmer, E. (2015b). Introduction: The 2030 agenda. *Journal of Global Ethics, 11*(3), 262–269. doi:10.1080/17449626.2015.1119928

Pang, E., Wong, M., Leung, C. H., & Coombes, J. (2019). Competencies for fresh graduates' success at work: Perspectives of employers. *Industry and Higher Education, 33*(1), 55–65. doi:10.1177/0950422218792333

Panich, W. (2013). *Enjoy learning in the 21st century*. Sodsri-Saritwong Foundation.

Park, Y., & Kim, Y. (2018). A Design and Development of micro-Learning Content in eLearning System. *International Journal on Advanced Science, Engineering and Information Technology, 8*(1), 56. Advance online publication. doi:10.18517/ijaseit.8.1.2698

Partnership for 21st Century Skills-P21. (2011). *Framework for 21st century learning*. http://www.p21.org/storage/documents/1._ p21_framework_2-pager.pdf

Patel, K. K., & Patel, S. M. (2016). Internet of things-IOT: definition, characteristics, architecture, enabling technologies, application & future challenges. *International journal of engineering science and computing, 6*(5), 6122-6131.

Patole, S. (2012). A study of auto components industry in India: Issues and challenges. *South Asian Journal of Marketing & Management Research, 2*(7), 155–160.

Pattanapichet, F. (2009). *A development of the competency-based English oral communication course for undergraduate public relations students* [Unpublished doctoral dissertation]. Chulalongkorn University Intellectual Repository (CUIR).

Patton, M. Q. (1999). Enhancing the quality and credibility of qualitative analysis. *Health Sciences Research, 34*(5 Pt 2), 1189–1208. https://www.ncbi.nlm.nih.gov/pmc/articles/PMC1089059/

Pavlovic, D., Zubovic, J., & Zdravkovic, A. (2016). Youth expectations in job search in Serbia. *Industry*, 44(4). https://www.researchgate.net/publication/314027512_Youth_expectations_in_job_search_in_Serbia#fullTextFileContent

Pavlovic, D., Bjelica, D., & Domazet, I. (2019). What Characteristics in the Youth Labor Market of Serbia Are Likely to Result in Employment? *Stanovnistvo*, *57*(2), 35–47. doi:10.2298/STNV190823006P

Pavlovic, D., Zubovic, J., & Obradovic, V. (2017). Relationship of the Youth Unemployment and Determinants of the Labour Market in the Balkan Countries. *Industry*, *45*(4), 153–166. doi:10.5937/industrija45-15155

Pawlewski, P. (2019). Interactive layout in the redesign of intralogistics systems. *Intelligent Systems in Production Engineering and Maintenance*, 462–473.

Pedersen, R. (2009). Empirical research on empathy in medicine—A critical review. *Patient Education and Counseling*, *76*(3), 307–322. doi:10.1016/j.pec.2009.06.012 PMID:19631488

Pedraza, J. M. (2021). The micro, small, and medium-sized enterprises and its role in the economic development of a country. *Business and Management Research*, *10*(1), 33. doi:10.5430/bmr.v10n1p33

Pejic-Bach, M., Bertoncel, T., Meško, M., & Krstić, Ž. (2020). Text mining of industry 4.0 job advertisements. *International Journal of Information Management*, *50*, 416–431. doi:10.1016/j.ijinfomgt.2019.07.014

Perifanis, N.-A., & Kitsios, F. (2023). Investigating the influence of artificial intelligence on business value in the digital era of strategy: A literature review. *Information (Basel)*, *14*(2), 85. doi:10.3390/info14020085

Perron P(1989), The great Crash, the oil price shock and the unit root hypothesis, *Econometrica*, Vol-57, No-2, PP-1361-1401

Pettersson, F. (2018). On the issues of digital competence in educational contexts – a review of literature. *Education and Information Technologies*, *23*(3), 1005–1021. doi:10.100710639-017-9649-3

Pham, L. T., & Hoang, H. V. (2019). The relationship between organizational learning capability and business performance: The case of Vietnam firms. *Journal of Economic Development*, *21*(2), 259–269. doi:10.1108/JED-10-2019-0041

Picatoste, J., Pérez-Ortiz, L., & Ruesga-Benito, S. M. (2018). A new educational pattern in response to new technologies and sustainable development. Enlightening ICT skills for youth employability in the European Union. *Telematics and Informatics*, *35*(4), 1031–1038. doi:10.1016/j.tele.2017.09.014

Pirani, Z., Molvizadah, V., Sayyed, M. A., & Sasikumar, M. (2013). E-learning framework for learning disabled children. *International Journal of Computer Applications*, *63*(19), 38–42. doi:10.5120/10577-5703

Pirzada, K., & Khan, F. N. (2013). Measuring Relationship between Digital Skills and Employability. *European Journal of Business and Management*, *5*(24), •••. https://www.researchgate.net/publication/258205764_Measuring_Relationship_between_Digital_Skills_and_Employability

Pluemsamrankit, P. (2017). Factors affecting 21st century skills for learners. *Valayalongkorn Review Journal Humanities and Social Sciences*, *7*(3), 141–158.

Pollitt, D. (2001). A flexible workforce for the modern economy: The role of education and training. *Education + Training*, *43*(7), 345–376. doi:10.1108/EUM0000000005857

Pongpanich, K. (2018). Analysis of factors and competency indicators of the teaching and learning of kindergarten teachers using Thai as a second language. *Journal of Yala Rajabhat University*, *13*(1), 1–12.

Pouangplub, S., Suriyamegha, O., Ngamnikorn, T., Chatdee, P., & Khumsat, T. (2016). The enhancement of Thai language usage competencies in accordance with identity of study fields in Panyapiwat Institute of Management. *동남아연구*, *26*(1), 89—107.

Prahalad, C. K., & Hamel, G. (1990). The core competence of corporation. *Harvard Business Review*, *68*, 79–93.

Prasertphon, C., Kornpuang, A., Jansila, V., & Chatruprachewin, C. (2013). A model for developing learning management competency of new teachers at general education private schools. *Journal of Education Naresuan University*, *15*(5), 43–53.

Prozesky, D. R. (2014). Role of communication in teaching. Teaching Eye Health. University of Pretoria.

Puentedura, R. R. (2010). *SAMR and TPCK: Intro to Advanced Practice.* Retrieved from http://hippasus.com/resources/sweden2010/SAMR_TPCK_IntroToAdvancedPractice.pdf

Qi, Q., & Tao, F. (2018). Digital twin and big data towards smart manufacturing and industry 4.0: 360 degree comparison. *IEEE Access : Practical Innovations, Open Solutions*, *6*, 3585–3593. doi:10.1109/ACCESS.2018.2793265

Racsko, R., & Troll, E. M. (2020). Our digital education habits in the light of their environmental impact: The role of green computing in education. *Teaching Mathematics and Computer Science*, *18*(1), 69–86. doi:10.5485/TMCS.2020.0508

Raja, R., & Nagasubramani, P. C. (2018). Impact of modern technology in education. *Journal of Applied and Advanced Research*, *3*(1), 33–35. doi:10.21839/jaar.2018.v3iS1.165

Ramadhan, H. A., & Putri, D. A. (2018). Big Data, Kecerdasan Buatan, Blockchain, dan Teknologi Finansial di Indonesia. Direktorat Jenderal Apl. Inform. Kementeri. Komun. dan. *Inform (Silver Spring, Md.)*, 1–66.

Ramírez-Montoya, M. S., Mena, J., & Rodríguez-Arroyo, J. A. (2017). In-service teachers' self-perceptions of digital competence and OER use as determined by a xMOOC training course. *Computers in Human Behavior*, *77*, 356–364. doi:10.1016/j.chb.2017.09.010

Rao, K., Edelen-Smith, P., & Wailehua, C. U. (2015). Universal design for online courses: Applying principles to pedagogy. *Open Learning*, *30*(1), 35–52. doi:10.1080/02680513.2014.991300

Rapisarda, S., Ghersetti, E., Girardi, D., De Carlo, N. A., & Corso, L. D. (2021). Smart working and online psychological support during the covid-19 pandemic: Work-family balance, well-being, and performance. *International Psychological Applications Conference and Trends*, pp.301-306. DOI: 10.36315/2021inpact062

Rassameethammachot, S. (2008). *Guidelines for human potential development by competency-based learning* (6th ed.). Knowledge Compilation Division, Research Division, National Productivity Institute.

Ratten, V. (2012). Implementing cloud learning in an organization: A training perspective. *Industrial and Commercial Training*, *44*(6), 334–336. doi:10.1108/00197851211254761

Redecker, C. (2017). European framework for the digital competence of educators: DigCompEdu (No. JRC107466). Joint Research Centre.

Reisoğlu, İ., & Çebi, A. (2020). How can the digital competences of pre-service teachers be developed? Examining a case study through the lens of DigComp and DigCompEdu. *Computers & Education*, *156*, 103940. doi:10.1016/j.compedu.2020.103940

Rice, M. F. (2022). Special education teachers' use of technologies during the COVID-19 era (Spring 2020—Fall 2021). *TechTrends*, *66*(2), 310–326. doi:10.100711528-022-00700-5 PMID:35098256

Riemer, M. J. (2007). Communication skills for the 21st century engineer. *Global Journal of Engineering Education*, *11*(1), 89–100.

Rizos, V., Behrens, A., Van der Gaast, W., Hofman, E., Ioannou, A., Kafyeke, T., Flamos, A., Rinaldi, R., Papadelis, S., Hirschnitz-Garbers, M., & Topi, C. (2016). Implementation of circular economy business models by small and medium-sized enterprises (SMEs): Barriers and enablers. *Sustainability (Basel)*, *8*(11), 1212. doi:10.3390u8111212

Roberts, J. B., Crittenden, L. A., & Crittenden, J. C. (2011). Students with disabilities and online learning: A cross-institutional study of perceived satisfaction with accessibility compliance and services. *The Internet and Higher Education*, *14*(4), 242–250. doi:10.1016/j.iheduc.2011.05.004

Røkenes, F. M., & Krumsvik, R. J. (2016). Prepared to teach ESL with ICT? A study of digital competence in Norwegian teacher education. *Computers & Education*, *97*, 1–20. doi:10.1016/j.compedu.2016.02.014

Romagosa, F. (2020). The COVID-19 crisis: Opportunities for sustainable and proximity tourism. *Tourism Geographies*, *23*(3), 690–694. doi:10.1080/14616688.2020.1763447

Rosenberg, M. (2006). *Beyond e-learning: Approaches and technologies to enhance organizational knowledge, learning, and performance*. Pfeiffer.

Royal Academy. (2013). *Royal Institute Dictionary 2011*. Royal Academy.

Rubita, I. (2012). Sustainability Issues in Tourism: A Case study of Yuksam Village, Sikkim [AJTS]. *Atna Journal of Tourism Studies*, *7*(2), 117–129. doi:10.12727/ajts.8.7

Rupcic, N. (2023). Means to improve organizational learning capability. *The Learning Organization*, *30*(1), 101–109. doi:10.1108/TLO-01-2023-289

Rüßmann, M., Lorenz, M., Gerbert, P., Waldner, M., Justus, J., Engel, P., & Harnisch, M. (2015). Industry 4.0: The future of productivity and growth in manufacturing

Saenthong, N. (2007). *Let's get to know competency*. HR Center.

Saha, S. K. (2022). Bibliometric Analysis of Published Literature on the Pharmaceutical Supply Chain. [IJAL]. *International Journal of Applied Logistics*, *12*(1), 1–18. doi:10.4018/IJAL.309088

Saha, S. K., & Jha, A. (2018). Role of information technology on supply chain management of pharmaceutical industry. [IJAL]. *International Journal of Applied Logistics*, *8*(2), 39–68. doi:10.4018/IJAL.2018070103

Saha, S. K., & Jha, A. (2022). Impact of reminder on medication adherence: A structural equation model, based on study in Sikkim, India. *Bioscience Journal*, *38*(e38014), 1981–3163. doi:10.14393/BJ-v38n0a2022-59087

Saha, S. K., & Jha, A. (2023). Correlation Between Evaluative Beliefs of Patients, Reminder and Medication Adherence. [IJABIM]. *International Journal of Asian Business and Information Management*, *13*(1), 1–16. doi:10.4018/IJABIM.316174

Saha, S. K., Saha, S., & Jha, A. (2018). Use of Information Technology in the Supply Chain Management of the Pharmaceutical Industry: A Literature Review. *Supply Chain and Logistics Management: Concepts, Methodologies, Tools, and Applications*, *8*(2), 673–706.

Saha, S., Saha, S. K., Pandey, J. R., & Jha, A. (2022). Employee motivation for training and development: A study of a pharmaceutical company in Sikkim. In *Handbook of research on developing circular, digital, and green economies in Asia* (pp. 370–386). IGI Global. doi:10.4018/978-1-7998-8678-5.ch018

Salaebing, M., & Nuchmee, K. C. (2017). Result of a podcast learning to promote Malay language communication skills for undergraduate students. *Al-Hikmah Journal. Fatoni University*, *7*(13), 37–44.

Salas Pilco, S. Z. (2013). Evolution of the framework for 21st century competencies. *Knowledge Management & E-Learning*, *5*(1), 10–24.

Sandrone, S. (2022). Medical education in the metaverse. *Nature Medicine*, *28*(12), 2456–2457. doi:10.103841591-022-02038-0 PMID:36385158

Sangrà, A., Vlachopoulos, D., & Cabrera, N. (2012). Building an inclusive definition of e-learning: An approach to the conceptual framework. *International Review of Research in Open and Distance Learning*, *13*(2), 145–159. doi:10.19173/irrodl.v13i2.1161

Santos-Roldán, L., Canalejo, A. M. C., Berbel-Pineda, J. M., & Palacios-Florencio, B. (2020). Sustainable Tourism as a Source of Healthy Tourism. *International Journal of Environmental Research and Public Health*, *17*(15), 1–14. doi:10.3390/ijerph17155353 PMID:32722271

Santos-Vijande, M. L., López-Sánchez, J. Á., & Trespalacios, J. A. (2012). How organizational learning affects a firm's flexibility, competitive strategy, and performance. *Journal of Business Research*, *65*(8), 1079–1089. doi:10.1016/j.jbusres.2011.09.002

Sarrab, M., Elgamel, L., & Aldabbas, H. (2012). Mobile Learning (M-Learning) and Educational Environments. *International Journal of Distributed and Parallel Systems*, *3*(4), 31–38. Advance online publication. doi:10.5121/ijdps.2012.3404

Sattabut, T. (2012). A study of competency needs of human resource staff for industry sector. *Veridian E-Journal*, *5*(2), 426–448.

Schein, E. (1993). On Dialogue, Culture, and organizational learning. *Organizational Dynamics*, *22*(2), 40–52. doi:10.1016/0090-2616(93)90052-3

Schilling, M. A. (2008). *Strategic management of technological innovation* (2nd ed.). McGraw-Hill Education.

Schumacher, A., Erol, S., & Sihn, W. (2016). A maturity model for assessing Industry 4.0 readiness and maturity of manufacturing enterprises. *Procedia CIRP*, *52*, 161–166. doi:10.1016/j.procir.2016.07.040

Seabra, C., & Bhatt, K. (2022). Tourism Sustainability and Covid 19 Pandemic: Is there a Positive Side? *Sustainability (Basel)*, *14*(14), 1–14. doi:10.3390u14148723

Seale, J. K. (2013). *E-learning and disability in higher education: accessibility research and practice*. Routledge. doi:10.4324/9780203095942

Seman, N. A. A., Govindan, K., Mardani, A., Zakuan, N., Saman, M. Z. M., Hooker, R. E., & Ozkul, S. (2019). The mediating effect of green innovation on the relationship between green supply chain management and environmental performance. *Journal of Cleaner Production*, *229*, 115–127. doi:10.1016/j.jclepro.2019.03.211

Semivolos, A., & Anshari, M. (2023). The Application of Artificial Intelligence to the Process of Logistic During COVID-19. In *Handbook of Research on Artificial Intelligence and Knowledge Management in Asia's Digital Economy* (pp. 1–11). IGI Global.

Senge, P. (1990). The Leader's New Work: Building learning organizations. *Sloan Management Review*, *32*(1), 7–23.

Shaheen, M. A., Hayyat, A., Aslam, S., & Shahbaz, M. A. (2023). Developing a culture of continuous learning for sustainable job performance: TEVT sector as learning organization, Pakistan. *Journal of Social Sciences Review*, *3*(2), 817–827. doi:10.54183/jssr.v3i2.321

Sharma, A., & Sharma, D. (2018). Impact of GST on Automobile Industry in India. *International Journal of Mnagement Science and Technology*, *9*(3), 145–154.

Shater, A., A. J. AlMahdawi, and M. A. S. Khasawneh. "The Digital Learning of Disabled Students: Perceptions of Teachers in Public Schools." (2023).

Shater, A., AlMahdawi, A. J., & Khasawneh, M. A. S. (2023). The Digital Learning of Disabled Students: Perceptions of Teachers in Public Schools.

Sheng, X., Chen, L., Yuan, X., Tang, Y., Yuan, Q., Chen, R., Wang, Q., Ma, Q., Zuo, J., & Liu, H. (2023a). Green supply chain management for a more sustainable manufacturing industry in China: A critical review. *Environment, Development and Sustainability*, *25*(2), 1151–1183. doi:10.100710668-022-02109-9

Shukla, S., & Prajapti, D. (2019). A Study on Various Schemes for Msmes: With Special Reference to SIDBI. *Adhyayan: A Journal of Management Sciences, 9*(02), 27-30.

Siddiq, F., Hatlevik, O. E., Olsen, R. V., Throndsen, I., & Scherer, R. (2016). Taking a future perspective by learning from the past–A systematic review of assessment instruments that aim to measure primary and secondary school students' ICT literacy. *Educational Research Review*, *19*, 58–84. doi:10.1016/j.edurev.2016.05.002

Sigala, M. (2020). Tourism and Covid 19: Impacts and Implications for Advancing and Resetting Industry and Research. *Journal of Business Research*, *117*, 312–321. doi:10.1016/j.jbusres.2020.06.015 PMID:32546875

Singh, J., Malhotra, M., & Sharma, N. (2022). Metaverse in Education: An Overview. In D. Bathla & A. Singh (Eds.), *Applying Metalytics to Measure Customer Experience in the Metaverse* (pp. 135–142). IGI Global. doi:10.4018/978-1-6684-6133-4.ch012

Sisodia, S., & Agarwal, N. (2019). Examining Employability Skills of Healthcare Services in India: A Descriptive Literature Review. *The International Journal of Social Science, Management Engineering and Technology (IJSSMET), 10* (3).

Sisodia, S., & Agarwal, N. (2017). Employability Skills Essential for Healthcare Industry. *Procedia Computer Science*, *122*, 431–438. doi:10.1016/j.procs.2017.11.390

Skantz-Åberg, E., Lantz-Andersson, A., Lundin, M., & Williams, P. (2022). Teachers' professional digital competence: An overview of conceptualisations in the literature. *Cogent Education*, *9*(1), 2063224. doi:10.1080/2331186X.2022.2063224

Skaraki, E. (2023). Designing educational material to teach Braille to adult educators through the method of distance learning. *Advances in Mobile Learning Educational Research*, *3*(1), 602–609. doi:10.25082/AMLER.2023.01.009

Ślusarczyk, B. (2018). Industry 4.0: Are we ready? *Polish Journal of Management Studies, 17.*

Smith, S. J., & Basham, J. D. (2014). Designing online learning opportunities for students with disabilities. *Teaching Exceptional Children*, *46*(5), 127–137. doi:10.1177/0040059914530102

Soliman, E. (2021). Planning requirements to enhance digital green education skills for students of applied technology schools (future vision). *Journal of Education*, *91*, 2959–3017.

Stanojevic, D., & Pavlovic, D. (2022). A Profile of Youth not in Employment, Education or Training (NEET) in Serbia. Monography "Human Capital and Welfare", published by University of Antwerp, Antwerp, Belgium.

Starkey, L. (2020). A review of research exploring teacher preparation for the digital age. *Cambridge Journal of Education*, *50*(1), 37–56. doi:10.1080/0305764X.2019.1625867

Starkey, L., & Yates, A. (2021). Do digital competence frameworks align with preparing beginning teachers for digitally infused contexts? An evaluation from a New Zealand perspective. *European Journal of Teacher Education*. Advance online publication. doi:10.1080/02619768.2021.1975109

Statistical Office of the Republic of Serbia. (2021). *Labor Force Survey in The Republic of Serbia, 2020*. Bulletin. Retrieved from https://publikacije.stat.gov.rs/G2021/PdfE/G20215671.pdf

Stentoft, J., Adsbøll Wickstrøm, K., Philipsen, K., & Haug, A. (2021). Drivers and barriers for Industry 4.0 readiness and practice: Empirical evidence from small and medium-sized manufacturers. *Production Planning and Control*, *32*(10), 811–828. doi:10.1080/09537287.2020.1768318

Steyn, G. M. (2002). The changing principalship in South African schools. *Educare (San José)*, *31*(1-2), 251–274.

Stofkova, J., Poliakova, A., Stofkova, K. R., Malega, P., Krejnus, M., Binasova, V., & Daneshjo, N. (2022). Digital Skills as a Significant Factor of Human Resources Development. *Sustainability (Basel)*, *14*(20), 13117. doi:10.3390u142013117

Stratigea, A., & Giaoutzi, M. (2000). Teleworking and virtual organization in the urban and regional context. *NETCOM*, *14*(3-4), 331–357. doi:10.3406/netco.2000.1479

Sun, C. (2015). The cultivation of cross-cultural communication competence in oral English teaching practice. *English Language Teaching*, *8*(12), 7–10. doi:10.5539/elt.v8n12p7

Sureeyatanapas, P., & Yang, J.-B. (2021). Sustainable manufacturing and technology: The development and evaluation. *Pursuing Sustainability: OR/MS Applications in Sustainable Design, Manufacturing, Logistics, and Resource Management*, 111–140.

Susan, A., & Novianti, W. (2019). *Benefits of Technology for Business.* ***IOP Conf. Series: Materials Science and Engineering***, S.662, 032036. Doi:10.1088/1757-899X/662/3/032036

Suvarnaphaet, K. M., & Suvarnaphaet, P. (2023). Use of podcasts to develop students' speaking skills. Humanities. *Arts and Social Sciences Studies*, *23*(1), 1–8. doi:10.14456/hasss.2023.1

Swarbrooke, J. (1999). *Sustainable Tourism Management*. CAB International. doi:10.1079/9780851993140.0000

Talib, F., & Rahman, Z. (2013). Current health of Indian healthcare and hospitality industries: a demographic study.

Tamang, K., & Solanki, V. (2020). Tourist Perception and Satisfaction Level: A Study of East Sikkim Tourism Industry. *Journal of Xidian University*, *14*(4), 2559–2575.

Tamesberger, D., & Bacher, J. (2014). NEET youth in Austria: A typology including socio-demography, labormarket behaviour and permanence. *Journal of Youth Studies*, *17*(9), 1239–1259. doi:10.1080/13676261.2014.901492

Tamesberger, D., Leitgöb, H., & Bacher, J. (2014). How to combat NEET? Evidence from Austria. *Inter Economics*, *49*(4), 221–227. doi:10.100710272-014-0503-1

Tang, T. A., Mhamdi, L., McLernon, D., Zaidi, S. A. R., & Ghogho, M. (2016, October). Deep learning approach for network intrusion detection in software defined networking. In 2016 international conference on wireless networks and mobile communications (WINCOM) (pp. 258-263). IEEE. doi:10.1109/WINCOM.2016.7777224

Tan, J. P. L., Choo, S. S., Kang, T., & Liem, G. A. D. (2017). Educating for twenty-first century competencies and future-ready learners: Research perspectives from Singapore. *Asia Pacific Journal of Education*, *37*(4), 425–436. doi:10.1080/02188791.2017.1405475

Tapalova, O., & Zhiyenbayeva, N. (2022). Artificial intelligence in education: AIEd for personalised learning pathways. *Electronic Journal of e-Learning*, *20*(5), 639–653. doi:10.34190/ejel.20.5.2597

Tarconish, E., Lombardi, A., & Taconet, A. (2023). Disability Awareness & Inclusive Teaching Online Training Videos for College Instructors Featuring Students with Disabilities. *Journal of Postsecondary Education & Disability*, *35*(4). Cooc, N. (2023). National Trends in Special Education and Academic Outcomes for English Learners With Disabilities. *The Journal of Special Education*, •••, 00224669221147272.

Tarconish, E., Lombardi, A., & Taconet, A. (2023). Disability Awareness & Inclusive Teaching Online Training Videos for College Instructors Featuring Students with Disabilities. *Journal of Postsecondary Education and Disability*, *35*(4).

Tárraga-Mínguez, R., Suárez-Guerrero, C., & Sanz-Cervera, P. (2021). Digital teaching competence evaluation of pre-service teachers in Spain: A review study. *IEEE-RITA*, *16*(1), 70–76. doi:10.1109/RITA.2021.3052848

Tatongjai, N., In Chanjarain, Weeraphan Sririrth. (2015). Teacher's Competency in Vocational Institutions in the Upper Northern Thailand after Completing the Graduate Diploma Program in Teaching Profession. Journal of Education. *Mahasarakham University*, *9*, 138–152.

Teece, D., Pisano, G., & Shuen, A. (1997). Dynamic Capabilities and Strategic Management. *Strategic Management Journal*, *18*(7), 509–533. doi:10.1002/(SICI)1097-0266(199708)18:7<509::AID-SMJ882>3.0.CO;2-Z

Thapliyal, M., Ahuja, N. J., Shankar, A., Cheng, X., & Kumar, M. (2022). A differentiated learning environment in domain model for learning disabled learners. *Journal of Computing in Higher Education*, *34*(1), 60–82. doi:10.100712528-021-09278-y

The above 12 marketing mix elements have become the solutions and recommendations when they are combined with the consumers' feedback that has been clustered by the digital Linguistic Inquiry and Word Count (LIWC) into 12 words and transited to 12 marketing strategies.

The Lancet (2020) COVID-19: Protecting healthcare workers. The Lancet 395(10228): 922.

Thiem, K. C., & Dasgupta, N. (2022). From precollege to career: Barriers facing historically marginalized students and evidence-based solutions. *Social Issues and Policy Review, 16*(1), 212–251. doi:10.1111ipr.12085

Thongart, C. (2018). Thai language learning management using 21st century competencies-based. *Journal of Education Studies, 46*(1), 171–184.

Thongkaew, T. (2007). *Competency: Principles and practices.* Mahawatiyaraya Suan Dusit Rajabhat University.

Thuaamklang, K. (2012). *Development of educational media/innovations to promote academic standing.* Sathaporn Books.

Todhunter, F. (2015). Using principal components analysis to explore competence and confidence in student nurses as users of information and communication technologies. *Nursing Open, 2*(2), 72–84. doi:10.1002/nop2.19 PMID:27708803

Tokovska, M., Zat'kova, T. S., & Jamborova, L. (2022). Digital Competencies Development in Higher Education Institutions: A Mixed Methods Research Study. *Emerging Science Journal, 6*(Special Issue), 150–165. Advance online publication. doi:10.28991/ESJ-2022-SIED-011

Tomczyk, Ł. (2019). What do teachers know about digital safety? *Computers in the Schools, 36*(3), 167–187. doi:10.1080/07380569.2019.1642728

Tomlinson, B. (2012). Materials development for language learning and teaching. *Language Teaching, 45*(2), 143-—179. doi:10.1017/S0261444811000528

Tømte, C. E. (2013). Educating teachers for the new millennium? Teacher training, ICT and digital competence. *Nordic Journal of Digital Literacy, 10*(Jubileumsnummer), 138-154.

Tondeur, J., Aesaert, K., Prestridge, S., & Consuegra, E. (2018). A multilevel analysis of what matters in the training of pre-service teacher's ICT competencies. *Computers & Education, 122*, 32–42. doi:10.1016/j.compedu.2018.03.002

Tondeur, J., van Braak, J., Guoyuan, S., Voogt, J., Fisser, P., & Ottenbreit-Leftwich, A. S. (2012). Preparing student teachers to integrate ICT in classroom practice: A synthesis of qualitative evidence. *Computers & Education, 59*(1), 134–144. doi:10.1016/j.compedu.2011.10.009

Tong, D. H., Uyen, B. P., & Ngan, L. K. (2022). The effectiveness of blended learning on students' academic achievement, self-study skills and learning attitudes: A quasi-experiment study in teaching the conventions for coordinates in the plane. *Heliyon, 8*(12), e12657. doi:10.1016/j.heliyon.2022.e12657 PMID:36643330

Tran, L. T. B., Ho, N. T., & Hurle, R. J. (2016). Teaching for creativity development: Lessons learned from a preliminary study of Vietnamese and international upper (high) secondary school teachers' perceptions and lesson plans. *Creative Education, 7*(7), 1024–1043. doi:10.4236/ce.2016.77107

Trilling, B., & Fadel, C. (2009). *21st century skills: Learning for life in our times*. John Wiley & Sons.

Tshabalala, M., & Ndeya-Ndereya, C. (2015). How far are we with the implementation of blended learning? Implementation guidelines on offer. In D. Nwaozuzu, & S. Mnisi (Ed.), *South Africa International Conference on Educational Technologies* (pp. 27-36). Pretoria: African Academic Research Forum.

Tsitskari, E., Goudas, M., Tsalouchou, E., & Michalopoulou, M. (2017). Employers' expectations of the employability skills needed in the sport and recreation environment. *Journal of Hospitality, Leisure, Sport and Tourism Education, 20*, 1–9. doi:10.1016/j.jhlste.2016.11.002

Tugano, M. S., Tria, J. Z., & Tonio, J. Z. (2022). Modular learning amidst COVID-19 pandemic: Satisfaction among students in a higher education institution. ***International Journal of Professional Development, Learners and Learning, 4(2)***, ep2206.

Tumpa, T. J., Ali, S. M., Rahman, M. H., Paul, S. K., Chowdhury, P., & Khan, S. A. R. (2019). Barriers to green supply chain management: An emerging economy context. *Journal of Cleaner Production, 236*, 117617. doi:10.1016/j.jclepro.2019.117617

Tynjälä, P., & Häkkinen, P. (2005). E-learning at work: Theoretical underpinnings and pedagogical challenges. *Journal of Workplace Learning, 17*(5/6), 318–336. doi:10.1108/13665620510606742

Ulak, N. (2020). A Preliminary Study of Novel Corona Virus Disease (Covid 19) Outbreak: A Pandemic Leading Crisis in the Tourism Industry in Nepal. *Journal of Tourism and Hospitality Education, 10*, 108–131. doi:10.3126/jthe.v10i0.28763

UN. (2022). *UN Special Rapporteur on the Right to Education launches new report into the impact of digitalisation.*

UNEP & UNWTO. (2005). *Making Tourism More Sustainable - A Guide for Policy Makers*. UNEP and UNWTO.

UNESCO. (2010). *World data on education 7th edition 2010/11*. http://www.ibe.unesco.org

UNESCO. (2011). *UNESCO ICT competency framework for teachers*. https://unesdoc.unesco.org/images/0021/002134/213475e.pdf

UNESCO. (2018). *A Global Framework of Reference on Digital Literacy Skills for Indicator 4.4.2*. [Online]Available at: http://www.uis.unesco.org [Accessed: July 10, 2023].

UNESCO. (2018). *A Landscape Review: Digital Inclusion for Low-skilled and Low-literate People.*

UNWTO. (2007). Tourism and Least Developed Countries: A Sustainable Opportunity to Reduce Poverty. https://www.unwto.org/sustainable/doc/tourism-and-ldc.pdf, Accessed on 10th March, 2023 at 11.32 pm.

UNWTO. (2022): Impact Assessment of the Covid-19 Outbreak on International Tourism, UNWTO. www.unwto.org. Accessed on 11th March, 2023 at 10.20 pm.

Vaidya, S., Ambad, P., & Bhosle, S. (2018). Industry 4.0–a glimpse. *Procedia Manufacturing*, *20*, 233–238. doi:10.1016/j.promfg.2018.02.034

Vaisakh, J., Swarna Abirami, L., Krishnachandran, C., & Arun, K. (2020). Tourism Marketing Through Social Media After Covid 19 in Kerala. *International Journal of Advanced Science and Technology*, *29*(7), 11481–11486.

Van Deursen, A. J. A. M., Helsper, E. J., & Eynon, R. (2014). *Measuring Digital Skills. From Digital Skills to Tangible Outcomes project report*. www.oii.ox.ac.uk/research/projects/?id=112

Van Deursen, A. J. A. M., & van Dijk, J. A. G. M. (2009). Using the Internet: Skill related problems in users' online behavior. *Interacting with Computers*, *21*(5-6), 393–402. doi:10.1016/j.intcom.2009.06.005

Van Deursen, A. J. A. M., & van Dijk, J. A. G. M. (2010). Measuring Internet skills. *International Journal of Human-Computer Interaction*, *26*(10), 891–916. doi:10.1080/10447318.2010.496338

Van Heerden, D., & Goosen, L. (2023). The Implications of COVID-19 for Fostering Diversity and Inclusion Through Curriculum Transformation: Teaching Open and Distance E-Learning Students. In C. Tabane, B. Diale, A. Mawela, & T. Zengele (Eds.), *Fostering Diversity and Inclusion Through Curriculum Transformation* (pp. 73–100). IGI Global., doi:10.4018/978-1-6684-6995-8.ch005

Van Wyk, N. (2004). School governing bodies: The experiences of South African educators. *South African Journal of Education*, *24*(1), 49–54.

Vasic, V. (2019). *Skills Mismatch Measurement in Serbia*. Turin: European Training Foundation. (Report). https://www.etf.europa.eu/sites/default/files/2019-07/Skills%20mismatch%20measurement_Serbia_0.pdf

Vehbi, B. O. (2012). A Model for Assessing the Level of Tourism Impacts and Sustainability of Coastal Cities https://cdn.intechopen.com/pdfs/35710/InTech-A_model_for_assessing_the_level_of_tourism_impacts_and_sustainability_of_coastal_cities.pdf Accessed on 9th March, 2023 at 9.12 pm.

Vera, D., & Crossan, M. (2004). Strategic leadership and organizational learning. *Academy of Management Review*, *29*(2), 222–241. doi:10.2307/20159030

VERBI Software. (2021). MAXQDA 2022 [computer software]. Berlin, Germany: VERBI Software. Available from maxqda.com.

Versiani, Â., & Fachin, R. (2007). Avaliando aprendizagem em simulações empresariais. Cadernos EBAPE.BR, Rio de Janeiro, 5, 1-13.

Vijaya, P. (2011). Prospects of Clean Technologies for Sustainable Business. *Review of Management, 1*(2).

Vij, P., & Sharma, M. (2014). The effect of human capital management practices on employee positive behavioural outcomes. *International Journal of Management Practice, 7*(3), 204–221. doi:10.1504/IJMP.2014.063594

Vladisavljevic, M., Zubovic, J., Djukic, M., & Jovanovic, O. (2021). Inequality-Reducing Effects of Tobacco Tax Increase: Accounting for Behavioral Response of Low-, Middle-, and High-Income Households in Serbia. *International Journal of Environmental Research and Public Health, 18*(18), 9494. doi:10.3390/ijerph18189494 PMID:34574420

Vladisavljevic, M., Zubovic, J., Jovanovic, O., Djukic, M., Najdovska, N. T., Pula, E., ... Gjika, A. (2022). Tobacco tax evasion in Western Balkan countries: Tax evasion prevalence and evasion determinants. *Tobacco Control, 31*(Suppl 2), s80–s87. doi:10.1136/tobaccocontrol-2021-056879 PMID:35022328

Vorster, J., & Goosen, L. (2017). A Framework for University Partnerships Promoting Continued Support of e-Schools. In J. Liebenberg (Ed.), *Proceedings of the 46th Annual Conference of the Southern African Computer Lecturers' Association (SACLA)* (pp. 118 - 126). Magaliesburg: North-West University.

Vuorikari, R., Kluzer, S., & Punie, Y. (2022). DigComp 2.2: The digital competence framework for citizens - With new examples of knowledge, skills and attitudes, EUR 31006 EN, Publications Office of the European Union, Luxembourg.

Vuorikari, R., Punie, Y., Carretero, S., & van den Brande, L. (2016). *DigComp 2.0: The Digital Competence Framework for Citizens. Update Phase 1: The conceptual reference model*. Luxembourg Publication Office of the European Union.

Wahyuningtyas, R., & Anggadwita, G. (2017). The perspective of Managing Talent in Indonesia: Reality and Strategy. In: Ordoñez de P, Rablos, P. and Tennyson, R.D. (eds.), Handbook of Research on Human Resources Strategies for the New Millennial Workforce. IGI Global.

Wahyuningtyas, R., Fajriani, V.M, Chaudhuri, M.R. and Yusoff, Y.M. (2021). Gap analysis of multigeneration digital competence. *Proceedings of Synergizing Management, Technology, and Innovation in Generating Sustainable and Competitive Business Growth*. Taylor Francis Group.

Wahyuningtyas, R., Alamsyah, A., & Diliana, N. A. (2021). Mapping Digital Talent Based on Competency using Social Network Analysis.*2nd International Conference on ICT for Rural Development (IC-ICTRuDev)*, Jogjakarta, Indonesia, 2021, pp. 1-5, 10.1109/IC-ICTRuDev50538.2021.9656521

Wald, M. (2009). E. A. Draffan, and Jane Seale. "Disabled learners' experiences of e-learning.". *Journal of Educational Multimedia and Hypermedia, 18*(3), 341–361.

Wamba-Taguimdje, S.-L., Wamba, S. F., Kamdjoug, J. R. K., & Wanko, C. E. T. (2020). Influence of artificial intelligence (AI) on firm performance: The business value of AI-based transformation projects. *Business Process Management Journal*, *26*(7), 1893–1924. doi:10.1108/BPMJ-10-2019-0411

Wang, M., Chen, Y., & Khan, M. J. (2014). Mobile cloud learning for higher education : A case study of Moodle in the cloud. *International Review of Research in Open and Distance Learning*, *15*(2), 255–267. doi:10.19173/irrodl.v15i2.1676

Wang, Q. (2008). A generic model for guiding the integration of ICT into teaching and learning. *Innovations in Education and Teaching International*, *45*(4), 411–419. doi:10.1080/14703290802377307

Wang, S., Christensen, C., Cui, W., Tong, R., Yarnall, L., Shear, L., & Feng, M. (2023). When adaptive learning is effective learning: Comparison of an adaptive learning system to teacher-led instruction. *Interactive Learning Environments*, *31*(2), 793–803. doi:10.1080/10494820.2020.1808794

Wang, Y., Jiang, S., Wu, C., Cai, X., & Wang, F. (2022). Impact of the Global Megatrends, COVID-19, and Digital Economy on Professional Career Management Transformation in Asian Countries. *Sustainability (Basel)*, *14*(17), 10981. doi:10.3390u141710981

Wannewitz, M., Petzold, J., & Garschagen, M. (2023). What makes people adapt together? An empirically grounded conceptual model on the enablers and barriers of collective climate change adaptation. *Frontiers in Climate*, *5*, 1213852. doi:10.3389/fclim.2023.1213852

Warinin, N. (2014). *The model of teachers' competency development in learning management for primary schools in KamphaengPhet primary educational service area office 2*. [Unpublished doctoral dissertation]. Nakhon Sawan Rajabhat University.

Watkins, K., & Marsick, V. (1993). *Sculpting the Learning Organization*. Jossey-Bass.

Werner, A., & Schneider, J. M. (1974). Teaching medical students interactional skills: A research-based course in the doctor-patient relationship. *The New England Journal of Medicine*, *290*(22), 1232–1237. doi:10.1056/NEJM197405302902206 PMID:4825853

Wichit, W. (2009). *Podcast with teaching and learning*. http://www.wittaya.net

Wiley (2023). Digital Skills Gap Index 2021. Available at: https://dsgi.wiley.com/, accessed at June 21st, 2023

Won, J. Y., & Park, M. J. (2020). Smart factory adoption in small and medium-sized enterprises: Empirical evidence of manufacturing industry in Korea. *Technological Forecasting and Social Change*, *157*, 120117. doi:10.1016/j.techfore.2020.120117

World Bank. (2021). Beyond Unicorns: Harnessing Digital Technologies for Inclusion in Indonesia. World Bank. [Online], Retrieved July 10, 2023, from.

World Bank. (2023). Database on youth unemployment, retrieved from the International Labour Organization. "Labour Force Statistics database (LFS)" ILOSTAT. ilostat.ilo.org/data. Available at: https://data.worldbank.org/indicator/SL.UEM.1524.NE.ZS, accessed at June 21st, 2023

World Economic Forum (WEF). (2022). *Travel and Tourism Development Index 2021: Rebuilding for a Sustainable and Resilient Future*. Geneva, Switzerland.

World Economic Forum. (2016). The Future of Jobs Report 2016, Retrieved from https://www3.weforum.org/docs/WEF_Future_of_Jobs.pdf

World Economic Forum. (2020). The Future of Jobs Report 2020. Retrieved from https://www.weforum.org/reports/the-future-of-jobs-report-2020

Wright, G., Cilliers, L., Van Niekerk, E., & Seekoe, E. (2017). The next stage of development of elearning at UFH in South Africa. *International Conference on E-Learning* (pp. 32-38). Lisbon: International Association for Development of the Information Society.

Wu, L.-Y. (2010). Applicability of the resource-based and dynamic-capability views under environmental volatility. *Journal of Business Research*, *63*(1), 27–31. doi:10.1016/j.jbusres.2009.01.007

Xu, J., & Pero, M. E. P. (2023). A resource orchestratipon perspectiv of organizational big data analytics adoption: Evidence from supply chain planning. *International Journal of Physical Distribution & Logistics Management*, *53*(11), 71–97. doi:10.1108/IJPDLM-04-2022-0118

Yassin, J., Aralas, S., & Basa, D. E. K. (2021). Tourism Density Effect on Environmental Performance Index: Evidence in ASEAN Countries, *Applied. Environmental Research*, *43*(1), 93–101.

Yodyingyong, K. (2009). *Innovative organizations: Concepts and processes*. Chulalongkorn University Book Center.

Yoon, J. S., Shin, S. J., & Suh, S. H. (2012). A conceptual framework for the ubiquitous factory. *International Journal of Production Research*, *50*(8), 2174–2189. doi:10.1080/00207543.2011.562563

Yoon, S. H. (2022). Gender and digital competence: Analysis of pre-service teachers' educational needs and its implications. *International Journal of Educational Research*, *114*, 101989. doi:10.1016/j.ijer.2022.101989

Zailani, S., Govindan, K., Iranmanesh, M., Shaharudin, M. R., & Chong, Y. S. (2015). Green innovation adoption in automotive supply chain: The Malaysian case. *Journal of Cleaner Production*, *108*, 1115–1122. doi:10.1016/j.jclepro.2015.06.039

Zdravkovic, A., Djukic, M., & Bradic-Martinovic, A. (2017). Impact of FDI on unemployment in transition countries: Panel cointegration approach. *Industrija*, *45*(1), 161–174. doi:10.5937/industrija45-13548

Zhang, X., Meng, Y., Ordóñez de Pablos, P., & Sun, Y. (2019). Learning analytics in collaborative learning supported by Slack: From the perspective of engagement. *Computers in Human Behavior*, *92*, 625–633. doi:10.1016/j.chb.2017.08.012

Zhao, Y., Llorente, A. M. P., & Gómez, M. C. S. (2021). Digital competence in higher education research: A systematic literature review. *Computers & Education*, *168*, 104212. doi:10.1016/j.compedu.2021.104212 PMID:36568577

Zhao, Z., Li, X., & Zhou, X. (2020). Distribution route optimization for electric vehicles in urban cold chain logistics for fresh products under time-varying traffic conditions. *Mathematical Problems in Engineering*, *2020*, 1–17. doi:10.1155/2020/9864935

Zubovic, J., Zdravkovic, A., & Pavlovic, D. (2015). Effects of regulation on youth unemployment: Evidence from European countries. *Industrija*, *43*(2), 129–144. doi:10.5937/industrija43-8483

About the Contributors

Patricia Ordóñez de Pablos is a professor in the Department of Business Administration in the Faculty of Business and Economics at The University of Oviedo, Spain. She completed her education in The London School of Economics, UK. Her teaching and research interests focus on the areas of strategic management, knowledge management, organizational learning, human resource management, intellectual capital, information technologies, with special interest in Asia (Bhutan, China, Laos, Myanmar). She is Editor-in-Chief of the International Journal of Learning and Intellectual Capital (IJLIC) and International Journal of Asian Business and Information Management (IJABIM), respectively. She has edited books for IGI Global, Elsevier, Routledge, and Springer. In 2021, she earned placement on Stanford University's "Ranking of the World Scientists: World's Top 2% Scientists" list.

Haji Mohammad Nabil Almunawar is currently an associate professor at the School of Business and Economics, Universiti of Brunei Darussalam (UBDSBE), Brunei Darussalam. He was the former dean of UBDSBE. He received his bachelor's degree in 1983 from Bogor Agricultural University, Indonesia, master's degree (MSc) from the Department of Computer Science, University of Western Ontario, London, Canada in 1991, and Ph.D in Computer Science/Information Systems from the University of New South Wales in 1998. Dr. Almunawar has published more than 150 papers in refereed journals, books, book chapters, encyclopedias, and international conference proceedings. He has more than 30 years of teaching experience in the area of information systems. His overall research interests include applications of IT in management, e-business/commerce, digital marketplace/platform, digital business ecosystem, health informatics, information security, and cloud computing. Currently, he focuses his research on digital transformation, digital marketplace, digital platform, and digital business ecosystem.

Anshari has been named on the Stanford University List of Top 2% World Researchers in 2022 for the categories of ICT, Business & Management. Currently, he serves as Deputy Director Institute of Policy Studies and Academic Staff at School of Business & Economics, Universiti Brunei Darussalam. His professional experience started when he was IT Business Analyst at Astra International. Research Fellowship from The Government Republic of China (Taiwan) at National Taiwan University (Jan-Dec, 2014). Research Fellowship from King Saud University - the Kingdom of Saudi Arabia 2009. Senior Associate Researcher of Informatics Department, Universitas Islam Negeri Yogyakarta, Indonesia.

* * *

Hani Gita Ayuningtias is a full-time Lecturer of the Business Management of Telecommunications and Informatics Department, School of Economics and Business, Telkom University, Indonesia. She teaches courses in Organizational Behavior, Human Resource Management and Interpersonal Comunication Business. Her research interests include HR Analytics/ People Analytics, Wellbeing, Future of Work and Woman Leadership. She is an active researcher who has published in leading international and national journals as well as international proceedings.

Tuğba Bahçekapılı Özdemir is a researcher at the Department of Computer Education and Instructional Technology, Trabzon University, Turkey. Her research interests are technology integration, technological leadership and mentoring, innovative use of technology in k-12 classrooms.

Ayça Çebi is an associate professor at the Department of Computer Education and Instructional Technology, Trabzon University, Turkey. Her research interests include adaptive educational system, educational data mining, technology integration, and digital competence.

Canan Çolak is an assistant professor at the Department of Educational Sciences, Giresun University, Turkey. She has some research interest; Teacher education, instructional design, digital literacy, 3D virtual worlds

Tushar Das, former research fellow of Netaji Subhas Open University, now working as head Master of Sadananda Mission High School. He has published research papers in edited books and journals. his area of interest is macroeconomics and international trade.

Janhavi Ravindra Deshmukh is MBA graduate from Christ University, Bangalore. She has specialized in Lean Operations and Systems.

Mihajlo Djukic graduated from the Faculty of Economics of the Belgrade University, Serbia (Bachelor and Master studies) and obtained PhD degree in finance at the Faculty of Economics, Technical University of Kosice, Slovakia. He has been employed as a researcher with the Institute of Economic Sciences (IES) since 2010. His research focuses on economic policy, development and evaluation of public policies. Mihajlo has participated in several projects commissioned by the state institutions at the national and local level, as well as by international organizations (SDC, GIZ, European Commission), which have dealt with the issues related to drafting and evaluation of strategic documents and policy analyses in various fields – local development, youth policy, health economics, etc.

Leila Goosen is a full professor in the Department of Science and Technology Education of the University of South Africa. Prof. Goosen was an Associate Professor in the School of Computing, and the module leader and head designer of the fully online signature module for the College for Science, Engineering and Technology, rolled out to over 92,000 registered students since the first semester of 2013. She also supervises ten Masters and Doctoral students, and has successfully completed supervision of 43 students at postgraduate level. Previously, she was a Deputy Director at the South African national Department of Education. In this capacity, she was required to develop ICT strategies for implementation. She also promoted, coordinated, managed, monitored and evaluated ICT policies and strategies, and drove the research agenda in this area. Before that, she had been a lecturer of Information Technology (IT) in the Department for Science, Mathematics and Technology Education in the Faculty of Education of the University of Pretoria. Her research interests have included cooperative work in IT, effective teaching and learning of programming and teacher professional development.

Sumaira is a faculty in management discipline with Ph.D. in entrepreneurship from Centre for Management Studies, Jamia Millia Islamia, New-Delhi. She has done MBA in marketing and finance from Centre for Management Studies, Jamia Millia Islamia, New-Delhi and BBA from Islamic University of Science and Technology, Kashmir in 2013 and 2011 respectively. She has qualified UGC NET/JRF/SRF exams. She is a gold medallist in graduation from Islamic University of Science and Technology, Kashmir. She has authored more than 11 Scopus indexed publications in journals of national and international repute. She has also authored an edited book and published several book chapters as well. Her research areas include entrepreneurship, women entrepreneurship, strategic entrepreneurship, motivation etc.

Asim K. Karmakar is Assistant Professor (Stage-3), Department of Economics, Jadavpur University, Kolkata and Life Member, Indian Society of Labour Economics. He is presently the Executive Member of the Indian Economic Association (IEA) and Life Member of The Indian Econometric Society (TIES); formerly the Managing Editor (2011-2014) of the Quarterly Referred Journal Artha Beekshan of Bengal Economic Association. He has completed three minor research projects sponsored by the CAS, Jadavpur University and the UGC. He is the recipient of the prestigious VIFRA 2015 LIFETIME ACHIEVEMENT AWARD for his outstanding contribution in the field of Indian Economy, conferred by Venus International Foundation, Chennai amongst Afro-Asian nationals. Under his guidance, 2 students have obtained their Ph.D degrees and 3 have obtained their M.Phil Degree.

Peter Kisembo is an IT enthusiast with a track record with regards to the use of IT in teaching and learning after having completed a degree in IT at Mbarara University of science and technology. I taught Mathematics, Physical Science and Computer Application Technology at a South African public school for over 8 years. Also worked as the ICT coordinator for the different schools in my cluster where I was tasked with assisting educators with regards to coming up with different strategies on how to integrate ICT into the classroom as well as how to be able to effectively use technology to achieve the required teaching and learning out comes. I am a proactive learner, and I ensure that I stay up-to-date with industry trends in Technology. I am always looking for ways to enhance my knowledge by exploring emerging technology trends that will enable me to contribute to the success of software development initiatives. I take ownership of my work with ethical considerations, and I demonstrate a strong commitment to delivering results. I use my excellent collaborative skills to thrive in a fast-paced environment that is driven by innovation. With the experience gained over the years, I started my own IT company Appline technologies mainly dealing with developing eLearning content to be used by educators across the African continent. I also went ahead to pursue an Honours qualification in Computing with the University of South Africa to widen my expertise further in technology especially relating to eLearning.

Siva Kumar Pujari received his PhD from Indian Institute of Technology Roorkee, India. He obtained his B.Tech. (mechanical engineering) from Acharya Nagarjuna University, Guntur, Andhra Pradesh, and M.Tech. from National Institute of Technology, Trichy, India. He has published several research papers in journals of international repute. His areas of interest include SC Resilience, supply chain disruption management and risk management.

Ramakrishnan N is a Professor and Head of Lean Operations and Systems specialization at the School of Business and Management, CHRIST (Deemed to be University), Bangalore. He has 34 years of experience which includes 23 years in senior roles including Global Head responsibilities in leading IT organizations and 11 years in full-time roles in academia. His PhD is in the area of software engineering, and has an M.S. in Software Systems from Birla Institute of Technology & Science, Pilani, after an Engineering degree and a post-graduate diploma in Business Administration. He has certification in 'Global Business Leadership' from U21 Global, Singapore and Harvard Business School Publishing, and is a certified Internal Quality Auditor.

Dejana Pavlović holds a Doctorate in Labor Economics from the University of Belgrade, Faculty of Organizational Sciences in Management. Her Doctorate focused on the impact of social and economic factors in the Youth Labor Market of the Republic of Serbia. She also holds both Bachelor and Master degrees from the same institution. From January 2013 to June 2015, she worked at the Economic Institute in Belgrade as a Research Assistant. Since June 2015, she has been working at the Institute of Economic Sciences, as a Research Associate. During her career, she participated in several projects on labour market research analysis, gaining experience in project drafting and becoming an expert on the current state of labor in the Serbian market. From March of 2019 to January of 2020, she served as a consultant to the Social Inclusion and Poverty Reduction Unit of the Government of the Republic of Serbia. She has been involved in a plethora of projects bearing international and national importance. Moreover, she has attended numerous scientific conferences and has published several scientific research papers within the field of the youth labour market.

Ilknur Reisoğlu is an associate professor in the Department of Computer Education and Instructional Technologies, Recep Tayyip Erdoğan University, Turkey. She received her PhD in computer education and instructional technology from Atatürk University, Turkey. Her research interests include information seeking, digital competence and 3D virtual worlds.

Rashmi S is MBA graduate from Christ (Deemed to be University) in Lean Operations and Systems. She has completed bachelor's degree in mechanical engineering. She is interested in Supply Chain Management, its trends and challenges.

Saibal Kumar Saha (UGC NET Qualified) holds a Ph.D in Management Studies, a first-class master's degree in MBA and a first-class bachelor's degree in Electronics and Communication Engineering. He has 12+ years of experience and has worked in MNCs like Cognizant Technology Solutions and Tata Aig Life. He has served National Institute of Technology - Silchar, Jyotirmoy School of Business – Kolkata, University of Technology and Management - Shillong, Sikkim Manipal Institute of Technology and is currently working as Associate Professor at Christ University Bangalore.

Ayşe Meriç YAZICI holds a bachelor of public administration, a master degree in business administration from Istanbul Aydin University and a PhD in business administration from Istanbul Aydin Univeristy. She holds a postdoctoral degree from Lund University, Sween. She is also Visiting Scholar at the Blue Marble Space Institute of Science, Washington D. C., U. S. A. Currently at Istanbul Gelisim University, faculty member.

Index

L

M

N

O

P

R

S

T

U

V

Y

www.ingramcontent.com/pod-product-compliance
Lightning Source LLC
LaVergne TN
LVHW080604300926
849162LV00005B/1041
* 9 7 8 1 6 6 8 4 9 0 8 9 1 *